Latino Education in
the United States

D0059251

Additional Praise for Victoria-María MacDonald's *Latino Education in the United States*:

MacDonald has provided a comprehensive, readable, and provocative guide for those interested in the historical evolution of Latino education in the United States. The combination of shrewd introductory essays, carefully selected readings, and extensive bibliography should make *Latino Education in the United States* one of the preferred reference books in the fields of Latino studies, education, and history. . . . One of the most impressive elements of *Latino Education in the United States* is that it provides an outlet to the many voices associated with Latino educational issues: voices of oppression, hope, discrimination, opportunity, and relentlessness. MacDonald's scholarly command of the historiography of Latino education, combined with extensive archival research, make this book a must-read for those interested in Latino and educational issues.

—*Félix V. Matos-Rodríguez, Ph.D. and Director, Centro de Estudios Puertorriqueños, Hunter College*

Latino Education in the United States

A Narrated History from 1513–2000

Victoria-María MacDonald

First published 2004 by
PALGRAVE MACMILLAN™
175 Fifth Avenue, New York, N.Y. 10010 and
Houndmills, Basingstoke, Hampshire, England RG21 6XS.
Companies and representatives throughout the world.

PALGRAVE MACMILLAN is the global academic imprint of the Palgrave
Macmillan division of St. Martin's Press, LLC and of Palgrave Macmillan Ltd.
Macmillan® is a registered trademark in the United States, United Kingdom and
other countries. Palgrave is a registered trademark in the European Union and
other countries.

ISBN 1–4039–6086–0 hardback
ISBN 1–4039–6087–9 paperback

**Library of Congress Cataloging-in-Publication Data Available from the
Library of Congress**

MacDonald, Victoria-María, 1960-
Latino education in the United States: A narrated history from 1513-2000 /
    written and edited by Victoria-María MacDonald.
    p. cm.
Includes bibliographical references and index.
ISBN 1-4039-6086-0 (hardcover)—ISBN 1-4039-6087-9 (pbk.)
1. Hispanic Americans—Education—History. 2. Hispanic
Americans—Education—History—Sources. I. Title: Latino Education
in the United States:
A narrated history from 1513-2000. II. Title: Latino Education in the United States:
A narrated history from 1513-2000. III. Title.

LC2670.M33 2004
371.829'68073—dc22

2004043362

A catalogue record for this book is available from the British Library.

Design by Autobookcomp.

First edition: November 2004
10 9 8 7 6 5 4 3 2 1

Printed in the United States of America.

With gratitude and love to my parents,
George Ewen MacDonald and Leonor Posada MacDonald,
who instilled in me a love for learning and gave me the gift
of a bicultural heritage.

Contents

Acknowledgements

It is a pleasure to acknowledge the assistance and support for this project. My editor, Amanda Johnson at Palgrave Macmillan, originally encouraged me to submit a book proposal on this topic. Many thanks for her steadfast encouragement, enthusiasm, and support for me personally and for this project. The Spencer Foundation provided three rounds of funding for my research on Latino educational history between 1997 and 2000. The financial assistance and affirmation of the significance of this work were critical to its conception and creation. A Collaborative Research Grant from the National Academy of Education also permitted additional archival research, enriching the range of documentary sources I was able to include in this volume.

An outstanding team of graduate students at Florida State University worked during the last year to ensure the completion of this manuscript. Research assistant Mark R. Nilles was invaluable. His determination to see this project finished, good humor through many adventures writing this book, and terrific editing skills are greatly appreciated. Kristina M. Goodwin pursued copyright releases and tracked archival sources with zeal and resoluteness. Elizabeth (Lisa) Hoffman Clark's outstanding research and editing skills pulled the book together into one coherent volume. To these students and many others who provided moral support and encouragement, particularly Eleanore Lenington, I am deeply grateful.

Former chair of the Department of Educational Foundations and Policy Studies and Professor Emeritus of Florida State University, Cynthia Wallat, patiently and consistently supported me over the years, understanding that research in a new field is time-consuming, requiring thought and a need to "simmer." I would also like to thank colleague and friend Emanuel Shargel, who has always been an enthusiastic supporter of my work. Many thanks also to staff members Amy McKnight and David Darío Anderson, who provided terrific help in juggling my several responsibilities in the department while finishing the manuscript. Jen Simington provided superior copyediting and skillfully took the book through the final production stages.

I owe a tremendous debt of gratitude to my doctoral advisor and mentor, Joel Perlmann, of Bard College. He opened the world of research to me, particularly the examination of ethnicity, and has provided continuous guidance in my development as a scholar. The pathbreaking work of Guadalupe San Miguel, Jr., and Rubén Donato in the field of Chicano educational history provided a foundation for this work. Their encouragement and support were very helpful. Many years ago, former dean of the Harvard Graduate School of Education, Patricia A. Graham, recommended that I explore the possibility of studying Hispanic education as a research topic. Although I did not follow her advice for many years, I hope she will be pleased with the publication of this book. My thesis advisor in the Department of History at Wellesley College, Jacqueline Jones (now at Brandeis University), provided a model of a scholar who was building a new field in American history while also raising a family. To her and the many women in academia who forged a path for my generation, many thanks.

Archivists and reference librarians across the country patiently handled inquiries, photocopying requests, and visits. In particular, I would like to recognize the librarians and staff at The Cushing Memorial Library, Texas A and M University; the Library of Congress; the Wisconsin Historical Society; Gladys Dratch and Maryléne Altieri at Special Collections, Monroe C. Gutman Library, Harvard Graduate School of Education; El Centro de Estudios Puertorriqueños at Hunter College; and the Rockefeller Archives, Sleepy Hollow, New York.

Family and friends also prodded me and provided support for the completion of this project. My three sons, David Marshall Huntzinger, Michael George Huntzinger, and John MacDonald Huntzinger, have grown up hearing about my book projects. I thank them for their love and comic relief from the rigors of academia. I hope that with this book they will learn something about their Latino heritage. They can also breathe a sigh of relief that it is *finally* done! My brothers, Robert George MacDonald, Leslie Charles MacDonald, and Charles (Carlos) Kenzie MacDonald, have always cheered on their little sister over the years. I am particularly grateful to Charles for necessary support during the final stages of the manuscript. My grandmother, María del Carmen de Contreras Posada (1900–2003), gave me the gifts of the Spanish language, courage, a love for life, and a feisty spirit. She passed away during the writing of this book, but I know that she is always with us. To my loving parents, George Ewen MacDonald and Leonor Posada MacDonald, I dedicate this book.

Introduction

The role of a socially committed historian is to use history, not so much to document the past as to restore to the dehistoricized a sense of identity and possibility. Such "medicinal" histories seek to re-establish the connections between peoples and their histories, to reveal the mechanisms of power, the steps by which their current condition of oppression was achieved, through a series of decisions made by real people to dispossess them; but also to reveal the multiplicity, creativity and persistence of resistance among the oppressed.

—*Aurora Levins Morales*, The Historian as Curandera[1]

When my oldest brother arrived home from kindergarten in the 1950s, he excitedly reported, "Mommy, Mommy, everyone speaks like Daddy!" With his entrance into the New York public schools he was making his debut into one of American society's most formative cultural institutions, its schools. At his new school he was expected to speak a language different from his own language of birth and his mothers' native tongue, in addition to learning a culture and history distinct from that of his father. But, after all, one of the main purposes of public schools in the United States has been to Americanize the diverse immigrant peoples that have arrived on its shores voluntarily or who have become members through colonization or annexation.

Born in 1960, I came of age alongside the dramatic growth of the Latino population in the late-twentieth-century United States, but had little understanding of the long and rich history of Hispanic peoples in North America. My Colombian-born mother critiqued the Anglocentric history presented in our schools. All I knew growing up was that my three brothers and I were usually the only Spanish-speaking children in the public schools of East Williston, Long Island, and Potomac, Maryland. How and why Latino peoples became a part of the United States was not mentioned in the curriculum. The history of Latinos in the United States was awaiting a new generation of scholars. The young girl with dark eyes, brown skin, and brown hair who was often excluded from the social circles of "American" girls with fair skin and blonde hair used the schools to become one of those scholars. *Latino Education in the United States* is thus both an academic work designed to fill a large gap in our knowledge and a personal labor of love to integrate the lived experience of Latinos' educational experiences with the American historical record.

WHO ARE LATINOS?

The shifting terminology utilized for peoples of Hispanic descent in the United States has caused conflict and confusion.[2] (See documents 8.6a and 8.6b.) The enormous diversity of Latino peoples, ranging from those who have deep roots in New Mexico dating from the late 1500s to recent arrivals from Central America, traverse social class, generation, nationalist identity, and Spanish language variation. In this volume, the terms "Latino" and "Hispanic" are used interchangeably to denote those individuals born in the United States and in Latin America, Central America, the Spanish-speaking Caribbean, and Spain who are of Hispanic heritage, inclusive of Native American and African ancestry. "Latino" and "Hispanic" are thus umbrella terms that attempt to synthesize and clarify the historical experiences of an extraordinarily diverse group of peoples.

Latino Education in the United States intersects three fields of scholarship: American history, education, and Latino studies. In each of these fields, Latino educational history has been barely explored or superficially examined. As explained in a historiographic essay, Latino educational history is like a rich, unearthed archeological site awaiting the work of researchers.[3] In American history, the study of Latino peoples began slowly in the early twentieth century but did not flourish until after the Chicano Civil Rights era of the 1960s and 1970s. Among educationists, scholars focus attention on the impact of linguistic and government policies on Latino children but offer only a handful of monographs that are strictly historical studies. Furthermore, the available histories are localized and do not offer a synthesis of the diverse Latino experiences. The growing field of Chicano and Latino studies has always identified education as a central theme in the Latino experience but awaits further research integrating schools with the broader sociohistorical context of American history.[4]

MAJOR THEMES IN THE DOCUMENTS

Historian Peter Novick has described the initial steps in the evolution of a new field as the "brick and mortar" phase.[5] In this stage, primary sources are identified, collected, and organized. This collection of documents thus brings to light the Latino educational experience through a diversity of sources. In some cases, traditional government documents are utilized to provide the reader with an understanding of the stated policies of Spain, Mexico, and the United States toward education. First person narratives—from diaries, letters, and even fiction—balance the records of local, state, and federal agencies. In many cases, stated policies were ignored at the time, and social historians have utilized personal narratives, census records, and birth and death records to construct a more adequate portrait of the historical experience. For example, within the Spanish missions, institutions created to teach and Christianize Native Americans, an extraordinarily high death rate nearly eradicated numerous tribes. However, letters from priests in the missions to their superiors often painted a different, and far more optimistic, portrait of mission conditions (see document 2.2).

The several educational equity legal cases contained in this volume are viewed as key to portraying the Latino experience as one not solely of oppression but of creative agency. The lengths to which communities must go in order to marshal the resources necessary for a legal case (lawyers fees, petitions, plaintiffs, etc.) reveal the value of education to Latinos. As the year 2004 celebrates the fiftieth anniversary of the landmark Supreme Court school desegregation case *Brown v. Board of Education*, scholars have also begun to examine and compare school desegregation in Hispanic communities. In this volume, the critical and lesser known school desegregation cases *Mendez v. Westminster (CA) School District* (1946) and *Delgado v. Bastrop (TX), Independent School District* (1948) are included as examples of pre-*Brown* school desegregation efforts among Mexican Americans in the Southwest. Although more accessible to the public, more recent U.S. Supreme Court cases such as *Lau v. Nichols* (1974) and *Plyler v. Doe* (1982) are included because of their profound influence in shaping the educational opportunities of late-twentieth-century Hispanic immigrants.

Photographs are also utilized in this collection because of their ability to put the reader in direct contact with the material. Even though the intended purpose of the original photographs and their selection here reflect biases, photographs can nonetheless permit the reader to personally analyze situations. For example, the expressions of Cuban refugee children caught in the political limbo of resettlement camps are difficult to capture in words (see document 6.5). In addition, photographs such as those of Anglo children sitting side by side with Mexican migrant workers in Wisconsin public schools in the early 1960s (document 5.12) retrieve from obscurity rarely known experiences.

The arrangement of these documents is primarily chronological, but occasionally overlap, reflecting the varied experiences of Latinos in the Southwest, Midwest, East Coast, and U.S. territories. The first chapter, "The Colonial Era: Schooling under Spanish Rule, 1513–1821," covers three hundred years in a succinct manner. The boundaries of the United States had not yet extended, even through colonies, to Santa Fé, New Mexico, or even Texas at the time of this Spanish colonial era. However, U.S. colonies are included because they illuminate the deep roots of the Hispanic American legacy. Furthermore, the documents in the first three chapters reveal the depth and primacy of the close link between Latinos, Catholicism, and education. Subsequent chapters proceed chronologically through the nineteenth and twentieth centuries, but readers will note overlaps, such as the conquest of Puerto Rico (1898) while segregated "Mexican" schools were simultaneously being created in the Southwest. The turbulent civil rights era of the 1950s and 1960s originally received attention from historians as the fight for equity among the African American community. In this large section of the book, the struggles of Mexican American and Puerto Ricans to also redress inequities in the search for educational opportunity is brought to light. During the post-1965 era the Latino community expanded dramatically both numerically and through the inclusion of Caribbean, Central, and Latin American peoples. The school experiences of these post–Civil Rights era children have occurred against the backdrop of significant political debates. The status and rights of English Language Learners, political refugees, legal and illegal immigrants shape their future prospects, as illuminated in the documents in the latter portion of this volume.

In most cases, documents were selected from various archives across the country because they reveal an aspect of Latino history that is difficult for the average student or researcher to access. In many cases, students in my undergraduate and graduate courses at Florida State University previously utilized these documents fruitfully for analysis and reflection on the Latino schooling experience. The selection of documents was also made to purposefully expand the prevailing narrative of the history of education. During the last few decades the history of schooling was enriched through the addition of the experiences of women, African Americans, and Native Americans. The primary documents in this collection are intended to open a window into the Hispanic American encounter with both private and public schools and institutions of higher education. Researchers will also benefit from exposure to the diversity of records—school reports, diaries, legislative acts, photographs, interviews, and public hearings—that can be utilized to construct a history.

As a preliminary work, future scholars will undoubtedly fill in the gaps in this work. More material on the Midwest, migrant workers, and the new immigration pathways to the American South will help balance the current selection. Voices from the first generations of Latina teachers are also difficult to find and warrant future inclusion. Furthermore, this volume devotes considerable attention to the twentieth-century experience, particularly the decades of the 1950s through 1970s. Against the backdrop of the larger social revolutions of this era, the leadership and activism of Puerto Rican and Mexican American youth transformed the political landscape. Federal and state governments, higher education institutions, and foundations became responsive to Latino demands for bilingual education, equitable resources, Spanish-speaking teachers and administrators, and a curriculum that was inclusive of the Latino experience. Although the late twentieth century has witnessed encroachments upon some of the gains of the Civil Rights era, these decades permanently opened doors of opportunity for Latino children and college youth.

Since the liberalizing of immigration laws in 1965, the United States has witnessed the second greatest immigration wave in its history. Latinos are now the largest minority group in the United States and a very young population. However, the current educational achievement of Latino children and youth lags considerably behind that of their white and African American peers. High drop-out rates, exclusion from honors and advanced placement classes, language difficulties, and low participation at the college and university level are among some of the challenges facing the Latino community. *Latino Education in the United States* brings together a collection of documents intended to illuminate the complex historical legacies that have shaped the contemporary Latino educational experience and provide guidance for the future.

CHAPTER ONE

The Colonial Era

Schooling under Spanish Rule, 1513–1821

INTRODUCTION: NEW SPAIN, 1513–1821

I have determined that schools be established where they do not exist as ordered by law and statutes; that the parents be induced, by the gentlest means and without the use of coercion, to send their children to the said schools . . . that the Presidentes and Audiencias look after the election of efficient teachers and the assignment of their salaries according to the population and conditions of the settlements . . . and that they ask the priests to persuade their parishioners with the greatest gentleness and affability, of the advantage and expediency of their children learning Spanish for their better instruction in the Christian doctrines and polite intercourse with all persons.

—*Royal Orders from King Charles III,*
November 5, 1782[1]

Formal and nonformal education in northern New Spain occurred within the context of Spanish exploration, conquest, and settlement. Conquistadores carried out these activities under the name of both the crown and the church. As historian David Weber explained, the explorers believed they could "serve God, Country, and themselves at the same time."[2] The search for gold and other riches was no less a part of Spain's intent as it rose to international power in the fifteenth and sixteenth centuries. Although Spaniards who pushed into the American Southwest were disappointed in their quest for material wealth, they created a permanent imprint upon Native American culture. The Spanish imposed both their language in verbal and written forms and the beginnings of formalized European education. The collision of cultures, languages, and religions over three centuries produced a new people who are the ancestors of today's Southwestern Latinos.

Conquest in the Caribbean islands and Latin America began in the late 1400s and early 1500s. Exploration in the modern-day United States dates back to Juan Ponce de León's arrival in Florida in 1513. Financed at first by the Spanish government, and later with private funds, explorers spanned into the New World. Upon reaching a settlement of Native Americans, the Spanish would issue a *requerimiento*, or notification. The requerimiento ordered natives to accept Spanish rule and Christianity or risk their lands, lives, and liberty.[3]

The Spanish government provided sustained royal financial and military support for the religious conversion of Native Americans, a chief vehicle for the imposition of Spanish language and culture. By 1526 at least two priests were required to accompany any exploration parties to New Spain. Catholic orders, chiefly from the Franciscan and Dominican orders created a network of missions that extended northward up California's Pacific coast, westward through modern-day Texas, and inward to New Mexico and Arizona. A short-lived series of missions were also created in the Southeast, starting with the founding of St. Augustine, Florida in 1565. The missionaries began the intertwined tasks of fulfilling their duty to the church to Christianize the natives and their duty to the Spanish empire to teach them the Spanish language and loyalty to the Crown.

During the long period of Spanish reign, local conditions and Native American rebellion shaped colonial policies. The Council of Indies, located in Spain, was the governing body for all colonial activities. The council issued reforms such as the 1573 Royal Orders for New Discoveries, prohibiting explorers from conquering and physically harming Native Americans. However, the distance from authorities in Spain, opportunities for exploitation, and religious zeal limited the effectiveness of these reforms.[4]

The Spanish conquest had a devastating impact on Native American populations. Infectious European diseases (particularly smallpox), abuse of the *encomendero* (a system in which Natives involuntarily worked Spaniard's lands), enslavement, and forced relocation into missions permanently altered the demographics of the Americas. Close contact between Europeans and Native Americans in the mission compounds often resulted in alarmingly high rates of Native American mortality.[5]

Furthermore, colonial policies dictated the land, civil, and political rights of individuals in New Spain based upon their skin color, race, ethnicity, and national origin.[6] European-born *peninsulares* occupied the highest legal and social status in New Spain. The second tier consisted of *criollos*, individuals born in New Spain of pure Spanish parentage. Legally, Native Americans occupied the third tier. In return for accepting colonial rule, they were accorded land rights and the church became their legal protector.[7] Socially, Indians were below both African slaves and free blacks.[8]

Adding to the complexities, however, of long centuries of settlement, many soldiers and settlers entered into unions (legitimately or illegitimately) with Indian women. The children of these unions were called *mestizos*. Mestizos occupied a nebulous legal and political status in the eyes of the Spanish crown. Depending upon factors such as the status of the father and whether the parents were married in the Church, some mestizos eventually "passed" into society as criollos.[9] In the status-conscious New Spain these legal, social, and political distinctions impacted educational opportunities.

During the 1600s and 1700s, Spanish explorers pushed northward from central Mexico into lands that form the contemporary United States. Both settlers and missionaries carried out formal and nonformal educational activities after the rudiments of living were established. Three discernible forms of education—settlers' schools, missions, and nonformal education—emerged simultaneously, each reflecting the hierarchical nature of Spanish colonial society. Settlers' schools, charged with the preservation of Spanish language, culture, and religion in the New World, represented the first tier. Run by either secular teachers or under the auspices of the missions, this formal education for cultural transfer was initially reserved only for the children of Spanish settlers, civil leaders, and military officers. Over the course of time, these schools also included children born in the New World, many of whom were mestizos.

The second and most prevalent type of formal education occurred in the missions. Missions were generally enclosed compounds run by priests and lay brothers to Christianize and "civilize" the Native Americans, as will be described in detail later in the chapter. The educational function of the missions was reserved exclusively for Native Americans as colonial policy dictated the separation of whites from Indians in the missions.[10] Unlike the social purpose of cultural *transfer* in settlers' schools, mission education was purposefully designed to *replace* Native American languages, religions, dress, and other cultural attributes with the Spanish language, Roman Catholic faith, and European mores and customs. Although some priests learned native languages, the missions' role in the deculturalization of Native Americans was extensive.[11]

Nonformal education occurred throughout New Spain and represents the third form of learning evident in the colonial era. Similar to any frontier situation, settlers brought books, pamphlets, and other means of noninstitutional learning to share with their families and fellow settlers. The copious letters between officials in Spain and New Spain preserved in the royal archives reveal a high level of literacy among the officers and elite explorers. Although the Spanish Inquisition attempted to censor books considered heretical, target works such as Miguel de Cervantes's 1605 *Don Quixote* were favorites listed in the registers of ships headed for the New World.[12] The type of nonformal education seventeenth-century nun Sor Juana Inés de la Cruz pursued—studying "not only without a teacher but also without fellow students with whom to compare notes" (document 1.2)—occurred throughout New Spain but of course is less documented due to the noninstitutionalized nature of those experiences.

SETTLERS' SCHOOLS IN THE COLONIES

Colonial and diocesan policy required the residential separation of Native Americans and Spanish populations, particularly during the first centuries of exploration. In some cases, children of soldiers, civil officers, and settlers were taught in the missions, but they received instruction separately from Native American children. Schooling, although prized among the Spanish elite, was often subsumed as a priority to finding and acquiring the basic necessities of living. Frequent attacks from hostile natives, illness, lack of arable land, and other factors were serious obstacles in the formation of permanent settlements where schools were desired. Despite these deterrents, the historical record reveals continuous attempts to create schools.

The curriculum of the settlers' schools, a combination of classical learning and the Roman Catholic catechism, reflected the Spanish culture's close association of education with religion. In this sense the history of Latino education in the Southwest and Southeast is similar to that of subjects of the British colonies. The 1647 Old Deluder Satan Law, one of the earliest laws for schools in Massachusetts, was titled such because of its intent: "It being one chief project of that old deluder, Satan, to keep men from the knowledge of the Scriptures, as in former times by keeping them in an unknown tongue . . . it is therefore ordered, that every township in this jurisdiction, after the Lord hath increased them to the number of fifty householders, shall forthwith appoint one within their town to teach all such children as shall resort to him to write and read."[13] Schooling in the colonial era, whether under British or Spanish rule, had as its central purpose knowledge of the Scripture.

In the Spanish colonies, exhortations and governmental decrees to regulate education far exceeded schools' compliance, similar to New England.[14] As early as 1778 the king of Spain issued decrees for the establishment of schools to teach Spanish. The orders were repeated again in 1782, 1793, and 1815.[15] Localized conditions in the colonies shaped the timing of individualized responses to these mandates. Settlers' schools in Spanish Florida, for example, began in the 1500s and ended in the 1700s. New Mexico settlement began as early as 1598 in Santa Fé, but Native American rebellion stalled permanent schools until after 1800. Permanently settled later than New Mexico, schools in Texas and California generally did not appear until the 1700s.

Settlers' Schools in La Florida

Claimed for King Philip II of Spain by *adelantado* (self-financed explorer and representative of the king) Pedro Menéndez de Avilés in 1565, St. Augustine is the oldest city of European origin in modern-day United States. Settlers from Spain began arriving to St. Augustine in the late 1500s. Although they had arrived to create missions for the Indians, the Franciscan priests extended their skills to children of the settlers. As early as 1606 the Franciscans established a classical school and preparatory seminary in St. Augustine.[16] In 1634 the Royal Crown granted a *cedula* for the opening of another school in St. Augustine. The duration of both of these schools is unrecorded but, like many frontier institutions, were most likely short-lived.[17]

Although the foundation of settlers' schools was recorded in St. Augustine in 1736 and again in 1785, by Franciscans, the prospects for permanent Spanish settlement in Florida was disappearing by the 1780s.[18] Aggressive British invaders spurred Indian rebellions and the massacre of friars, women, and children. A lack of riches—such as the gold and silver found in Mexico—curbed enthusiasm for Florida as a New Spain colony. One symbol of Spain's decreasing influence in Florida was the Crown's licensure of an English language school in St. Augustine in 1805. Previously, only Spanish language schools were permitted in the colonies.[19] Florida eventually fell under British and French rule before restoration to the United States in the Adams-Onis Treaty of 1819.[20]

Settlers' Schools in New Mexico

New Mexico underwent two eras of conquest. In 1598 adelantado Don Juan de Oñate settled Santa Fé, New Mexico (see document 1.1.). Native Americans resisted the exploitative practices of Spanish settlers, particularly the encomendero system and the denigration of native religions. Native Americans formed an alliance and staged the Pueblo Revolt of 1680. An estimated four hundred out of three thousand Spaniards in New Mexico were killed. The revolt included the destruction of churches and schools, and the slaughter of 21 of the 33 Franciscan priests. It was one of the deadliest insurrections against Europeans in North American history.[21] Reconquest or *la reconquista* of the far northern New Spain colonies did not begin again for over a decade in this far and remote northern frontier.

Recorded interest in schooling did not appear again in New Mexico until the early 1700s. In 1717 and 1721 residents of Santa Fé discussed the creation of schools, but no concrete results followed.[22] Bernardo P. Gallegos has found in the correspondence and wills of eighteenth-century New Mexicans references to private tutors receiving payments for teaching youth. For instance, New Mexican Francisca de Misquia indicated in her will that she owed Antonio Durán "12 pesos for the instruction of my children." Beyond private tutors, little evidence of established schools in the early 1700s has surfaced.[23]

Early-nineteenth-century New Mexico witnessed the beginning of formalized schools. In 1805 the colonial government sent two teachers from Mexico City to Santa Fé to start a trade school in the craft of weaving.[24] Between 1805 and 1821 several schools were opened not only in the larger settlement of Santa Fé but also in Albuquerque, Taos, Belen, San Miguel, and Santa Cruz. According to Gallegos, schools were created primarily for the children of soldiers and officers of presidios but were also attended by children from the surrounding areas. These children were called *vecinos*, or "neighbors." The curriculum at the Santa Fé Presidio school consisted of the Catholic doctrine, writing, reading, and counting. Average monthly attendance at this school varied from as low as forty to as high as 201 pupils between 1808 and 1820.[25]

Resources for teacher salaries varied by locality but were limited everywhere. Teacher salaries in New Mexico were comparable to a soldier's pay. A Spanish soldier in the early 1800s received 450 pesos per year, although some of that amount was paid through goods sold to soldiers at highly inflated prices. In 1812 New Mexico a teacher in Santa Fé received five hundred pesos per year and teachers in Albuquerque and Santa Cruz were paid three hundred pesos per year.[26]

Despite exhortations from women intellectuals such as Sor Juana Inés de la Cruz to allow women to teach, little evidence of women teachers in New Spain has been found. Within the religious convents, women taught each other in addition to receiving instruction from priests. The equivalent of New England colonial "dame schools," schools taught by widows or older women with children in New Spain has been referred to as *amigas* (friends) schools (see document 1.2).[27] Far more research is needed to construct an adequate portrait of the New Spain colonial teacher.

Children in New Mexico or the other far northern colonies who wanted higher education, and had the means available, were sent to either Mexico City, or even Europe, for further instruction.[28] Sixteenth-century formal higher education in European Spanish society was limited to men of the elite classes and the clergy. Universities had existed in Spain from the Middle Ages. The most famous was Salamanca, founded by the king of León in 1218 and granted privileges to confer degrees by the pope in 1255. In the Americas the Spanish founded universities in Santo Domingo, Hispaniola, in 1538, Lima, Peru, and Mexico City in 1551, and Bogotá, Colombia, in 1580. Furthermore, the Jesuits and Dominicans founded seminaries (called colleges) for their orders in virtually all of the colonies.[29] Although access to higher education required travel and money, families of means relied upon the universities in the interior of New Spain to educate their children beyond the settler's schools.

Settlers' Schools in Texas

Educational conditions in one of northern New Spain's largest settlements, San Fernando de Bexar (modern-day San Antonio, Texas) open a window into the difficulties of creating a frontier school. Settlers' schools were first recorded there in 1746 and again in 1789. Petitions from schoolteachers of this late-eighteenth-century era illuminate the frustrations of local educators. For instance, local communities often would not or could not provide teachers a living wage. Teachers such as Don José Francisco de la Mata criticized parents for allowing their children to "[run] about as vagabonds engaged only in pernicious pursuits . . . and other idle entertainment which lead only to perdition" (document 1.4).[30] Several years later, San Antonio's governor, Juan Bautista Elquezabal, issued an 1802 proclamation that town officials would be held responsible for requiring parents to place their children in school or face "severe penalties." In response to this order the town hired Francisco Ruiz as schoolmaster.[31]

In 1812 San Antonio residents donated funds to create a schoolhouse. In 1815 the *cabildo* (town council) requested a teacher to instruct in the "rudiments of our Holy Religion and the Primary Branches (Primeras Letras)."[32] The 1819 contribution toward the teacher's salary was paltry, consisting of 55 pesos, 4 reales, and a fanega (about two bushels) of Indian corn for the entire school term.[33] The irregular curriculum, poor teacher preparation, and low teacher salary evident in this era of San Antonio was not particular to New Spain but also characteristic of rural, nineteenth-century district schools of the northern and midwestern United States.[34]

Settlers' Schools in Alta California

Similar to New Mexico, California experienced two phases of settlement. California was first discovered in 1542 by Juan Rodríguez Cabrillo. Settlement did not begin in Alta California until over one hundred years later. Between 1697 and 1767 missionaries from the Society of Jesus (Jesuits) were granted almost exclusive rights to settle Baja California as long as they relied not on royal funds but on private benefactors.[35] Subsequently, Jesuits created a string of 17 missions in Baja California, from Loreto northward several hundred miles (but still south of modern-day California).[36] During the period of the Jesuits, mention is made of a private school that a Father Ugarte ran for settlers, and a petition to the Crown was made in 1717 for the "founding of a general school," but little other evidence of formal education has surfaced.[37]

In a surprising and politically motivated move, King Charles III expelled Jesuit priests from all of New Spain in 1767. Historians generally conclude that Charles III viewed the increasing political and financial power of Jesuits in particular, and the Roman Catholic Church in general, as a threat.[38] The Jesuits were replaced with the Franciscans, an order viewed as friendlier to the Crown's political aims. The Franciscans moved northward to modern-day California and brought with them soldiers and settlers.

According to historian Hubert Howe Bancroft, literate settlers arriving to California after 1769 often taught their own children. In addition, a woman called an *amiga* (friend) could open her home, where she "instructed not only her own children but those of her neighbors, or even an ambitious soldier who aspired to be a corporal."[39] Governor Diego de Borica (1794–1800) is credited with the spread and enforcement of settlers' schools in Alta California. The small towns of San José and Monterey opened public schools in 1795. In San José retired sergeant Manuel Vargas started the school in a public granary. In 1796 32 pupils were instructed in Santa Barbara under the sailor José Manuel Toca. The town of Santa Barbara was ordered to pay Toca $125 for instruction of the children. Each soldier was required to pay him one dollar for instruction, as literacy was required for promotion.[40]

Schools in San Diego (1796) and San Francisco (1797) served the small but growing population of mestizos who settled Alta California.[41] Corporal Manuel Boronda, a carpenter, offered to teach the San Francisco school free of charge. One key to the success of the California public schools was Governor Borica's system of accountability. He ordered, and received, copies of reports and students' copybooks every few months.[42] Settled relatively late in the Spanish colonial era, California's settlers' schools reflected the more secular aims of King Charles III's Bourbon reign. Instead of missionary priests extending classes to settlers, teachers were more often lay people from the military.

Summary of Settlers' Schools

During the last decades of Spanish rule, New Mexico, California, and Texas began the creation of formal schools that persisted unevenly through the Mexican Era (1821–1848 for New Mexico and California; 1821–1836 for Texas).[43] Literacy rates remained steady despite the lack of formal schools. For example, one-third of male settlers and soldiers between 1692 and 1821 in colonial New Mexico could sign their names, a strong indicator of literacy.[44] The schooling available to children of settlers and civil and military leaders served the colonial era's rudimentary need for literacy to communicate with Spanish officials, conduct trade, and, in the case of priests, record baptisms, marriages, and deaths. Spanish settlers' schools thus represented a continuation and affirmation of the Spanish religious and linguistic heritage.

MISSION SCHOOLING AND CULTURE

> *I compiled a large catechism . . . in the language of these Natives. By dint of effective patience, we succeeded in having nearly all the men learn by heart the large catechism, and the very aged the little one. With the women, because after all they are women, teaching the catechism did not proceed so well. Those who already know the said number of prayers in their own language learn it in perfect Castilian, which costs them hard labor, because they are even greater blunderers than I.*

> —*Report of Father Marian Payeras, Mission Purísima Concepción, Alta California, January 13, 1810*[45]

The most systematic and well-documented form of education available in northern New Spain occurred within the vast network of Catholic missions. It was within the missions that Native Americans experienced direct alteration of their culture, politics, economy, and demography. Once viewed by historians as paternalistic communities, the missions have been examined in recent years from the point of view of Native Americans, using new techniques in ethnohistory, social history, and anthropology.[46] Native Americans who spent time in the missions were often cast out of their former tribes and as a result remained in the mission. The missions' function of deculturalization powerfully shaped the Euro-Indian relationship. Native Americans attempted to preserve elements of their culture even within the constraints of highly circumscribed mission life.[47] However, the combination of military enforcement from soldiers armed with superior firearms and cavalry and the weakened state of tribes from disease and malnutrition placed Native Americans at a severe disadvantage to fight the Spanish conquest.

The missions' key role in forging and pacifying the northern frontier received official backing from the Crown. The state financed travel expenses, salary, and supplies for the priests and lay brothers. The missions were originally financed for ten years—the time considered sufficient to convert and "civilize" a Native American community. After that period Native Americans were supposed to be allowed self-governance and granted small land parcels. In practice, most missions lasted longer than ten years, particularly in the seventeenth and eighteenth centuries.

Missions could be extensive in their size and land holdings. A comprehensive mission included a large compound of dwelling houses, workshops, a chapel, storehouses, classrooms, a kitchen, and servants' (and sometimes slaves') quarters. Walls with gates enclosed many of the missions and were erected to protect residents from hostile Native Americans and to monitor the traffic of Indians entering and exiting the mission. Beyond the compound were cultivated fields, ranches for livestock, and orchards.

In order to fulfill their royal charge of converting and civilizing the Native Americans, missionaries employed various strategies. In some cases, Florida and New Mexico in particular, missions were built where Indians already had settlements. This strategy then allowed priests to entice Indians with gifts, food, and a show of elaborate church decorations and vestments.[48] The level of coercion used to keep Indians as residents of the mission varied. The most zealous priests exacted severe punishments of confinement and whippings on those who escaped. Others relied on paternalistic goodwill and moral coercion to keep their charges within the mission. The source of many uprisings against Spaniards stemmed, in part, from the often overzealous desire of friars to eradicate all forms of Native American religions, including the role of the powerful medicine men as healers.[49]

Once within the confines of the mission, the Catholic orders, Franciscan, Dominican, or Jesuit, started and regulated the day with the call of the bells, a concept novel to Native Americans, who had previously lived free from rigid schedules. At a typical mission, the bells first rang at dawn; adults and children proceeded to the chapel for mass and catechism. The mass was in Latin and the catechism was typically in Spanish, with interpreters for non-Spanish speakers when necessary. Afterwards, the new converts, called "neophytes," breakfasted and began their tasks. Adult men were generally sent to the fields or to train and work in the blacksmithing, carpentry, or brick-making shops. Soldiers were often sent to supervise the agricultural laborers in the fields. Women worked at the looms, made pottery and baskets, and prepared meals for the fieldworkers. Children, particularly the boys, attended school and performed chores.

In the evening, the entire community was brought together again to pray the rosary, recite other prayers, and receive more religious instruction. In some missions, residents were allowed time each day or week to leave the compound and hunt, fish, or visit relatives. In more restrictive situations (especially for Native Americans younger than 18 and unmarried), residents were locked in each evening.[50]

The educational goals of the missions were clearly articulated by the Crown. The actual practices that occurred were negotiated activities between Native Americans and missionaries. For instance, each missionary was required to learn the language of the particular tribe among which he labored. The laws even required that the friar be examined in that language before he was allowed to proceed to the particular tribe or nation.[51] In practice, the enthusiasm for learning native languages was greatest during the 1500s and 1600s.

Records from missionaries in Florida and Georgia, for example, discuss the work of Dominic Augustine Baez who, in 1568, prepared a grammar book and catechism in the Indian's native language.[52] According to historian David Weber, by 1572 Franciscans had produced "at least eighty literary works in native languages—catechisms, grammars, and dictionaries (usually copied by hand rather than printed)."[53] Francisco de Pareja prepared a bilingual catechism and confessional in Spanish and Timucuan (a language largely used by Native Americans in Florida). Pareja's bilingual text was first published in 1612 in Mexico City and reprinted several times during the colonial era.[54] The difficulty of mastering so many different languages, and the more widespread use of Spanish after initial conquest and settlement, contributed to less emphasis on learning Native American languages toward the end of the colonial era.

The teaching of the Spanish language was a key objective of the Crown. The laws of the Council of Indies stated: "We beg and urge the archbishops and bishops to instruct and order the priests and missionaries in charge of Natives to incline and direct the Indians, through the gentlest and kindest means, to the study of the Spanish language in order that they may be taught our Christian doctrine in this tongue."[55] The teaching of Spanish was thus closely linked to evangelization. Missions relied on catechisms such as Fray Alonso de Molinas's 1546 *Doctrina Christana*. Rote memorization, the recitation of prayers, and basic reading and writing was carried out to varying degrees.[56]

The missionaries focused their attention on training boys as future evangelists and leaders. Girls' education was not entirely neglected, but it reflected European gendered roles of the colonial era. For instance, in the California missions of the 1790s, boys were educated in the industrial arts such as tanning, carpentry, and blacksmithing. For academics, they were taught to "read, write, and sing by note." The girls "learned to sew, spin, weave, and do household work."[57] Older girls were kept protected by a matron, often the wife of a soldier. In the eighteenth-century diary of a Franciscan missionary, he observed that the matron "by day kept them [the girls] with her, teaching them to sew, and other things, and at night locked them up in a room, where she kept them safe . . . and for this they were called the nuns."[58] As illustrated in the document of Sor Juana Inés de la Cruz, found at the end of this chapter, girls who sought higher education found convent life one of the few options available. Although it appears that some Native American girls were sent to lower Mexico to become trained as nuns, the historical records of explorers and missionaries focus almost exclusively on the educability of men; this deserves further exploration.

The selection of male youth to receive advanced instruction in the Catholic faith and Spanish language emerged as a key strategy for the missionaries. Discovering that adults were far more resistant to cultural change, the priests targeted their efforts toward future leaders, the young boys. In Ramón Gutiérrez's analysis of New Mexico missions, he viewed the priests as "engaging in a war for the hearts and minds of Indian children."[59] In the missions, the young men were kept separate from their parents from the age of seven until marriage.[60] The result was to "turn them against their living and past culture . . . [and] make them outsiders to the world they now [had] to convert."[61]

The utilization of education and Catholic indoctrination for the younger generation disrupted traditional patterns of generational authority. For instance, as early as 1523, missionary Peter of Ghent asked Hernán Cortés for permission to separate sons of Indian elites from their families for educational and assimilation purposes.[62] Historian Bernard P. Gallegos found in New Mexico that these children were called *doctrinarios* and sometimes included girls. The doctrinarios became cultural brokers or "agents of literacy" between the Spanish and Indians.[63] An additional class of children, *genizaros* (orphaned Indians residing in Spanish households as slaves) also crossed cultures. Owners were required to educate the genizaros, thus spreading the Spanish language and culture back into the Native pueblos when the slaves were freed at maturity.[64]

Toward the end of Spain's long reign (1513–1821) King Charles III ushered in liberal reforms that improved the legal and civil status of nonwhites and increased calls for secularization of the missions. Several political and economic forces—including increased competition from Russia, England, and France for New Spain lands; rebellion among the criollos in the colonies; and a strained royal treasury—spurred these reforms.[65] The missions had always been intended as transitional enterprises in the exploratory era, not permanent entities. In the early 1800s, the Crown called for the missions to secularize—to divide the lands among the Native Americans and allow secular clergy (brothers and priests who did not belong to a specific order) to take charge of the parish—but the Church resisted these measures.[66] Furthermore, the vast holdings of the Franciscan priests along the coast of California had aroused envy and greed among newer settlers. In most places, it was the independence of Mexico from Spain that finally forced secularization.

Missions played a key role in the frontier development of New Spain. As an educational institution, they were extremely organized and had highly educated teachers. The curriculum brought to Native Americans was a conservative force. Its role was to transfer and replace native languages with the language of the colonizing Spanish power. For the first generation of children who learned to read and write Spanish (often called *ladinos*), these skills could become assets in a world that was permanently transformed economically, politically, and demographically. However, few opportunities for independent living were permitted prior to independence from Spain. In some cases, educated Indians formed rebellions and became leaders of warring tribes.[67] At the end of the colonial era, literacy skills were more practically useful when the missions were secularized and Native Americans were granted limited land rights and governance in the new pueblos. The intent of the missions, however, had never been to raise a generation of youth ready to challenge the status quo.

NONFORMAL EDUCATION

Colonial communities such as those in New Spain, often isolated geographically and with few provisions, relied chiefly upon the family and Church for education.[68] Literacy increased in the western world during the 1400s and 1500s. The invention of the printing press, the Protestant Reformation, which encouraged direct study of the Bible, and increased trade, requiring both numeracy and literacy, were all contributing factors. Ballads, broadsides, pamphlets, and other materials were read aloud at street corners or in homes to family members.[69] In New Spain, literate settlers provided education for family members and fellow colonists and their children when schools were unavailable. Furthermore, those who were already educated brought with them books to maintain Spain's cultural attachment to the arts, and for teaching purposes. Expensive to reproduce, books were also symbols of wealth, signifying elite status.

Hundreds of books were transported across the Atlantic in the trunks of the elite, learned clergy, and faculty of the new universities.[70] The availability of centers of learning and the publication of numerous scholarly works on the New World in the interior colonies contributed to a cultured environment in the growing cities. According to historian Harold Perkin, "The last century of Spanish rule, with cities larger and more splendid than any in contemporary North America, was the golden age of Latin American higher education, with chairs of Indian languages as well as arts, philosophy, theology, and medicine."[71] This stratified focus on higher learning spread throughout the colonies as settlers brought with them poetry, literature, and religious works.

The Spanish Inquisition regulated this type of nonformal education in both Spain and New Spain. Implemented in 1478 by King Ferdinand V and Queen Isabella, the Inquisition was a branch of government designed to stamp out heretics. As a result of the Inquisition, all written materials leaving for the colonies were inspected for evidence of heresy and confiscated if not approved. One benefit of the Inquisition for modern-day scholars is that port inspectors were required to keep lists of acceptable and nonacceptable books. Records show that the *Life of St. Francis* was approved but the *Life of Julius Caesar* was not and would be confiscated.[72] One study of books brought to New Spain revealed fiction as the most popular genre, followed by "chivalries and histories." Not surprisingly, most of the titles were religious in nature. The *Historia pontifical y católica* was often exported, followed by the lives of saints. Histories of the New World were also favored in both Spain and the colonies. *Conquista del Perú* and Ribadeneira's *Descubrimiento de las Indias* (Discovery of the Indies) were frequently found in the homes and small libraries of the elite as well as in the missions. Collections of poetry and ballads were also quite common. Lope de Vega, Juan de Mena, and Montemayor were authors whose works found their way across the ocean. Classic literature listed in the ship's registers included the *Odyssey* and the *Illiad*.

CONCLUSIONS

During the late colonial era, Latin American–born individuals (criollos) grew increasingly dissatisfied with Spain's rigid and hierarchical policies. Specifically, Spanish-born citizens (peninsulares) were routinely placed in privileged church and state positions, marginalizing those born and educated in Latin America. The Spanish government, responding to both a French invasion on the European continent and discontent from its colonies, began a series of reforms in the 1800s.

Criollos began to gather secretly and speak of independence in the late 1700s and early 1800s. One famous figure of the pre-revolutionary period was Catholic priest Miguel Hidalgo y Costilla. In 1810 he delivered his famous "Grito de Dolores" (cry of Dolores), demanding independence for the colonies.[73] In response to the revolutionary spirit brewing in the colonies, the liberal faction of the Spanish Cortes (parliament) inaugurated a new era in Spanish political history. The Cortes created a liberal constitution in 1812, requiring the king to become more responsive to the parliament; granting citizenship to all Spanish subjects, including Indians; and abolishing the Inquisition.[74] These radical reforms were threatened when Ferdinand VII regained the throne in 1814 and put a stop to these liberal measures. Eventually, the Spanish military revolted and restored the 1812 constitution. Despite these rapid political upheavals, the egalitarian principles of the constitution reached the far-northern Mexican colonies in the early 1800s and, according to historian David Weber, the effects "continued to be felt long after independence."[75] The official independence of Mexico from Spain in 1821 inaugurated a tumultuous 25-year period of political, economic, and social change, with more rhetorical support for schools than actual provisions.

DOCUMENT 1.1

Memorial for the Viceroy about the Missions in New Mexico, 1595

In the following document, you will see evidence of the Crown's official protective stance toward Indian service. The Crown took measures to limit abuse and exploitation of Indians in order to dissuade rebellion. Note the use of an Indian woman as translator and cultural broker, in addition to the orphan boys, who will be indoctrinated into Christianity and taught the Spanish language; assimilation was a key strategy in the civilization and conquest of New Spain.

From George P. Hammond and Agapito Rey, eds. *Don Juan de Oñate: Colonizer of New Mexico, 1595–1628*. Volume 5 of Coronado Cuarto Centennial Publications, 1540–1940 (Albuquerque: University of New Mexico Press, 1953), pp.77–80. Original preserved in the Archives of the Indies, Seville, Spain. Italicized words represent the viceroy's actions in response to Oñate's proposed modification of his contract.

It is requested that his lordship increase to twelve the number of friars who are to go, so that there may be enough to distribute throughout the land that has been won and conquered, for otherwise the natives recently converted and baptized would lack ministers to maintain them in Christianity, for some friars may become sick and die on the road or after reaching there.

Let the father commissary general consult with me on this matter.

In the capitulations [contract] *five priests and one lay brother are granted. Let what is prescribed in article 26 of the ordinances be observed.*

It is asked that his lordship issue a cedula or royal order that neither the governor nor his royal officials may impede the erection of convents, hospitals, churches, or chapels wherever and in what manner might seem proper to the friars, as ordered by their prelate, but that they shall render all possible aid and support in this cause.

Let the governor, in the name of his majesty, and the prelate, in the name of the friars, jointly proceed to give an account to the viceroy and, meanwhile, carry out the plans on which they are in agreement.

This is not discussed in the capitulations.

It is asked that his lordship order the governor to issue instructions that the Indians who are scattered may be gathered into towns and suitable places where it may be possible to preach and administer the sacraments to them with greater facility, and that this be done with the approval of the friars.

Although it seems that this point may be very useful, it is more appropriate for the future when the land shall be more explored and subdued. Let the governor and the prelate be mindful from the start of what may be done in this matter and report to the viceroy.

This is not dealt with in the capitulations.

It is asked that his lordship instruct the governor and his royal officials not to determine or order, of their own accord, the places where churches may conveniently be located and built for the teaching of the faith, but, rather, in consultation with the prelate and the friars, in order that, with fuller counsel, they may be better located.

The same answer as to paragraph two; and in conformity with it, the governor shall not be required to establish churches without the approval of the prelate.

Not in the capitulations . . .

It is asked that his lordship order that there be moderation in imposing personal services on the Indians, and, if possible, that none be required of them now in the beginning in order not to exasperate and disturb them or to arouse some hatred against the Christians, as the natives are not accustomed to such forced services.

Let the governor be strongly urged to observe this moderation and avoid it [personal service] entirely at present, if possible. In the instructions to Don Juan [de Oñate], he is forbidden to allot any Indians for work at all, particularly in the mines.

It is asked that in the beginning the Indians in the pueblos shall not be counted [taxed] with the exactness that is being employed in New Spain.

If the levying of tribute should be necessary, it must be as light as possible.

Not in the capitulations . . .

It is asked that his lordship allow the marriage of an Indian woman of New Mexico who is in a convent of those in seclusion so that she may go with the army as the first light and torch of the language, for since she came from that land as a child and may therefore have forgotten her mother tongue, she should be able to learn it again with ease when she finds herself among her own people. But, unless she first marries some honorable man, his lordship should not permit her to go at all.

Examine well to learn if this Indian has forgotten the language, and thus, as she would be of no value, she need not go, but if she must go, let the governor see to it that she is married.

The terms of the contract stipulate that this Indian woman shall be given to him if she is alive.

It is asked that his lordship order that letters of recommendation be sent to the governors and alcaldes of some of the chief pueblos of New Spain, requesting them to find some orphan boys and to give them to us, because they will learn the language of the people of New Mexico more easily and thus make the preaching more effective.

Let letters be given to the alcaldes mayores, corregidores, and chiefs of five or six Indian pueblos, asking this of the chieftains and charging their magistrates to aid the friars in getting them to agree to do it kindly, of their own will, and without causing any offense to anyone.

DOCUMENT 1.2

Sor Juana Inés de la Cruz, Reply to Sor Philothea, 1691

Sor Juana Inés de la Cruz (1651–1695), a Jeronymite nun, is one of colonial Spain's most gifted poets and writers. Although she did not live in northern New Spain, but near Mexico City, her autobiographical comments are included here because they illuminate both the form of schooling available to young women in colonial Spain and the difficulties faced by young women who desired more schooling. Sor Juana advocated that women become teachers for the benefit of girls. This particular passage is an excerpt from a longer response Sor Juana wrote when she was challenged by a male cleric to defend her secular tendencies in her writings. The reference to a "most illustrious Lady" is fictional, as the challenger was the Bishop of Puebla, Manuel Fernández de Santa Cruz.

Reprinted by permission of the publisher from *A SOR JUANA ANTHOLOGY*, translated by Alan S. Trueblood, pp. 205, 210–213, 216–217, 232–233, 243, Cambridge, Mass.: Harvard University Press, Copyright © 1988 by the President and Fellows of Harvard College.

My most illustrious Lady:

. . . To go on with the account of this strong bent of mine, about which I want you to be fully informed, let me say that when I was not yet three, my mother sent a sister of mine, older than I, to learn to read in one of those establishments called Amigas [girls' elementary schools], at which point affection and mischievousness on my part led me to follow her. Seeing that she was being given lessons, I became so inflamed with the desire to learn to read, that I tricked the mistress—or so I thought—by telling her that my mother had directed her to give me lessons. This was not believable and she did not believe me, but falling in with my little trick, she did give me lessons. I continued attending and she went on teaching me, no longer as a joke, since the event opened her eyes. I learned to read in so short a time that I already knew how when my mother found out, for the mistress kept it from her in order to give her a pleasant surprise and receive her recompense all at one time. I kept still, since I thought I would be whipped for having acted on my own initiative. The person who taught me is still alive (may God preserve her) and can attest to this.

I remember that at this period, though I loved to eat, as children do at that age, I refrained from eating cheese, because someone had told me it made you stupid, and my urge to learn was stronger than my wish to eat, powerful as this is in children. Afterward, when I was six or seven and already knew how to read and write, along with all the sewing skills and needlework that women learn, I discovered that in the City of Mexico there was a university with schools where the different branches of learning could be studied, and as soon as I learned this I began to deluge my mother with urgent and insistent pleas to change my manner of dress and send me to stay with relatives in the City of Mexico so that I might study and take courses at the university. She refused, and rightly so; nevertheless, I found a way to read many different books my grandfather owned, notwithstanding the punishments and reproofs this entailed, so that when I went to the City of Mexico people were astonished, not so much at my intelligence as at the memory and store of knowledge I had at an age at which it would seem I had scarcely had time to learn to speak.

I began to study Latin, in which I do not believe I had twenty lessons in all, and I was so intensely studious that despite the natural concern of women—especially in the flower of their youth—with dressing their hair, I used to cut four or five fingers' width from mine, keeping track of how far it had formerly reached, and making it my rule that if by the time it grew back to that point, I did not know such-and-such a thing which I had set out to learn as it grew, I would cut it again as a penalty for my dullness . . . I became a nun because, although I knew that that way of life involved much that was repellent to my nature—I refer to its incidental, not its central aspects—nevertheless, given my total disinclination to marriage, it was the least unreasonable and most becoming choice I could make to assure my ardently desired salvation. To which first consideration, as most important, all the other small frivolities of my nature yielded and gave way, such as my wish to live alone, to have no fixed occupation which might curtail my freedom to study, nor the noise of a community to interfere with the tranquil stillness of my books . . .

I also felt that being a nun and not a lay person, I should, because of my ecclesiastical status, make a profession of letters . . .

What I might point out in self-justification is how severe a hardship it is to work not only without a teacher but also without fellow students with whom to compare notes and try out what has been studied. Instead I have had nothing but a mute book as teacher, an unfeeling inkwell as fellow student, and, in place of explanation and exercises, many hindrances, arising not only from my religious duties (it goes without saying that these occupy one's time most profitably and beneficially) but also from things implicit in the life of a religious community—such as when I am reading, those in a neighboring cell take it upon themselves to play music and sing. Or when I am studying and two maids quarrel and come to me to settle their dispute. Or when I am writing and a friend comes to visit, doing me a great disservice with the best of intentions, whereupon I not only must put up with the bother but act gratefully for the injury . . .

Oh, how much harm would be avoided in our country if older women were as learned as Laeta and knew how to teach the way Saint Paul and my Father

Saint Jerome direct! Instead of which, if fathers wish to educate their daughters beyond what is customary, for want of trained older women and on account of the extreme negligence which has become women's sad lot, since well-educated older women are unavailable, they are obliged to bring in men teachers to give instruction in reading, writing, and arithmetic, playing musical instruments, and other skills. No little harm is done by this, as we witness every day in the pitiful examples of ill-assorted unions; from the ease of contact and the close company kept over a period of time, there easily comes about something not thought possible. As a result of this, many fathers prefer leaving their daughters in a barbaric, uncultivated state to exposing them to an evident danger such as familiarity with men breeds. All of which would be eliminated if there were older women of learning, as Saint Paul desires, and instruction were passed down from one group to another, as is the case with needlework and other traditional activities.

For what drawback could there be to having an old woman, well-versed in letters and pious in conversation and way of life, in charge of the education of maidens? And what harm in preventing the latter from going to waste either for lack of instruction of from having it imparted to them through the dangerous medium of male masters? . . .

From this convent of Our Father Saint Jerome of the City of Mexico, the first day of the month of March of the year sixteen hundred ninety-one. Your most favored servant kisses your hands,

Juana Inés de la Cruz

DOCUMENT 1.3

Report on Mission Rosario—Father Solís, Texas, 1768

Fray Gasper José de Solís was sent to inspect various missions in New Spain in the late 1760s. In this account of Mission Rosario he provides his opinion of the prosperity of the missions and reveals the often paternalistic spirit of priests. His negative attitudes toward the Native Americans are clearly expressed and reflect prevailing sentiments about their culture.[76]

From Frederick Eby, compiler, *Education in Texas: Source Materials*, University of Texas Bulletin no. 1824 (Austin: University of Texas, 1918),1–3.

The opinion which I have formed of this mission of Nuestra Señora del Rosario is as follows: As to material wealth it is in good condition. It has two droves of burros, about forty gentle horses, thirty gentle mules, twelve of them with harness, five thousand cattle, two hundred milch cows, and seven hundred sheep and goats. The buildings and the dwelling, both for the ministers and for the soldiers and the Indians, are good and sufficient. The stockade of thick and strong stakes which protects the mission from its enemies is very well made . . .

The mission was founded in 1754. Its minister, who, as I have already said, is Fr. Joseph Escovar, labors hard for its welfare, growth, and improvement. He treats the Indians with much love, charity and gentleness, employing methods soft, bland, and alluring. He makes them work, teaches them to pray, tries to teach them the catechism and to instruct them in the rudiments of our Holy Faith and in good manners. He aids and succors them as best he may in all their needs, corporal and spiritual, giving them food to eat and clothing to wear. In the afternoon before evening prayers, with a stroke of the bell, he assembles them, big and little, in the cemetery, has them say the prayers and the Christian doctrine, explains and tries to teach them the mysteries of our Holy Faith, exhorting them to keep the commandments of God and of Our Holy Mother Church, and setting forth what is necessary for salvation . . .

[The Indians] are all barbarous, idle, and lazy; and although they are so greedy and gluttonous that they eat meat almost raw, parboiled, or half roasted and dripping with blood, yet rather than stay in the mission where the *padre* provides them everything needed to eat and wear, they prefer to suffer hunger, nakedness, and other necessities, in order to be at liberty and idle in the woods or on the beach, giving themselves up to all kinds of vice, especially lust, theft, and dancing.

DOCUMENT 1.4

Petition to the Cabildo with Response, Don José Francisco de la Mata, Texas, 1786

In this account of a schoolmaster in a settlers' school, and the cabildo's official response, the community's ambivalent attitude toward mandatory education is revealed. Throughout the late eighteenth and nineteenth centuries in both the eastern and southwestern parts of the United States, zealous educators and reformers such as Don José Francisco de la Mata encountered apathy from the community. However, these early endeavors paved the way for more successful public schooling models later in the nineteenth century. Note that signatures at the end of this document and others are followed by the term "rubric." During this era, a rubric was a fancy flourish added to someone's signature to make it more distinctive and difficult to falsify.

From Frederick Eby, compiler, *Education in Texas: Source Materials*, University of Texas Bulletin no. 1824 (Austin: University of Texas, 1918), 8–10. Originals can be accessed at the Bexar Archives, a collection of official Spanish documents from the Spanish and Mexican eras. East Texas Research Center, R.W. Steen Library, Stephen F. Austin State University Campus, Nacogdoches, Texas.

Very Illustrious *Cabildo*.

I, Don José Francisco de la Mata, native of the villa of Saltillo and a resident of the capital of Texas from the year [17]86 to the present date, must and do appear in all due form before you with the most respectful submission and say that I have been much grieved at heart upon seeing the ignorance in this community as to what is the law of God, a knowledge of those principles was lacking among the children of this villa[.] [I have also been much grieved at heart] to see the children, as is well known, running about as vagabonds engaged only in pernicious pursuits such as [playing with] arrows [and] ropes, [and spending their time] in childish games and other idle entertainment which lead only to perdition[.] They have no respect for the officials, [and] no reverence for the aged or the distinguished[.] This I know from personal experience since most of them did not even know how to make the sign of the cross[.] Moved by love for God, our Lord, I decided to open a school of my own free will, as I did, that I might gather them all and give them an education, and teach them their letters, attainments which are essential to the perfect state necessary for the service of our Divine Mother, and for the obedience and respect which they should have for their parents so that they may be, as they are at the present time, an example to this community[.] In view of all I have said and what I shall say it is indispensable that I appeal to your tribunal and legal court in order that my gathering of children may be maintained with greater efficiency and zeal than heretofore attained by my teaching[.] You will, with due knowledge of the priest and vicar of this parish, kindly demand and insist that the said community, that is, the fathers and mothers of families, contribute a small sum annually for a fixed length of time[—]twelve *reales*[—]. In this it seems to me I do them a great justice (subject to your approval in the matter) considering the present misery as well as all other circumstances [for example] they have not appreciated in any manner, the importance of my work and its benefit to their children[.] I have been satisfied with the said small recompense, in order to draw to my teaching as many children as there may be in this neighborhood[.] And since this small pittance is barely sufficient for my subsistance [*sic*], I beg of your generous consideration that you do as I have said[—]this seems to me to be only just[.] At the same time [I beg that you will] authorize the establishment of my school in the name of his Majesty (may God keep him) so that the fathers and mothers of families when their children are subjected to such slight punishments as may be necessary shall not come, as they have come, into my presence to irritate me and insult me by words and threats, a thing which has been noticed even by their own children[.] Also [I beg that] they be not allowed to withdraw any child now included or who may be included in my flock on account of the slight punishments which may be necessary to inflict on them; for, from this, results, as has been proven, the ruin of the children [and] of their parents, and continual inconveniences to judges, but principally it seriously grieves our great God and Lord as the Holy Fathers in their preaching and doctrines teach us[.] Therefore, I beg that the consideration of this, my representation, be with due knowledge and approval of the Governor, so that you may determine what may lead to the end I seek, and thus, to secure the rules that may be suitable for the management of my school—Therefore I beg and plead that you be kind enough to accept this as framed for the lack of better facilities[.]

José Francisco de la Mata [Rubric]

Having seen and approved, and in consideration of what the applicant sets forth in the preceding document [i.e.] that the parents of children should pay twelve *reales* annually for each child they may have in school, and realizing that he is doing a work of charity, and this his petition is just and equitable, for it results in the welfare of the public; therefore justice shall be done to him as he requests[.] To this end we will ask on our part, that the *cura vicario* and ecclesiastical judge of this villa through his saintly admonitions, shall instruct the said parents in regard to the duty they have to see that their children do not fail to go to school, as well as to refrain from going to complain impudently to the teacher when their children are punished, with the understanding that they will bring upon themselves such punishment as may be fitting[.] They have no right to take their children from school for such a childish reason as it is our duty to see that all children receive instruction[.] No other person can give this instruction since he has offered to maintain the said school for six years counting from this day to the summer of [17]95, obliging himself, as he has done, to be subject to the full rigor of the law if he fails to comply[.] ... We thus decree, command, and sign in this *sala capitular* in the exercise of our rights, and witness in the *villa* of San Fernando, and *Precidio* [*sic*] of San Antonio de Bexar on the first day of May, one thousand seven hundred and eighty-nine[.]

Ignacio Calbillo [Rubric]
Julian de Rocha [Rubric]
Jose Antonio Saucedo [Rubric]
Marco de Zepeda [Rubric]
Angel Nabarro [Rubric]
Joachin Flores [Rubric]
Jn Felipe Flores [Rubric]
Joaquin de Orendain [Rubric]

San Antonio de Bexar
1" de Mayo de 1789

I approve the preceding proposition made by Don José Francisco de la Mata[.]

CHAPTER TWO

Education during the Mexican Era, 1821–1848

The education of youth has always been one of the most important bases for the felicity of Peoples, and the prosperity of their Government. The Mexican, who, unfortunately, groaned under the despotic and savage sway of the ambitious sons of Iberia, has never occupied himself in perfecting this most important institution, which would already have placed him on a level with the most cultured nations. The corrupt Government at Madrid only cared to suck up, by whatever means within its reach, the precious resources of the Americas, and studiously and craftily to retard the growth of enlightenment.

—*1821 School Ordinance, San Fernando de Bexar, Texas[1]*

INTRODUCTION

The social, economic, and political changes accompanying the independence of Mexico from Spain in 1821 profoundly affected schooling in the far northern colonies. Most significantly, Mexican independence ended the close relationship between education and religion that largely defined the colonial era. The end of state-sponsored religious missions, a new spirit of egalitarianism, and constitutional requirements for schooling combined to bring new importance to public schooling. Unlike before, public schools became a critical component of the creation of an educated citizenry. In this regard, the link between education and the republic echoed Jeffersonian principles articulated in the early Republican Era of the United States.[2] However, decades of political upheaval in the fledgling Mexican state and the unintended negative consequences of the closure of the missions prevented the widespread establishment of public schools.

Royal orders requiring public primary schools had increased during the late Spanish colonial era (1800–1821). In practice, however, government resources favored universities for the upper classes and missions for Native Americans. The Mexican Revolution altered this colonial system. Public education was thrust into debates over the shape and character of the new republic. Borrowing from the ideology of the American republic, the Mexican secretary of state argued, "without education liberty cannot exist."[3] Mexico's

1824 constitution adopted these ideals and required the provision of education for the masses. However, neither state nor federal funds were authorized for public schools, leaving local communities with the complete burden of generating resources. Similar to public schooling measures passed in the legislatures of the pre–Civil War American South that went unrealized, ambitious plans rarely resulted in permanent institutions outside of the largest and most prosperous communities.[4] In general, many parallels exist between the evolution of Mexico's public school system during these formative years and that of the United States immediately prior to the common school movement of the 1830s and 1840s.

POLITICAL, ECONOMIC, AND SOCIAL CHANGE IN MEXICO, 1821–1848

Although many scholars date the beginning of Mexican Independence to the 1810 "Grito de Dolores," criollo military officer Augustín de Iturbide officially declared Mexico independent from Spain in 1821.[5] The Mexican Revolution was chiefly a revolt of the Mexican-born criollos against Spanish domination, but egalitarian principles also stirred reformers.[6] Historians Michael Meyer and William Sherman describe Mexican history during the 1830s and 1840s as one which "constantly teetered between simple chaos and unmitigated anarchy."[7] The average presidential term during the 1830s and 1840s was seven and one-half months. Because of delays in communication and transportation, government officials in the far northern frontier often had difficulty knowing which regime's policies and laws they were following. As a result, local officials often created their own laws or used outdated policies.[8]

The revolution's 1821 Plan de Iguala (Equality) declared "the social and civil equality of Spaniards, Indians, and Mestizos."[9] The Mexican Era thus ushered in changes in social relations between the main populations of colonial society—Spanish, Indian, black—and the many variations (mestizos, Afromestizos, etc.) that three hundred years of intermingling had produced. In the new nation, racial classifications in public documents were forbidden and replaced with the more general cultural term, *gente de razón* (people of reason). Martha Menchaca asserts that by the time of the Mexican Era, "gente de razón" was a term referring to Catholics and the racially mixed heterogeneous population that "practiced Spanish-Mexican traditions," but the term excluded tribal Indians.[10] Despite revolutionary ideals disparaging Spain's three-century-long racial and ethnic caste system, a racial hierarchy that prized whiteness and devalued the darker hued Indian, black, and Afromestizo members of its population continued during the Mexican Era (1821–1848).[11] Furthermore, slavery was not abolished, although liberal plans were enacted for gradual emancipation. In fact, the new constitution of 1824 banned the slave trade, set slave children free at the age of fourteen, and required adult slaves to be set free after ten years.[12]

In addition to cultural changes among frontier Mexican society, economic shifts also impacted education. Historians generally agree that during the Mexican Era, trade and accumulation of capital increased stratification between social classes. The abolition of Spain's restrictive policies opened the door for trade with Americans, French, British, and Russians.[13] William Bucknell's opening in 1822 of the Santa Fé Trail, facilitated overland trade between the east and west coasts. Ships from the American northeast and Europe exchanged luxury goods and other manufactured products for raw products in the ports of California. Demand for cattle hides, tallow, and wool among manufacturers overseas contributed to the emergence of a wealthy Mexican ranchero class in Texas, California, and New Mexico, occupying the highest tier of Mexico's new social order.[14] This wealthy class was able to provide financial support to bolster local school initiatives.

The military presidio (fort), an institution central to the conquest and maintenance of Spain's far northern colonies, adapted to the new changes wrought by the revolution. The authority of military officers in the presidios, designed to protect settlements and missions, was subsumed to new town councils called *ayuntamientos*. However, the soldiers and officers from the presidios also donated their time and skill as teachers. In addition, they raised funds for the new public schools. Additionally, because Native Americans, often armed with guns and supplies from American traders, increased the frequency of their attacks upon settlements, education often took a back seat to the basic necessity for safety and protection.[15]

The Spanish missions had been agricultural and educational settlements for three centuries; they, too, came under fire in the new order. In the eyes of many reformers, the 1821 Plan de Iguala and 1824 constitution declaring Native Americans as citizens equal to those of Spanish descent rendered the missions antiquated institutions of feudal Spanish society. In addition, many people felt that missionaries had monopolized valuable land for settlement, resulting in their economic and political gain. These sentiments contributed to the government's decision to secularize the missions. Secularization, as described below, ended the formal provision of education to Native Americans. Whether social, economic, or political, the cumulative effect of Mexican independence disrupted traditional hierarchies. Education was no longer reserved only for elite members of society—nor was it simply a tool to impose Christianization and Hispanicization upon indigenous peoples.

DECLINE OF THE MISSIONS

As the reader may recall from chapter one, during Spanish colonial rule missions were the primary cultural institutions on the frontier. For example, mission libraries were often extensive and operated as informal lending libraries for the surrounding settlers. California's Mission San Loreto contained a library of 466 volumes and Mission San Jose de Comundú possessed 126 volumes.[16] In addition, priests were highly educated men who had studied at European universities or the new colleges in colonial Mexico. They often brought with them broad philosophical and religious ideas that even the Inquisition could not suppress. Whether or not modern historians evaluate the practices of the missionary priests as beneficial or harmful, it is fact that they taught the Native Americans and mestizos the Spanish language, recorded and translated Indian languages and histories, spread the Catholic religion, and introduced European industrial trades. Furthermore, in the most remote areas, they provided separate schools to settlers' children. Under Mexican independence, cultural, educational, and religious institutions suffered from the diversion of resources and energies to political and military matters.

The original role and purpose of the missions under Spanish colonial times became anachronistic after independence. Since the 1500s, missionaries from various orders, including the Franciscans, Dominicans, and Jesuits, were required to accompany all exploring parties to the New World. These priests were called "regular clergy" and they remained under the specific authority of their order. The other branch of clergy, called "secular clergy," were priests who reported to a bishop rather than a private order.[17] The crown supported regular clergy in the missions for ten years. After this period regular clergy were expected to transfer ownership to secular clergy, or "secularize." Unlike its modern definition, the term "secularization" in 1800s Mexico meant a transfer in governance from specific orders (e.g., Franciscans or Jesuits) to parish-supported priests (secular clergy). As a part of this process missions would become self-supporting parishes, no longer requiring government subsidies. Furthermore, land that the missionaries or Crown had held in trust for Native Americans was to be turned over to them as private property. In return, Native Americans were required to pay taxes and support the parish. Lastly, any surplus lands would become a part of the public domain, available for purchase.[18] Although secularization was, on paper, an evolutionary step in the development of Spanish frontier society, in practice it was rarely carried out before the 1800s. In 1813 the liberal Spanish Parliament demanded the immediate secularization of all missions ten years of age or older, but few carried out the orders.[19] It was not until the 1820s that secularization finally began in the northern frontier of Mexico.

Several factors contributed to the end of the missions by the 1820s in Texas, Arizona, and New Mexico and the mid-1830s in California. Ideologically, the paternalistic and condescending attitudes toward Native Americans on the part of Spanish priests clashed with the liberal ideals of the independent government. In addition, the fight for independence created anti-Spanish and anticlerical feelings among the colonists. As a result, few Spanish-born priests could be persuaded to work on the frontier.[20] Finally, both the Spanish government during the rebellion and the Mexican government in the 1820s and 1830s directed funds away from the missions to military or other needs. Missionary priests were thus unable to adequately provide for the needs of their "charges."[21]

Historians generally agree that missions in Arizona, New Mexico, and Texas had secularized by the 1820s. Obtaining an accurate count of Indians after Mexican independence becomes problematic because the government forbid enumeration using Spanish racial classifications. Native Americans baptized and still living in the missions were called "neophytes." When missionaries believed they were sufficiently converted to Hispanic and Christian culture, they were gradually allowed first to visit and then return to their villages, or *rancherías*, permanently. Governmental decrees for secularization sped this process as Native Americans themselves left the missions once they were no longer coerced to remain.[22] By 1820 in Arizona only 1,127 neophytes remained in the missions—9,200 Indians were living in their own rancherías. In Texas, Indians secularized in the 1800s were counted as part of the 7,800 Mexican gente de razón, making it difficult to estimate their population numbers. In the ethnically heterogenous population of New Mexico, an 1827 census revealed 43,433 citizens in and around Santa Fé. By 1840 it was estimated that about ten thousand Christianized Indians were living in communities apart from the gente de razón.[23] The most detailed information on secularization and Native American populations is based upon the later-settled and larger system of missions in California.

Extending from San Diego to Cape Mendocino and encompassing 14 million acres of fertile land, the Alta California missions were particularly well endowed.[24] Unlike the New Mexico, Arizona, and Texas missions, which were almost exclusively supported by the government, the California missions were required to find private benefactors.[25] Money from wealthy Spanish benefactors was placed into a financial entity called the "Pious Fund." From that fund the Franciscans invested in land, ranches, and livestock—assets that dramatically increased in value during the nineteenth century.[26]

Native Americans greatly outnumbered gente de razón in California. In 1821 only 3,200 settlers were estimated to be living in scattered towns and ranches. An estimated 21,000 neophyte Indians lived in the missions in the 1820s, a number that rapidly declined after the governor of California officially secularized the missions in 1834. According to historian George Phillips, as the missions secularized, Native Americans left to find work on ranches, headed to larger towns such as Los Angeles, or joined unchristianized tribes.[27]

Disease also decimated an extraordinary number of neophyte Indians in California. Historians Paul Farnsworth and Robert Jackson detail the devastating effect of gathering large numbers of Native Americans in the close, poorly ventilated quarters of the missions. Smallpox, venereal disease, and other infectious European diseases contributed to high mortality rates. The missions brought new "uncivilized" Indians into the missions to keep their numbers up but could not maintain pre-1810 levels. For example, 1,361 neophytes were living in Mission San Juan Capistrano in 1812. By 1820, the number had decreased to 1,064 and by 1830 there were only 925.[28] High infant mortality and low life expectancy in the missions contributed to historians Erick Langer and Robert Jackson's conclusion that the California missions were "death camps" for Native Americans. In their study of Mission

Soledad in Alta California, for example, more than 90 percent of the children died before reaching the age of ten; hence, within one generation more than 90 percent of the future indigenous population of that mission had disappeared.[29]

Native Americans and missionaries held differing views on the decline of the missions. Unaccustomed to the notion of Indians as individuals capable of self-governance, missionaries expressed alarm at their fate under secularization. Reverend Gonzales Rubio, returning to California in the 1830s, wrote, "My Mission had already been secularized, and I had no resources. I could do nothing for the Indians, who were like boys of one hundred years. It is only with liberality you can draw them towards you: give them plenty to eat and clothes in abundance, and they will soon become your friends, and you can then conduct them to religion, form them to good manners, and teach them civilized habits."[30] What is also revealed in the priest's statements are the terms (food, clothing, etc.) under which Native Americans would choose to stay in the missions (see document 2.3).

Other contemporary observers illuminated the constrained nature of Indian education during the decline of the missions. British captain Frederick W. Beechey, a frequent visitor to the California missions in the 1830s, reflected, "there may be occasional acts of tyranny, yet the general character of the padres is kind and benevolent . . . It is greatly to be regretted that with the influence these men have over their pupils, and with the regard those pupils seem to have for their masters, the priests do not interest themselves a little more in the education of their converts, the first step would be in making themselves acquainted with the Indian language. Many of the Indians surpass their pastors in this respect, and can speak the Spanish language, while scarcely one of the padres can make themselves understood by the Indians. They have besides, in general, a lamentable contempt for the intellect of these simple people, and think them incapable of improvement beyond a certain point."[31] Comments such as these provide hints of the still-rigid beliefs Europeans held about the educability of Native Americans.

Historians today note the administrative barriers and unrealistic expectations hindering the smooth transition of Native Americans toward land ownership and self-governance. Specifically, Menchaca and others point out that the 1824 constitution's General Law of Colonization resulted in significant parcels of land being placed in the hands of a small number of settlers.[32] Christianized Native Americans often entered into unfair labor contracts with new landowners. Subsequently, without the paternalistic protection of missionaries, many Indians became trapped in a system of debt peonage.[33]

For three centuries missions had provided close, sustained acculturation not only toward Indians but also between Indians and the Spanish. According to Farnsworth and Jackson, the late mission years permitted "stabilized pluralism" in which "native Americans adopted a culture that was neither Native nor Spanish but exhibited elements of both."[34] Overall, the secularization of the missions in the 1820s and 1830s ended an important institution in the educational, religious, and cultural life of the Southwest. When Anglo settlers began moving into the Southwest in large numbers—to Texas in the 1820s and elsewhere by the 1840s—the abandoned and decrepit missions sent a misleading message that neither Spaniards nor Mexicans placed much emphasis upon cultural institutions in general and education specifically.

PATHS TOWARD MEXICAN PUBLIC SCHOOLS

Rule No. 2: School. No director or maestro will be appointed to the school who is not examined and approved by the departmental assembly in reading, writing, and counting. At the same time he will have to make evident the depth of his Christian fervor and patriotism, and he should show his comprehension of the republican, federal, representative, and popular system of government and the great advantages which accrue to our immortal nation through this system.

—Rules and Regulations for a Public School
Villa de Santa Cruz de la Cañada, New Mexico, 1827[35]

Under colonial rule, the purpose of public education was to transfer and maintain the Spanish culture, language, and the Catholic religion in the New World. For example, the Spanish king's 1782 proclamation requiring Spanish-language schools used "better instruction in the Christian doctrines and polite intercourse with all persons" as the justification for education. This had little to do with Enlightenment ideals of liberty and citizenship.[36] After 1821, the role and purpose of public schools shifted ideologically to reflect the spirit and purpose of independence. Growth in the actual number of public schools in the Mexican Era was less significant than the population's changing perception of the role of education in a republic. Historian Richard J. Altenbaugh found a similar pattern in U.S. history, arguing that schooling "underwent a profound transition, not so much institutionally but conceptually, during the early days of the [American] republic."[37]

The implementation of public schools during the Mexican Era was uneven and depended upon the energies and resources of local communities. However, it was very characteristic of the "district school" stage of educational development in the U.S. frontier of the early 1800s. During this district era, rural communities often obtained their own resources for schools through donations of money, firewood, and food. Furthermore, the qualifications of local teachers were often meager and communities hired itinerant schoolmasters with few formal qualifications.[38]

The Mexican nation's antimonarchical republican spirit was reflected in both the rhetoric and administrative structure created to support education. The new secretary of state's 1823 view that "without education, liberty cannot exist" received formal backing in the 1824 constitution.[39] Unlike the U.S. Constitution, which did not mention education specifically but declared in the tenth amendment that education is the responsibility of the states, the Mexican constitution specifically required education in Article 50, No.1.

Within Article 50, both primary and university schooling were mentioned. It required the establishment of colleges for the "Marine, Artillery and Engineer Departments" and for teaching "the natural and exact sciences, the political and moral sciences, the useful arts and languages." Furthermore, the General Congress could "regulate the public education in their respective states," as long as Congress did not "prejudice" the "rights which the states possess."[40] Despite the broad-minded and progressive attitude embodied in the constitution, the absence of a predetermined federal funding mechanism (e.g. taxes or land grants) weakened the implementation of the public school system.

The administration of Mexico's public education was top-down. Congress passed laws at the federal level that the various departments (states) were required to implement. At the departmental level, laws were then passed on to the local level. Power over local schools regarding finances, teacher hiring and firing, and other administrative matters rested with the local political body called the *ayuntamiento*. The ayuntamiento held considerable power over the fledgling public schools, similar to the role of mid-nineteenth-century school boards in U.S. history.

At least two major pieces of educational legislation were passed in the National Congress between 1821 and 1848. In 1833 Congress issued sweeping and detailed legislation regulating the public schools. (See document 2.3 for entire School Act of 1833.) Among the 19 articles in the School Act were requirements for the creation of normal schools and the appointment and pay of school inspectors to visit public schools. Furthermore, it was ordered that primary schools be opened at each of the six national colleges. Tensions between the Catholic Church and the state appear to underlie Article 9, which ordered fines for parishes or religious orders who were requested to open schools and "fail[ed] to do so." The 1833 decree further required annual public examinations at local schools and provisions for children who, "due to their poverty, deserve to be helped." Primary education for girls was also specifically addressed. In addition to the prescribed curriculum of "reading, writing, arithmetic and the political and religious catechisms," it was stated that girls "shall be taught to sew, embroider, and other useful occupations of their sex."[41]

Three years later, in 1836, Congress ordered departments to establish public schools in each pueblo (village). The ayuntamientos were also named as the chief administrative units, a function they fulfilled de facto for numerous years.[42] The legislative acts of the 1830s, designed to be a blueprint for the constitution's educational mandate, were implemented with varying success across the new Mexican nation. In large urban areas such as Mexico City, public primary and secondary schools were numerous, well-supported, and in the close vicinity of institutions of higher education. In rural Mexico, most notably the far northern departments of California, New Mexico (which included Arizona), and Texas, the ability to find qualified teachers, adequate supplies, and even buildings presented a formidable challenge. Of course, this should come as no surprise since remote, rural schools in any nation are the hardest to establish and support.

EARLY TEXAS SCHOOLS IN THE MEXICAN ERA

The evolution of San Fernando de Béjar (modern-day San Antonio), illuminates the conditions under which public education attempted to secure a permanent footing in Mexican society. Béjar, as the town was commonly called, attempted to carry out in good faith the state of Texas-Coahuila's constitutional requirements for education. On March 11, 1827, the Congress of Texas-Coahuila ratified its new constitution. Title VI, Article 215 required: "In all the towns of the state a suitable number of primary schools shall be established, wherein shall be taught reading, writing, arithmetic, the catechism of the Christian religion, a brief and simple explanation of this constitution, and that of the republic."[43] A commissioner appointed by the state legislature issued a ruling in September of that year further requiring new towns to set aside lands for a market square, jail, burial grounds, and "a school and other edifices for public instruction."[44] Although educational historians have often noted that both constitutional and rhetorical demands for schools during formative eras were often ignored, Béjar possessed the resources to carry out these legislative acts.[45]

In 1828 the ayuntamiento of Béjar issued an ordinance for public schools. Victoriano Zepeda, secretary of the ayuntamiento, and five other members articulated their views of schooling. They linked education with the recent revolution, stating that Mexicans had "broke finally the ominous chain which bound us, elevating us to the rank of free men, independent of any other." (Refer to document 2.1 for entire text.) Furthermore, the ayuntamiento pointed out that an annual gift of six hundred dollars for four years from local citizen General Anastacio Bustamente would allow Béjar to create permanent schools, despite being "one of the most distant [towns] from the center, of the least populous, of the poorest in moneyed citizens, and finally, vexed by the terrifying hostilities which it has suffered from the savages through long periods of time."[46] The school was created and benefited from some state assistance. For example, in 1828 the State of Texas-Coahuila "brought one hundred charts, thirty six catechism and other supplies" for the school children of Béjar.[47]

San Fernando de Béjar was not the only town in Texas-Coahuila to create schools or maintain those schools begun during the colonial era. Early in 1829 the Congress of Texas-Coahuila issued Decree No. 92. The decree required schools based upon the Lancaster Plan to be established in the capital of each department. The Lancaster Plan, originating in England, was a monitorial system of education that gained popularity in Europe, the United States, and Mexico in the early 1800s. Under this plan, advanced students could efficiently train or "monitor" those students below them, thus saving money for the schools.[48] The scope of Decree No. 92 was broad and ambitious for the times. It required towns to pay teachers 800 dollars per annum and expected each school to have a minimum of 150 scholars. In addition, although allowance was made for children whose parents were "unable to pay," tuition was charged to the parents. Parents were required to pay 14 dollars a year for each primary-school child and 18 dollars for each student receiving secondary instruction.[49] The city of Leona Vicario carried out this directive, establishing a Lancastrian school. In 1829 the school had 57 pupils, the teacher was paid his full eight hundred dollars, including a house rented for him, and money was spent on slates, table, benches, and paper. Indeed, the city paid out $1,250.00 for its school that year, many of those funds collected from tuition.[50]

The State of Texas-Coahuila's plan for Lancastrian schools was quickly scaled back to one that was more financially feasible. In April of 1830, the congress formally recognized "the obstacles that have arisen for strictly fulfilling the decree No. 92." Instead of renewing the original plan, they asked for "six public primary schools," a teacher salary that was only five hundred dollars per year, and other measures that revised the original decree, creating more realistic goals.[51]

The town of Nacogdoches also actively worked to create a school. Similar to nineteenth-century associations in the United States that called themselves "friends of education," the citizens of Nacogdoches created a Board of Piety on January 16, 1831, to raise funds for both a new church and school. The group posted and distributed a circular in March of 1831 in both the English *and* Spanish languages, reflecting the rapid increase in the number of Anglo settlers who came to Texas in the 1820s. The circular emphasized that a church was necessary to "worship the Gospel agreeable to the Roman Catholic Religion which is professed by the Mexican Nation." The primary school was, in their opinion, "of absolute necessity in civilized society attended by any form of government wherever." The community's response illuminated the range of frontier resources in early-nineteenth-century Texas. Two colonels were able to donate one hundred pesos each; other donations ranged from Don Patricio Torres's "month's service of a hired laborer," to the "yearling calf" from Don Concepción Ybarba and the five pesos from Don Jesús Santos (see document 2.4 for entire list of contributors). The State of Texas-Coahuila further boosted funding to the new schools with an 1833 land grant of "four sitios of land," whose proceeds would be "appropriated entirely and exclusively as a fund of the primary schools."[52] In Texas, existing records document the initiative of several communities to create public schools. The pattern in California reflected another strategy in the Mexican Era—utilizing mission properties to open public schools.

PUBLIC SCHOOLING IN CALIFORNIA DURING THE MEXICAN ERA

The two principal colonial institutions—the mission and the presidio—continued to adapt and change (or be changed) under Mexican governance. Unlike in Texas, where local communities created and funded schools, the governors of Mexican California provided more top-down direction regarding public schools.

In San José, the estimated population of whites and Indians was still small at the time of independence. For instance, the total population in 1822 was about 300 and increased to 600 by 1830.[53] The public school (preserved today as a historical site in downtown San José), had two teachers in 1821—brothers Joaquin and Antonio Buelna. In 1822 a "one-legged soldier named Labatida" was appointed as the teacher, presumably as a replacement. José Antonio Romero became the teacher in 1823 and apparently remained in that position for several years.

The salaries for the San José teachers were 15 dollars per month.[54] The average teacher salary in the Mexican Era—between 15 and 20 dollars per month—compared poorly to that of other public servants. The average pay for a presidial soldier was 30 dollars monthly; in some cases, soldiers drew dual pay from their roles as both soldiers and teachers.[55]

The scattered records from Monterey indicate that a school was in operation in the 1820s but closed temporarily in 1829, "for want of a teacher."[56] Manuel Crespo and Antonio Buelna (apparently the same from San José) taught for 15 to 20 dollars a month. The city's school inventory in 1829 consisted of "a table, one arithmetic, and copy-books."[57]

Further south in San Diego, the public school was held at the presidio. In 1829, 18 students were enrolled. Reflecting Mexico's official recognition of the Catholic Church as the church of the State, a "Padre Menendez" was reported as the teacher and he received from 15 to 20 dollars a month from the city funds.[58]

The governors of California issued numerous public pronouncements and decrees for public schools during the 1820s and 1830s. In 1822 Governor Pablo Vicente de Sola recommended that since no schools existed at the mission, "for a small sum the padres might hire teachers and do great good."[59] In 1824 a "hospicio de estudios" was proposed in the California legislature with four members voting in favor, but no further mention is found of the proposed school, and it is uncertain whether four members were sufficient for a passing resolution.

Under Governor José María Echeandía (1825–1831) California became more involved in the creation of school policy. In 1827 the legislature recommended the federal government send teachers to a small "colegio ó academia de gramática, filosofía, etc." (secondary school, or academy of grammar and philosophy).[60] The same year, in 1827 Echeandía ordered the establishment of public schools in the missions, but his orders were not favorably received at the struggling missions. Resistance may have stemmed from his policy requiring that "masters should be employed and all expenses paid by the missions, but that the schools should be under control of the ayuntamientos or other authorities."[61] Nonetheless, by 1829 Echeandía reported that seven of the southern missions had established schools—with student populations ranging from eight to forty-four.[62]

The patchwork school system of the 1820s improved in the 1830s with assistance from the federal government. In 1833 the California legislature passed an act requiring all towns to establish primary schools. The following year the central government sent 20 teachers to carry out this mandate. The teachers arrived in San Diego as part of the larger Hijar-Padres Colony, which totaled 239 people and included several professionals and skilled craftsmen. According to Cecil Hutchinson, the teachers then migrated to Los Angeles, San José, and Santa Barbara.[63]

PUBLIC SCHOOLING IN NEW MEXICO AND ARIZONA

Educational progress in New Mexico (which included Arizona during the Mexican Era) appears to have been the slowest among the new nation's northern settlements in terms of formal institutional development. The work of teacher and priest Antonio José Martínez dominates the narrative of New Mexican education. A native of Taos, Martínez spent six years in advanced schooling in central Mexico before returning to establish a boys seminary in 1823. He soon expanded the Taos seminary to boys wishing to receive higher-level lessons in grammar, morals, rhetoric, logic, reading, writing, and arithmetic.[64] An additional educational contribution was Martínez's purchase of a printing press; in 1835 he brought a press and a printer named "Baca" to Taos. The press produced urgently needed schoolbooks, religious materials, and newspapers.

Historians know more about the qualifications desired of New Mexico's public school teachers than the individuals themselves. In 1827, for example, Villa de Santa Cruz de la Cañada adopted detailed rules for their teacher or "maestro." Political values of the new nation were blended with progressive European pedagogical views of the late eighteenth and early nineteenth centuries. The maestro was ordered to "take great care in teaching the children to treat citizens with the kind of courtesy and politeness which ought to shine forth in a son of our glorious Republic." The close Church-State relationship is illuminated in Rule No. 11, which required teachers to assemble the entire school after each Sunday mass for a visitation/inspection by the ayuntamiento. Students were required to "present for review their books, copy books, etc."[65] This type of public accountability was certainly a feature of many rural schools in the United States of the nineteenth century, often taking the form of public spelling bees or award presentations.[66] The close link of these activities with Catholic religious practices distinguishes it from the Pan-Protestant American experience.

NONFORMAL EDUCATION DURING THE MEXICAN ERA

Outside of the publicly recorded measures taken toward public schools in the Mexican-Era Southwest are the numerous informal arrangements for schooling held in various settings, called nonformal education. While documented less frequently, this was most likely the most prevalent form under which literacy was obtained in the frontier southwest. Furthermore, the opening of trade with other countries and the arrival of British, American, and other "foreigners" and their supplies spread the printed word and the need for and interest in at least rudimentary literacy.

Prior to Mexican Independence, government officials from the Spanish Inquisition monitored and censored reading materials brought to the colonies. The end of Spanish rule, the spirit of freedom imbued in the ideology of the new nation, and the weakened status of the Catholic Church encouraged the introduction of a greater number and variety of books and other print materials in northern Mexico after 1821.[67] In addition, the government provided formal protection of freedom of expression, further creating a more enlightened atmosphere concerning controversial ideas and literature. For instance, Article 12 of the 1827 constitution of Texas-Coahuila declared, "the state is also obligated to protect all its inhabitants in the exercise of the right which they possess of writing, printing and freely publishing their sentiments and political opinions, without the necessity of an examination, or critical review previous to their publication."[68]

The political and economic changes of the Mexican Era led to the creation of a wealthy *ranchero* class. This class was able to use its greater expendable income to purchase luxury items such as leather bound books.[69] Young men from this class also learned from local officials and professionals to expand their education. For example, Californio Juan Bautista Alvarado clerked for the Monterey settler Nathan Spear. Spear reported that he "was in the habit of imparting to him [Alvarado] when in his employ a good deal of information about other countries and governments."[70] Mexican Governor Pablo Vicente Sola tutored Mariano Guadalupe Vallejo in political and historical events and guided a reading of Cervantes's *Don Quixote*. For a brief period in the 1830s a small group of Californios were tutored together at the ranch of William E. P. Hartnell in Monterey.[71]

Although the education of girls was formally prescribed in the School Act of 1833, contemporary observers note that girls were usually taught at home or in local arrangements. For example, Apolinaria Lorenzano recalled her early experiences teaching in a small school "established by a widow, where gente de razón girls learned to read, pray, and sew." She was also employed at the San Diego mission teaching sewing and reading to girls.[72] Lorenzano's teaching experiences appear similar to the female-run "amiga" schools of the colonial era, which appear to have continued during the Mexican years.

The introduction of printing presses to the far northern Mexican states also expanded opportunities for literacy. An American entrepreneur began operating a press in San Antonio in 1823. His bilingual newspaper was short-lived, but it was the first in Texas. By the early 1830s, presses were operating in New Mexico and California producing newspapers, leaflets, and books to a receptive audience.[73]

CONCLUSION

The development of public schools in the far northern Mexico frontier began during 27 years of rapid political, economic, and social change. By 1844 Mexico as a whole had almost 60,000 students attending 1,310 public primary schools.[74] However, educational development was relatively slow in the frontier communities of New Mexico, Texas, and California. Most children were educated at home, at a neighbor's home, or in a private or public school. The elite were sent to the United States, England, France, or even the Sandwich Islands.[75] The resources of the Mexican government between the 1820s and 1840s were put toward maintaining inner political stability and protection from the warring Native Americans on the frontier. Daniel Tyler calculated that the Mexican government was allocating only a miniscule amount of funds toward education compared to the amount given to the military during those tumultuous decades.[76] The constitutional requirements for schools and subsequent legislative acts remained only partially fulfilled during the Mexican Era. As Anglo settlers began moving into Texas in the 1820s and ultimately declared independence in 1837, Tejanos had established, at least in concept, the idea of government-sponsored schooling. The U.S. victory in the Mexican-American War (1846–48) led to the subsequent subordination of Mexican laws and customs to U.S. goals. These changes were carried out in political, economic, and social arenas, including education.

DOCUMENT 2.1

Statement of the Ayuntamiento on the Purpose of Public Education

The nationalist tone of this document illuminates the desire of Mexicans to distance themselves from Spain's tight monarchical control over the colonies. Similar to resolutions for schools in the Republican Era of the United States, education is linked to the creation of a freethinking citizenry.

From I. J. Cox, "Educational Efforts in San Fernando de Bexar," *Texas Historical Association Quarterly* (July 1902): 62–63. Courtesy the Texas State Historical Association, Austin.

> *San Fernando de Bexar,*
> *13 of Mch., 1828.*

The foregoing Ordinance having been put under general discussion, it has been approved in its entirety by this *Ayuntamiento*.

> *Capitular Hall of San Fernando de Bejar,*
> *13 of March, 1828.*
> Ramon Musquiz
> Juan Martin de Beramendi
> José Maria de la Garza
> Manul. Flores
> Juan Angl. Seguin
> Victoriano Zepeda
> Sec'y ad Interim.

The education of youth has always been one of the most important bases for the felicity of Peoples, and the prosperity of their Government. The Mexican, who, unfortunately, groaned under the despotic and savage sway of the ambitious sons of Iberia, has never occupied himself in perfecting this most important institution, which would already have placed him on a level with the most cultured nations. The corrupt Government at Madrid only cared to suck up, by whatever means within its reach, the precious resources of the Americas, and studiously and craftily to retard the growth of enlightenment. Nothing, in truth, was more natural than this iniquitous behavior, since the first, increasing its riches, satisfied all the desires of its vain and haughty natural caprice; and the second secured it in the domination of the richest and most productive of its evil-acquired patrimonies, blinding us to the important knowledge of our Native rights.

Nevertheless, the natural empire of the reason, which some day comes to prevail, and the characteristic qualities of all the children of this soil, in union with other joint causes, broke finally the ominous chain which bound us, elevating us to the rank of free men, independent of any other.

In spite of this, and of the paternal beneficent institutions of our Present Government, to which belongs the establishment of primary Schools, the spirit of discord which still endures amongst us has impeded it from occupying itself with this, as with other matters that undoubtedly make for the aggrandizement of the Nation, all its efforts being employed in assuring our internal and external tranquility, which is doubtless the corner stone of the social edifice.

In spite of all, and in virtue of the ardent desires of the towns, there are already seen in most of them educational establishments for the youth who will form the future generation, which will come to secure completely Mexican Liberties; and among these, although one of the most distant from the center, of the least populous, of the poorest in moneyed citizens, and finally, vexed by the terrifying hostilities which it has suffered from the savages through long periods of time, [Béjar] has just made a heroic and extraordinary effort, stirred up by several of its citizens, and by that worthy citizen, General Anastacio Bustamente, to make a collection amongst all its citizens, amounting to six hundred dollars annually and lasting for four years, in order to carry to accomplishment the desire which in all time it has had for the education of its youth.

Yes, unfortunate Béjar, truly worthy of a better fate, you are the one which has just given so heroic a testimony of beneficence in spite of your notorious poverty; with difficulty do you commence to lift yourself from the abject state into which you had sunk, thanks to the presence of that philanthropic General and the aid of the Supreme Federal Government.

Be filled, then, citizens of Béjar, with the ineffable satisfaction which is produced by the important services directed to the good of your children, of society in general, and of the adored Country to which we belong, awaiting the glorious day in which you may either experience the fruit of your sacrifices for this pious establishment, or in which your ashes may receive a new being, through the eulogies which, without doubt, your posterity will lavish upon you.

DOCUMENT 2.2

Secularization of the Missions— Missionary Point of View, 1833

In this recollection of the demise of the missions, an active participant paints a bucolic scene of their former glory. The Native American point of view of this event has unfortunately received little documentation. Through the work of modern day archeologists, estimates of Indian illness and disease in the missions—as mentioned below—has received greater attention and inclusion in the historical narrative.

From Msgr. Francis J. Weber, as first published in *Documents of California Catholic History (1784–1963)* (Los Angeles, CA: Dawson's Book Shop, 1965), pp. 33–36.

GONZALES RUBIO TO JOSÉ JOACHIM ADAM, SANTA BARBARA, SEPTEMBER 1864.

On my landing in this country, which happened on the 15th of January, 1833, there were in existence from San Diego up to San Francisco Solano 21 Missions, which provided for 14,000 to 15,000 Indians. Even the poorest Missions, those of San Rafael and Soledad, provided everything for divine worship, and the maintenance of the Indians. The care of the neophytes was left to the Missionary, who, not only as Pastor, instructed them in their religion and administered the sacraments to them, but as a householder, provided for them, governed and instructed them in their social life, procuring for them peace and happiness.

Every Mission, rather than a town, was a large community, in which the Missionary was President, distributing equally burdens and benefits. No one worked for himself, and the products of the harvest, cattle and industry in which they were employed was guarded, administered and distributed by the Missionary . . .

In the inventory made in January, 1837, the result showed that said Mission [San José] numbered 1,300 neophytes, a great piece of land, well tilled; storehouses filled with seeds; two orchards, one with 1,600 fruit trees; two vineyards—one with 6,039 vines, the other with 5,000; tools for husbandry in abundance; shops for carpenters, blacksmiths, shoemakers, and even tanneries, and all the implements for their work.

The fields were covered with live stock: horned cattle, 20,000 head; sheep, 500; horses, 459. For the saddle 600 colts of two years, 1,630 mares, 149 yoke of oxen, 30 mules, 18 jackasses, and 77 hogs.

Twice a year a new dress was given to the neophytes, amounting in the distribution of $6,000. When the mission was secularized I delivered to the Mayor-domo then in charge some $20,000 worth of cloth and other articles which the store-house contained.

. . . The other Missions, called "the Northern," though having been already secularized, were in utter bankruptcy, and the same can be affirmed for the most part of those of the south, down to San Diego; for it was observed that as long as the Missions were in the hands of the missionaries everything was abundant; but as soon as they passed into the hands of laymen everything went wrong, till eventually complete ruin succeeded, and all was gone. Yet, we cannot say that the ambition of those men was the cause, since, though the Government in the space of four years, divided seven ranches to private individuals—the smallest of a league and a half—yet in spite of this cutting off of part of my Mission lands, the Mission was every day progressing more and more . . .

I was able to save only a small relic of these tribes during the pestilence of 1833, in which I collected together some 600 Indians. I would have saved more during the small-pox epidemic of 1839, but my Mission had already been secularized, and I had no resources. I could do nothing for the Indians, who were like boys of one hundred years. It is only with liberality you can draw them towards you: give them plenty to eat and clothes in abundance, and they will soon become your friends, and you can then conduct them to religion, form them to good manners, and teach them civilized habits.

Do you want to know who were the cause of the ruin of these Missions? As I was not only a witness but a victim of the sad events which caused their destruction, I have tried rather to shut my eyes that I might not see the evil, and close my ears to prevent hearing the innumerable wrongs which these establishments had suffered. My poor neophytes did their part, in their own way, to try and diminish my sorrow and anguish.

DOCUMENT 2.3

1833 Legislative Act for the Public Schools of Mexico

The federal government of Mexico actively passed several decrees regarding education during its first two decades. The individual states of Mexico were then required to implement these decrees at the local level. This central governmental involvement in education was different from the U.S. emphasis on state regulation of public schools. Note the details regarding teacher education and preparation. The first state-supported normal school in the United States was established by the State of Massachusetts in 1837, four years after this decree. Provisions for girls' education and children in impoverished circumstances distinguish this decree from similar measures passed in the colonial era.

> From Frederick Eby, compiler, *Education in Texas: Source Materials*, University of Texas Bulletin no. 1824 (Austin: University of Texas, 1918), pp. 85–88. Originals can be accessed at the Bexar Archives, a collection of official Spanish documents from the Spanish and Mexican eras. The documents are housed at the East Texas Research Center, R.W. Steen Library, Stephen F. Austin State University Campus, Nacogdoches, Texas.

OFFICE OF THE SECRETARY OF STATE
Department of the Interior
Decree No. 56

His Excellency, the Vice-President of the United Mexican States has been pleased to send me the following decree.

The Vice-President of the United Mexican States in exercise of the supreme executive power and the power granted him the 19th instant law of the general congress, decrees:

Art. 1. That a normal school be established for those preparing themselves to give primary instruction.

Art. 2. A similar school shall be established for women preparing themselves to give primary instruction.

Art. 3. A primary school for children shall be established in each of the six institutions of higher education, and they shall be separated from each other as much as circumstances will allow having if possible, a separate door of entrance; but they shall be under the supervision and care of the president or vice-president of the institution.

Art. 4. Reading, writing, arithmetic, and the political and religious catechisms shall be taught in these schools. The teacher shall enjoy a salary of seventy-five dollars a month, but no quarters will be furnished.

Art. 5. The Board of Directors shall establish in each parish of the federal city where schools of higher education do not already exist similar primary schools for children in which reading, writing, arithmetic and the two aforesaid catechisms shall be taught.

Art. 6. The same shall be done in each parish and sub-parish of the federal district.

Art. 7. The Board of Directors shall also establish subsequently in each parish of the different cities in the federal district a primary school for girls in which they shall receive the same instruction as outlined in article 4, and in addition they shall be taught to sew, embroider, and other useful occupations of their sex.

Art. 8. In addition to these primary schools which shall be supported from the funds appropriated for public instruction, the Board of Directors will be authorized to see that the obligations which some parishes and religious orders have contracted for the establishment of certain schools at their own expense be complied with, and these shall not be considered as free schools.

Art. 9. The Board of Directors shall have power to impose a fine of sixty dollars on each parish or religious order thus bound to support a school if it fails to do so, and the said sum shall be used to establish the said school in the place agreed upon, and which, in the judgment of the Board of Directors is best suited for the purpose.

Art. 10. The salary of the teachers of the two normal schools shall be one hundred dollars a month, with quarters furnished. The teachers shall give instruction in the mutual method of teaching, and they shall teach the Castilian grammar, elements of logic, elements of ethics, arithmetic, and both the political and religious catechisms.

Art. 11. The teachers of primary instruction shall receive a salary not exceeding sixty dollars a month, and they shall be furnished a school house.

Art. 13. The method of mutual instruction shall be put in practice in the primary schools established by the Board of Directors as soon as the necessary teachers can be secured.

Art. 14. In those supported by parishes and religious orders, all endeavors shall be made for the gradual adoption of the same method.

Art. 15. All the schools of the district except the institutions of higher learning shall be under the immediate supervision of a school inspector who will look after them, make frequent visits of inspection, and will report to the Board of Directors anything requiring their decision.

Art. 16. The school inspector shall be apointed [*sic*] by the government from a list of three submitted by the Board of Directors, and he shall receive a salary of 2,000 a year.

Art. 17. In each school there shall be an annual public examination presided over by the school inspector, and at such a time the prizes designated by the Board of Directors shall be awarded to those showing special advancement.

Art. 18. The teachers for the different schools shall be appointed this time by the Board of Directors at the proposal of the presidents of the schools, but in the future they shall be appointed by competitive examination.

Art. 19. The boys and girls who, due to their poverty, deserve to be helped with the necessary school supplies in order to be able to attend school, shall receive such help at the discretion of the said Board of Directors, and the previous recommendation of the inspector of schools.

Therefore I command this to be printed, published, circulated, and given its due observance. Palace of the Federal Government, Mexico City, October the 26th, 1833.—Valentin Gomez Farias—to D. Carlos Garcia.

And I transmit the same to you for your intelligence and subsequent action.

God and Liberty. Mexico, October 26, 1833.

DOCUMENT 2.4

Petition Written in Spanish and English to Raise Money for a School and Church, Nacogdoches, Texas, 1831

The fact that this circular was printed in both English (with several errors) and Spanish reveals the influence of Anglo settlers arriving to Texas in the 1820s. The petitioners' names also reflect the diverse ethnic (particularly Anglo and Mexican) population and the intertwined relationship of the Roman Catholic Church with the public schools. Donations of labor, animals, and even beans reflect both frontier conditions and community spirit.

From Frederick Eby, compiler, *Education in Texas: Source Materials*, University of Texas Bulletin no. 1824 (Austin: University of Texas, 1918), pp. 43–46. Originals can be accessed at the Bexar Archives, a collection of official Spanish documents from the Spanish and Mexican eras. The documents are housed at the East Texas Research Center, R.W. Steen Library, Stephen F. Austin State University Campus, Nacogdoches, Texas.

CIRCULAR

The Board of Piety of Nacogdoches, to the Settlers of this Frontier.

Fellow Citizens: A happy event of the most imperious, exquisite and irresistible circumstances, is, the necessity of two Establishments of piety usefull [*sic*] and necessary; the idea was brought forth by an assembly under the denomination herein mentioned.

A Church intended to celebrate and worship the Gospel agreeable to the Roman Catholic Religion which is professed by the Mexican Nation, and a primary School for the Education of the youth of this Circuit, they are both great and exclusive objects which the Junta expects to promote and raise by every legal, lawful and honorable means . . .

The building of a Church is of the greatest necessity for a christian and religious people. To maintain a School for the education of children is of absolute necessity in civilized society attended by any form of government whatever. If [eliments] elements constitute a man in his nature, religion and education insinuates moral principles in him . . . if you wish to have a School where to send your children to receive education and acquire the principles of

Religion and natural morality, make the sacrifice of a small portion of your interests, and deposit the same in the hands of the Treasurer appointed for the purpose; subscribe to the amount you think proper for so important an object, and be persuaded the funds shall not be disposed of conterary [contrary] to the aforesaid objects . . .

Jose de la Piedras, President—Pedro Elias Bean, V. President.—Adolfo Sterne (absent) Treasurer.—Frost Thorn.—Fr. Antonio Diaz de Leon, (temporary), Curate.—Manuel de Los Santos Coy, Alcalde.—J. Antonio Padilla, Secretary.

Nacogdoches, March the 10th, 1831.

List of the names of the persons who subscribed toward the construction of the church and school.

Colonel Dn José de las Piedras, on the part of the military contingent, and for the present, one hundred pesos.

Colonel Dn Pedro Bean, one hundred pesos.

Dn Adolfo Sterne, twenty-five pesos and one hundred pounds of nails.

Dn Patricio Torres, a month's service of a hired laborer.

Dn Juan Mora, ten pesos.

Dn Jesús Santos, five pesos.

Ynes Santaleón, a barrel of beans.

Dn Martin Ybarbo, a two-year-old steer and a barrel of corn.

Andres González, his person service with a yoke of oxen for eight days.

Dn Juan Lazarin, the same.

Dn Concepcion Ybarbo, a yearling calf.

D. Antonio Manchaca, ten pesos.

D. Bautista Chirino, a three-year-old steer.

Brígido Sineda, the service of a hired laborer for eight days.

Nacogdoches, January 18, 1831.
José de las Piedras. Treasurer
José Antonio Díaz de Leon, as Secretary.

This document is a copy of the original sent of the State Government. Bejar, 12 of February, 1831.

CHAPTER THREE

Americanization and Resistance

*Contested Terrain on the Southwest Frontier, 1848–1912**

INTRODUCTION

Beginning in the mid-nineteenth century the United States began an era of expansionism, supported ideologically by the notion of "Manifest Destiny." Journalist John O'Sullivan, who argued that Providence granted the United States a divine mandate to spread from coast to coast, coined this term in 1845. The ideology of the United States possessing a Manifest Destiny ultimately provided justification for the Mexican War of 1846–1848.

Since the early 1820s, increasing numbers of British, French, American, and Russian immigrants had begun settling in Mexico's frontiers. Mexico welcomed these settlers, particularly to Texas. The Mexican government offered inexpensive fertile land to settlers. In exchange, settlers were required to obey Mexican laws, learn the Spanish language, and convert to Catholicism. Furthermore, Mexico overlooked slave trade laws as an additional inducement to American slave owners.[1] The generous land distribution drew colonists, such as Stephen Austin, to Texas with hundreds of land-hungry families. The trickle into Texas became a flood, and by 1830 Americans overwhelmed Mexicans in Texas by 25,000 to 4,000.[2] The new settlers largely ignored the unenforceable laws regarding Hispanicization and Catholic conversion.

The independence of Texas in 1836 as a sovereign republic and its subsequent U.S. annexation in 1845 paved the way for the United States to spread to the West Coast. A weak Mexican military and government, unresolved border disputes between the United States and Mexico, resurgent Indian threats, demand for western lands, and President James K. Polk's fear

---

*The last section of this chapter is a reprint of Chapter Two from *The Majority in the Minority: Expanding the Representation of Latina/o Faculty, Administrators and Students in Higher Education*, edited by Jeanett Castellanos and Lee Jones (Sterling, VA, 2003) and reproduced by permission of the publisher. Copyright © 2003, Stylus Publishing, LCC.

of a British or Russian invasion of California contributed to the subsequent U.S. decision to declare war on Mexico.[3]

The defeat of Mexico, ratified in the Treaty of Guadalupe Hidalgo (1848), altered the political, economic, and social lives of Mexicans. The adjustment from Spanish to Mexican rule was less abrupt than that to American conquest. Spain and Mexico had at least shared the Spanish language and Catholicism as the official religion of the State. Before the war, Mexicans had not been immigrants to the region; after the war they became immigrants, colonized peoples on their former land. Articles VIII and IX of the Treaty of Guadalupe Hidalgo articulated the rights and responsibilities of 100,000 Mexicans who had been conquered. Under Article VIII individuals had one year to become Mexican citizens or seek U.S. citizenship. As the new territories entered statehood, their constitutions narrowed suffrage restrictions. According to Martha Menchaca the new constitutions only granted suffrage to Mexicans considered to be part of the "white" race. Mestizos, Indians, African Americans, and Afromestizos were denied political rights. For example, in 1849 California granted the vote to "every white, male citizen of Mexico who shall have elected to become a citizen of the U.S."[4] Blacks and Indians were excluded from this citizenship, although they had previously been protected under the Treaty of Guadalupe Hidalgo. Furthermore, statutes barred "nonwhite" populations from practicing law, becoming naturalized citizens, and, in many cases, marrying Anglos.[5] This racialization of Mexican peoples also extended to schooling. For instance, by the early 1860s, California's school code stipulated that "Negroes, Mongolians and Indians" be excluded from the regular public schools.[6]

Both citizenship rights and land rights granted under the Treaty of Guadalupe Hidalgo became the cause of considerable conflict between newly arriving Anglos and native-born Mexicans. During the California Gold Rush of 1849, for example, Mexicans were often labeled "foreigners" and thrown off claimed lands. Even Mexicans who carried their U.S. citizenship papers were forced out of gold mining extralegally and often with violence.[7]

Scholars of the Mexican experience have been highly critical of violations of the treaty that resulted in widespread land loss to Mexicans. As part of the treaty, Mexico ceded 500,000 square miles—including the contemporary states of California, New Mexico, Arizona, Utah, Nevada, and parts of Colorado and Wyoming—for only 15 million dollars. Mexicans owned much of this land in large tracts. Articles VIII and IX of the treaty protect the rights of Mexicans to continue ownership of land. "In the said territories, property of every kind, now belonging to Mexicans not established there, shall be inviolably respected. The present owners, the heirs of these, and all Mexicans who may hereafter acquire said property by contract, shall enjoy with respect to it, guaranties equally ample as if the same belonged to citizens of the United States."[8] Under pressure from Anglo settlers wishing to take title to Western land, the U.S. Congress passed the Land Act of 1851. The Land Act created boards of land commissioners in each new state and territory to adjudicate the validity of former Mexican land grant titles.[9] Between 1848 and 1900

Mexicans lost millions of acres in the Southwest. Historian Albert Camarillo identifies several factors leading to this tremendous loss of land. These factors include long and costly legal battles before the board of land commissioners; exploitation by lawyers and other unscrupulous Anglos; lack of Mexicans' English skills; "spendthrift practices" of the *Californio* elite; and land confiscation by squatters.[10] Californio is a term generally utilized to refer to the original Spanish land grantees in California that formed the middle and upper classes of Mexican society before United States conquest.

In general, historians agree that Anglo settlers in the states of Texas and California benefited more from "prompt and liberal adjudication" of land grants than did settlers in the territories of Arizona and New Mexico.[11] Whether in New Mexico, California, or Texas, the land loss among Mexicans significantly contributed to their diminished status of second-class citizens by the 1880s and 1890s.[12] The concomitant decline of the cattle industry, the Civil War, and economic downturns of the late nineteenth century exacerbated the diminished economic and social status of nearly all but a small group of elite Mexicans.[13] As will be explored, this declining social and economic status limited the abilities of formerly elite Californios and *Tejanos* to invest as highly in advanced education for their children.

The tangible aspects of the American conquest codified in citizenship and property law represented only some of the dramatic changes for Latinos in the nineteenth-century Southwest. Cultural conflict between the arriving Anglo-Protestant settlers and new Mexican Americans surfaced in muted terms during the 1830s and 1840s but escalated during the Mexican War and into the 1850s. Anglos arriving to Texas and California brought with them negative stereotypes of the character, religion, and racial composition of Mexicans. In general, Mexicans were disparaged as "greasers," immoral, sexually degenerate, indolent, "mongrels," "papists," and potentially subversive politicos.[14]

The belief in Anglo-Saxon Protestant superiority, which settlers brought to the Southwest in the mid-nineteenth century, resulted from a convergence of factors. Proponents of the Mexican War viewed Southwestern land as wasted in the hands of mongrel Mexicans. Californian settler T. S. Farnham, for instance, declared in 1840 that Californios were an "indolent, mixed race," and "the old Saxon blood must stride the continent."[15] The racial mixing between Spaniards, Mestizos, Native Americans, and African Americans over three centuries particularly offended Americans grappling with their own race questions over African American slavery and fears of miscegenation. As a slave state, Texas attracted Southern white migrants who viewed dark-skinned peoples with suspicion and suspected they may have been tainted with African blood.[16] As these white supremacist views were carried into the Southwest, they were often manifested as acts of violence, racial slurs, and blatant discrimination, contributing to the continued social and economic decline of former Mexican citizens.[17]

An additional factor shaping Anglo negativity toward Mexicans was the Roman Catholic religion. Anti-Catholicism resurfaced in the mid-nineteenth-century United States as thousands of immigrants, largely Irish Catholics, arrived in the 1830s and 1840s. On the East Coast, anti-Catholicism took the extreme form of convent and church burnings. The formation of the anti-Catholic, anti-immigrant political party—the Know-Nothings—culminated anti-Catholic hysteria in the 1850s. According to party members, Catholics were loyal only to the pope in Rome and thus represented a subversive threat.[18]

Mexicans who had traditionally combined Catholic religious feast days with municipal events often protested the attacks on Catholicism.[19] Further-more, as discussed in chapter two, public schools during the Mexican Era were often taught by priests or nuns or in Catholic Church buildings. The close alliance between the Catholic Church and public schools disturbed arriving Anglo-Protestants.

During the first decades after American conquest, Mexicans resisted the marginalization of their language, culture, and religion through several means. Varying by locality and time, Mexicans were able to retain some of their rights, and intermarried and assimilated with leading Anglo families as one strategy of survival. Others formed *mutualistas* (mutual aid societies) or participated in more militant and extralegal organizations such as Las Gorras Blancas (the White Caps). This famous resistance group sabotaged the introduction of barbed wire fences in New Mexico ranching areas.[20]

Eventually, Anglos overwhelmed Southwestern Latinos numerically, po-litically, and economically. The process of becoming residents of a nation with a separate and foreign linguistic, cultural, and religious heritage was often painful for many Latinos, whose roots in the new U.S. lands stretched back to the late 1500s. Proponents of public education, U.S. society's primary vehicle for Americanization among newly arrived immigrants to the East Coast, encountered unique challenges developing a secular educational institution within a historically Spanish Catholic culture.

EDUCATION AND NATION-BUILDING IN TEXAS, 1836–1900

The Americanization of Texas Mexicans, or *Tejanos*, began with the Texas Revolution of 1836.[21] The Anglo-led government of the Republic of Texas valued the potential benefits of a public school system. In fact, the republic's declaration of independence from Mexico referenced the lack of schools as one of its rationales for rebellion, stating, "It [the Mexican government] has failed to establish any public system of education, although possessed of almost boundless resources, (the public domain,) and although it is an axiom in political science, that unless a people are educated and enlightened, it is idle to expect the continuance of civil liberty, or the capacity for self-government."[22] Resolutions for public education were passed in the Congress of Texas, particularly the provision of land grants for schools, and newspaper editorials bemoaned the lack of genuine interest in supporting schools. However, citizens of the Republic of Texas encountered the same difficulties character-istic of Mexican Texas. Similar to most frontier areas, a combination of private, religious, and quasipublic institutions arose where enough students and a qualified teacher could be procured.[23]

In 1845, Texan citizens voted to be annexed by the United States, a step the United States was eager to approve. As residents of a newly acquired U.S. territory, Tejanos began to see their traditions, culture, and language come under fire. The establishment of American-style public schools after state-hood in 1845 brought the Tejanos into further conflict with the rapidly growing numbers of Anglo settlers.[24] In the decades following the Texas government's 1854 "Act to Establish a System of Common Schools," Tejanos wishing for public education experienced shifting attitudes toward Spanish language use, equal access, and the employment of Tejanos or other Latinos as teachers.

In the urban centers of the northeastern United States, public schools had long served the function of assimilating immigrants.[25] The southwestern experience differed because of pre-established Hispanic communities and the strong influence of German immigrants. Through law, the Texas Anglo–dominated legislature reinforced English as the public schools' proper lan-guage of instruction. Two years after Texas formally established public schools, an 1856 amendment stipulated that "No school shall be entitled to the [monetary] benefits of this act unless the English language is principally taught therein."[26] The amendment was approved again in Chapter 98, Section 9 in the 1858 legislature.[27] The state requirement of English-language instruction in Texas's nineteenth-century public schools was rarely fully adhered to in local communities; rural communities especially strayed from the requirement. Both German immigrants and Tejanos maintained their native languages in many public schools during the transitional decades of the 1850s through 1880s.

Tejanos did not so much reject the English language as attempt to preserve Spanish while also learning the language of their conquerors. Thus, for example, the Spanish-language newspaper in San Antonio, *El Bexareño*, advocated that public education be conducted in both languages.[28] Shifting state policies reflected the fluidity of Americanization measures. When the public school system in Texas was recreated in 1871 under Radical Rule during Reconstruction, a more flexible approach was pursued toward the language interests of Tejanos, Germans, and French. In his first annual report, Superintendent of Public Instruction J. C. DeGress stated that as a result of "the large proportion of citizens of German and Spanish birth and descent in our State," teachers would be permitted to teach the German, French, and Spanish languages, "provided the time so occupied should not exceed two hours each day"[29] (see document 3.1). Public school officials in other locales, such as among the German communities of the Midwest, had also compromised on language policies in order to keep immigrant children in the public school system.[30]

The presence of the Spanish language and use of Tejano teachers persisted during these transitional decades of Texan public schools. Arnoldo De León has documented the extensive number of Mexican-born or Spanish-named public school teachers in the counties of San Bexar, El Paso, Duval, and Nueces between 1850 and 1900.[31] State policy could not alter the ethnic heritage of Tejanos and other newcomers. School officials in the twelfth school district reported in 1872 that "of the population three-fifths are Mexicans, still speaking their own language and observing their own customs, and the remainder a mixture of all classes and creeds."[32] By 1886 the superintendent of public instruction indicated that although the law stated that "schools shall be conducted in the English language," many citizens complained that Spanish or German was being utilized as the language of instruction in public schools (see document 3.2).

The enormous size of Texas and its rural nature contributed to a variety of local arrangements, some of which incorporated the Tejano community members into the school systems. For example, on large ranches, owners created special schools for *vaqueros'* (cowboys) children. Children at the Randado Ranch in Zapata County and Los Ojuelos in Encinal County studied English and Spanish and took exams in three subjects at the end of each year.[33] The lines between public and private schooling were often blurred, particularly under the state's "community system." In the late nineteenth century, rural areas were permitted to use public funds for a school if the teacher could pass the Board of Examiner's test. Thus, the sisters of Nazareth Academy became certified by the state of Texas as community teachers and taught Mexican pupils in Spanish and English with public funds.[34]

In El Paso, a unique situation arose within the public school system. The city's public school system was established in 1883, yet only a "handful" of Tejano pupils attended. According to San Miguel, Jr., school officials' reluctance to teach non-English-speaking children led Tejano parents to establish a private school and hire Olivas V. Aoy, "an elderly Spaniard," to prepare Spanish-speaking children for the public schools.[35] The private school was called the Mexican Preparatory School and was formally created in 1887. In 1888 the El Paso school board incorporated the school into the city system. By 1899, the "Aoy" Mexican Preparatory School had twice been enlarged, enrolled three hundred students, and had signed additional children onto a waiting list. Many Mexican parents in El Paso sent their children to the Aoy School, which boasted the highest attendance of any elementary school (white or Mexican) in the city.[36]

Tejano parents seeking educational opportunities for their children also relied significantly upon the rapidly growing parochial schools sponsored by the Roman Catholic Church. According to Guadalupe San Miguel, Jr., Catholic schools were popular among Latinos in the Southwest for three reasons. First, Catholic schooling was seen as a form of preserving Latino identity because of the closely intertwined nature of religion with Latino culture. Second, the Catholic Church was willing to permit the speaking of Spanish in school and allow Mexican Americans to preserve other cultural traditions. Thus, instead of imposing "subtractive" measures upon Mexican children, measures that San Miguel, Jr., and Richard Valencia define as ones that not only "inculcate American ways, but also . . . discourage the mainte-nance of immigrant and minority group cultures," Catholic teachers permit-ted "additive" measures such as bilingual or trilingual language instruction.[37] Last, the recruitment of mostly female teaching orders provided an inexpen-sive method of staffing schools, and some of the sisters were native Spanish speakers from Spain or Mexico.[38] Both male and female teaching orders were also heavily recruited from France to the Southwest during this era, and they brought a liberal attitude toward the value of learning and teaching several languages.

Between 1848 and 1900 dozens of Catholic schools for boys and girls were established in Texas.[39] The Incarnate Word in Brownsville, Texas was founded in 1853 for girls between the ages of 5 and 18. Students came from Texas and Mexico to learn traditional subjects in addition to music, painting, sewing, and embroidery.[40] The Ursuline Sisters also established schools in both Galveston and San Antonio by the early 1850s. The Ursuline Academy of San Antonio (1851) taught traditional academic subjects and all students were required to learn Spanish, English, and French, "not only by theory, but by practice: the pupils were required to converse in these languages in the respective classes." Mother Joseph Aubert, a teacher at the school, wrote to her former superior in France in 1857 that nearly one hundred children from the "best" Spanish, German, French, and British families attended the school. Boarding students at the Ursuline Academy were reputed to come from the wealthier class of Hispanics who could afford tuition. However, the sisters also opened a "free day school principally for the benefit of Mexican children." Mother Aubert commented on this student population, describing that they consist "mainly of Mexicans who are now loyal American citizens. I love all these children . . . they are so affectionate and responsive to kindness."[41] The sisters were eventually asked to turn their day school over to another order that was not cloistered. One sister recalled, "It was with an aching heart . . . that the Ursuline Nuns beheld the 300 Mexican children of their free school that had been attached to the Academy pass from their care to that of the Sisters of the Incarnate Word."[42] The female teaching orders who came to Texas in the nineteenth century played a pivotal role in spreading education, particularly for Mexican girls and young women whose families preferred they attend single-sex schools rather than the emerging coeduca-tional public schools.

Texas Bishop Odin enthusiastically championed Catholic education for boys. The Brothers of St. Mary founded a school for young men in San Antonio in 1852. The academy quickly became St. Mary's University and offered advanced study that included Latin, history, algebra, and philosophy.

In 1888 the Brothers of St. Mary also opened the San Fernando Cathedral School as a preparatory school.[43] The Brothers of the Holy Cross came to Austin in the early 1870s and founded St. Edward's Academy for boys. The school expanded and was chartered by the state of Texas as St. Edward's College in 1885.[44] Schools for boys were also founded in Brownsville (1872), Houston (1899), and Waco (1899). Catholic orders of both men and women were recruited and brought from the northeastern United States and from Europe to work in Texas. Protestant denominations, particularly the Presbyterian Church, also saw Texas specifically and the Southwest in general as a missionary enterprise.

Although most missionary work among Latinos was conducted in New Mexico and Colorado, Tejanos enrolled their children in Protestant schools for several reasons: lack of alternative options (particularly during the frontier decades), perceived advantages for their children learning English from Anglo teachers; or anticlerical sentiments toward the Catholic church that dissuaded them from sending their children to parochial schools. One revealing description of the religious fervor and anti-Catholicism that missionary teachers brought to the Mexican field is recorded in the memoir *Twenty Years Among the Mexicans* (see document 3.3). Author Melinda Rankin, a New England–born Presbyterian missionary, arrived in Brownsville, Texas in 1852.[45] She established a successful school for girls called the Rio Grande Female Institute. She found that her English skills helped bring children from the local Catholic school: "I possessed one important advantage, namely, the Mexicans desired their children to learn English, and as that language was but imperfectly taught in the convent, many left and came to me on that account."[46] Miss Rankin sought funds from donors all over the country to build a permanent school in Brownsville. Upon its establishment in 1854, Rankin wrote satisfactorily, "A Protestant Seminary is reared in front of papal Mexico, and within its walls are gathered Mexican girls, whose improvement encourages me to hope that their consciences may become enlightened . . . this institution is one of the instrumentalities by which God intends to disenthrall benighted Mexico from the dominion of popery."[47] Miss Rankin remained in charge of the school until 1861, when Confederate sympathizers viewed her Yankee background as unpalatable, and, like many northern teachers in the South before and during the Civil War, she was driven out.[48] The school was apparently resurrected after Reconstruction as the Presbyterian Mission School in 1878. The wife of the Presbyterian church's pastor, Mrs. Hall, was the principal. A Mexican teacher, Luciana Media, was hired among the assistants who worked with more than 60 Mexican girls. Students attended free of charge and received instruction in both Spanish and English.[49]

As Texas changed its status from an independent republic to part of the United States, Tejanos experienced a political, social, and economic shift between 1848 and 1900. Tejanos attempted to seek the best educational alternatives for their children based on local circumstances. In some cases, parents viewed Americanization as the best strategy for adaptation and sent their children to the new public schools. Others selected Protestant schools that may not have charged fees, such as the Presbyterian Day School in 1840s San Antonio. Similarly, Anglo Protestants enrolled their children in Catholic parochial schools that appeared to offer a superior education compared to the limited public school alternatives. During the last half of the nineteenth century the Texas public school system began to impose harsher restrictions against the Spanish language and created separate Mexican schools. In response, some middle-class Tejano communities established their own private schools, such as El Colegio Altamirano in Jim Hogg County (1897–1958), that were free from the control of the Catholic Church or the state of Texas.[50]

FROM ALTA CALIFORNIA TO THE STATE OF CALIFORNIA: EDUCATION, AMERICANIZATION, AND THE CALIFORNIO POPULATION

From a want of any organized system of school instruction while California remained a Mexican province, it is not surprising that, in very many cases, the children of the older Californians have little or no education beyond that of repeating and a few reading the ceremonies and religius [sic] books of the Catholic church. It is true that there are exceptions to the position taken; but scarcely in sufficient numbers to form any considerable amount. This class of our population has heretofore been deprived of the advantages of schools; and now, since the parents of such children have been brought into contact with the Anglo Saxon race, the want of education becomes more apparent to them, and they are alive to the interests of this important subject.

—John G. Marvin, Superintendent of Public Instruction of the State of California, 1851[51]

Similar to the situation in Texas, a small number of Mexicans controlled a considerable portion of the land and political power in California at the time of American conquest in 1848. Historians estimate that these elites represented only about 5 percent of the Mexican population. In order to distance themselves from the negative connotations ascribed to Mexicans, they began using the name *Californio*.[52] However, even those of the elite Californio class found their Hispanic language, culture, and religion under assault by the state of California's new constitution and laws.

After U.S. conquest, the Gold Rush of 1849 brought thousands of Anglo settlers to the Golden State. By 1860, the Anglo and Mexican populations of California exceeded 380,000.[53] During this era of rapid change the public school system of California was viewed as a stable influence in molding the diverse groups of European immigrants, Californios, and Anglos into the future citizens of California. Unlike Texas, which experienced a long disruption in the public schools due to the Civil War and Reconstruction, California moved forward quickly in its public school development. A state superintendent was appointed in 1851 and schools were created in counties throughout the state. During the 1850s and 1860s Californios and Anglos clashed over language issues in the new public schools. Early public school reports illuminate the bilingual/bicultural environment present in many communities (refer to document 3.4). Similar to Tejanos, Californios did not reject the English language but wished also to preserve Spanish in both the home and the public domain.

The experience of Santa Barbara's Californios in the 1850s illuminates tensions over language issues throughout the Southwest as Mexicans found themselves becoming foreigners in their native lands. Because Californios comprised three-fifths of the population, they initially wielded considerable influence in the community. Spanish-language instruction was maintained in Santa Barbara's public schools in the early 1850s. Two male teachers from Chile taught geography, history, writing, and arithmetic.[54] However, the state of California began passing "subtractive" policies in 1855 that forbade the teaching of Spanish in public schools. The city's two Anglo school commissioners (two were Anglo and one was Mexican) called for English-only public schools. A temporary compromise was reached with the creation of a separate English school. However, the expense of maintaining two schools was prohibitive and the English- and Spanish-speaking public schools were combined into one bilingual school. Upset, Anglo parents withdrew their children from the bilingual school. The city's increasingly anti-Mexican newspaper, the *Gazette*, declared "the parents of American children unwilling that they should learn a confused jargon and gibberish, prefer to keep them at home." By 1858 the Anglo parents had won the battle for English-only instruction. Subsequently, many Mexican parents chose to enroll their children in a Catholic school that permitted Spanish.[55]

Mexicans in Los Angeles experienced similar adjustments. In 1850 the city council still called itself the ayuntamiento, the nomenclature from the Mexican Era. The city hired a Spanish-speaking retired soldier, Francisco Bustamente, who had formerly taught during the Mexican Era. Bustamente was contracted to "teach to the children first, second, and third lessons, and likewise to read script, and so much as [he] may be competent to teach them orthography and good morals."[56] Because of the new public school law requiring English in the public schools in 1855, the approximately 500 school-age children (majority Mexicans and Californios) were then schooled in English. In response, Antonio Jimeno del Recio offered to teach Spanish-speaking children at public expense. The town council (which still included one Latino member) approved the proposal and the Spanish class was taught for a brief period until funds ran out.[57] By the end of the 1850s, Anglo teachers, such as Miss Gertrude Hoyt, instructed Spanish- and English-speaking students in English. According to Pitt, "Yankee parents complained occasionally that she had difficult coping with the Spanish (and French) children to the detriment of Yankee children."[58]

Los Angelenos who wished for their children to learn Spanish and English, and could afford the fees, sent their children to one of California's many Catholic schools. During the first years of statehood in California, 1850–1855, Catholic schools continued to educate Mexican Americans and were often provided with public funds for their efforts. Section 10 of California's 1851 school law permitted religious schools to receive "compensation from the Public School fund in proportion to the number of its pupils, in the same manner as provided for district schools by this act."[59] By 1853, anti-Catholic sentiments were gaining ground in California and these liberal measures came under attack. Bishop Joseph S. Alemany asked the California superintendent of public instruction to continue supporting the Catholic schools, requesting, "I beg leave to ask you to aid us with your great influence, that the reported schools may not be altogether cut off from the public fund." According to Bishop Alemany, in 1853, 12 schools with a total of 579 pupils were being instructed in English, French, and Spanish languages in the Catholic schools throughout California.[60]

The 1855 revised school law permanently removed the option of public funds for private religious schools. The code included the stipulation that "no sectarian books should be used, and no sectarian doctrines should be taught in any public school under penalty of forfeiting the public funds." The superintendent of 1865 commented, "The stringent provision settled then, and probably forever, the question of an American system of public schools in this State, free from the bitterness of sectarian strife and the intolerance of religious bigotry."[61] One last effort was made to permit Catholic schools to receive funds in an 1860 bill introduced to the California Assembly by Hon. Zack Montgomery of Yuba County. The bill was argued vigorously in the Assembly but was ultimately defeated.[62] In the northeastern United States, fights over public funds for Catholic schools had engaged the energies of urban school officials during these same decades.[63] In the Southwest, the Catholic religion had been so closely intertwined with schooling for centuries that the separation of church and state represented a profound break in cultural traditions.

The racialization of Latino children after American conquest further shaped their educational circumstances in the post–Treaty of Guadalupe Hidalgo decades.[64] Most Mexicans in California and the Southwest were of varying degrees of *mestizaje*—the result of Native Americans, Spaniards, Africans, and Mexicans comingling over several centuries. Californios of darker complexion were placed at a distinct disadvantage if they classified themselves or were classified by the government as Native Americans (or blacks) instead of Mexicans. As a result, Menchaca suggests that many mestizos identified themselves to the government as Mexicans to avoid discrimination. For example, as early as 1858, "Negroes, Mongolians, and

Indians" in California were not allowed to attend schools for white children "under penalty of the forfeiture of the public school money by districts admitting such children into school."[65] The revised California school law of 1866 permitted some exceptions to this rule. Section 56 permitted a board of trustees by a majority vote to "admit into any public school half-breed Indian children, and Indian children who live in white families or under guardianship of white persons." In Section 57, "children of African or Mongolian descent, and Indian children not living under the care of white persons, shall not be admitted into public schools." Schools were required, however, to open public schools whenever at least ten parents of "such children" petitioned the school board.[66] Except for the very few children who may have had white guardians willing to petition for entry to white schools, the majority of mixed race children were placed in segregated schools. The description of one such school in Los Angeles underscores this point:

> There is also a small school of fifteen negro children of all the shades arising from blending all the primary colors of Spanish, American, Indian, and African parentage. They are engaged in the pursuit of knowledge under difficulties, as their little room ten by fifteen feet, has neither desks, blackboard, maps, charts, nor any kind of furniture, except a line of rough board seats without backs, around the walls.[67]

Anglo attitudes regarding the educability of Mexicans, Indians, and African Americans were commonly disparaging. John Swett, superintendent of public instruction for California from 1863 to 1868, often made comments such as "The boys' school, numbering, say forty scholars, held in a comfortable brick school-house, is attended mostly by children whose mother tongue is Spanish, and who are not remarkable either for order or scholarship."[68] In San Buenaventura, Swett drew the conclusion that since "the American residents there have established a private school and refuse to send their children to the public schools, where the 'native' children attend, we are led to suppose that its management is not the best in the world."[69] The decades of cultural conflict between Californios and Anglos resulted in the diminished political and economic status of Mexicans and their descendents. However, as historians Richard Griswold del Castillo and Albert Camarillo have found, Mexican communities increasingly relied upon themselves to preserve their language, culture, and identity.[70]

HISPANO SCHOOLING IN THE TERRITORIES OF NEW MEXICO, COLORADO, AND ARIZONA

Both demographics and geography shaped an educational history in the territories of Arizona (1848–1912), Colorado (1848–1876), and New Mexico (1848–1912) distinct from that of California and Texas. The rural and isolated nature of the land and majority Hispanic citizenry preserved the Spanish language and led to blended public-private schools for a longer duration than in California and Texas.

Arizona

Elhrenberg [AZ] was about one hundred and twenty-five miles up the river from Yuma . . . My school room was a building formerly used as a saloon . . . I had fifteen pupils not one of whom knew any English; and I knew nothing of Spanish. One of the trustees was of an old California family and was master of both English and Spanish. He helped me organize my school the first morning and told me that if I ever needed any help to send for him, as his store was quite near . . . I taught through the months of May, June, July and August. We had no electric lights in those days and a lighted lamp added too much to the heat and attracted too many insects. So I used to sit outside evenings and gather the children around me. It served me as a lesson in Spanish and them as one in English so that when I went to Yuma about October 1, I had a large vocabulary, probably thousands of words, all that were needed for ordinary conversation, but I knew nothing of the grammar."

—*Frank C. Lockwood*, Pioneer Days in Arizona: From the Spanish
Occupation to Statehood, *1932*[71]

After the Treaty of Guadalupe Hidalgo, the Southern portion of Arizona remained part of Mexico. In 1853 the United States purchased 30 thousand square miles from Mexico for 10 million dollars in the Gadsden Purchase. Until 1864, Arizona was part of the territory of New Mexico, and then it began its own territorial governance. Formerly called *Alta Pimeria*, the territory remained remote and home to hostile Indians, particularly the Apache. Ranching, long-distance freighting, and silver mining brought Anglo settlers to Arizona, but they were greatly outnumbered by Hispanics and Native Americans until railroads were built in the late 1880s.[72] As a result, the Spanish culture and language persisted, even in the new public schools.

The responsibility of creating Arizona's new school system was placed under the direction of the territorial governor, who was ex-officio superintendent of public instruction. Prior to the early 1870s public schools were sporadically created. For example, after the school law of 1868 records indicate that Augustus Brichta taught "fifty-five Mexican boys" for 6 months in a crudely constructed building.[73] A teacher named John Spring also taught an early school in Tucson in 1871. He "taught one hundred thirty eight boys, nearly all Mexicans, for fifteen months in a long, rudely furnished adobe building with dirt floor and dirt roof." The teacher tried bilingual education strategies; for instance, he "had to go over their lessons with them in Spanish before trying to teach them in English." The pupils had already received some schooling either "for brief periods in Mexico" or from private instruction. Attendance at Spring's school was high for that era, averaging 78 percent of enrollment over a 15-month period.[74] In Yuma, one former teacher recalled that in the first year of public schools in 1872, "There were pupils of all ages from married women to those just old enough to enter . . . only a very few knew any English, so the teachers just had to do the best they could."[75] The next year in Tucson, a "free public school" for girls was opened. A Mrs. L. C. Hughes taught the school in the old Pioneer Brewery.[76]

As Arizona formalized its school laws, Mexicans were incorporated into the power structure. For example, Governor Stafford identified Estévan Ochoa as a key supporter of public schools in his annual reports. He wrote of Ochoa, "Being a Mexican by birth, his devotion to the school system, and his clear, practical intelligence, have been of invaluable service in stimulating his people to sustain the public school system."[77] Ochoa was named as a member of the Tucson school board and elected state legislator in the 1870s.[78]

Regarding tolerance toward the Spanish language, Arizona's public school reports reveal a liberal attitude toward bilingualism not evident in Texas or California. As late as 1899, the Anglo superintendent of Apache County schools, John T. Hogue, critiqued public school teachers who worked among Hispanic populations and did not know the language. Hogue concluded his 1899 report, stating:

> I have one other suggestion to make. It has come under my observation that in school districts where the population is largely Spanish-speaking or Mexican, that the pupils, as a rule, do not progress well under the tutorage of teachers who have no knowledge of the Spanish language; in fact these pupils, the great majority of them, advance slowly and very unsatisfactorily in all cases where the teacher has no knowledge of the Spanish language. In view of this fact, and in consideration of the more significant fact that our Government has, as a result of the Spanish-American war, acquired an immense territory where the Spanish language is the dominant and prevailing language, I recommend that teachers in school districts where the inhabitants are largely Spanish-speaking people, should, in addition to being English scholars, also be Spanish scholars.[79]

In 1904 Hogue was still recommending the employment of bilingual teachers. (See document 3.5.)

The completion of railroad passages in the late nineteenth century, linking Arizona to California, and political developments in the Southwest contributed to an increase in ethnic tensions between Mexicans and Anglos in Arizona Territory. According to Thomas Sheridan, by the turn of the century, the gulf between Mexicans and Anglos had widened and "the segregation that characterized Texas and California took root in Arizona as well."[80] The Mexican community in Arizona maintained its bicultural identity through the publication of Spanish language newspapers such as *El Fronterizo* (1878–1914). Furthermore, mutualistas, such as the Alianza Hispano Americana, founded in 1894, provided legal and financial support and a social outlet for Arizona's Mexican community.[81]

New Mexico

New Mexicans also shared the relative tolerance toward the Spanish language displayed by Arizona school officials, but with one important difference—not only the Spanish language but also the Catholic Church was closely united with New Mexico's public schools through the nineteenth century. In contrast, Arizona secularized its public schools by the mid-1870s.[82] Indeed, one key example of the tensions between Mexicans in New Mexico, called

Hispanos, and Anglo settlers concerned the scope of Catholic involvement in public schooling and the centralized (often Anglo) territorial control of schools versus local control.

Established as a territory in 1850, New Mexico passed its first school law in the 1855–56 legislature. The law stipulated that the schools would be supported by a property tax and control would reside with the territorial government. Because they were accustomed to local control and holding large tracts of land, the new law was unpopular with Hispanos. Quickly, Hispanos repealed the law by a general vote of 5,016 against and 37 in favor. Anglos viewed the repeal as a rejection of education in general, instead of a rejection of less control and more taxes. The territorial governor William Pile condemned the vote, saying, "If more proofs of the present unfortunate condition of the mestizos were wanting, it may be shown that their indifference to education reaches not only hostility, but a hostility which has, perhaps, been expressed with more unanimity at the ballot-box than any similar instance in history." Governor Pile continued his diatribe, stating that the only explanation for such a vote was the "the people are so far sunk in ignorance that they are not really capable of judging of the advantages of education . . . the verdict shows that the people love darkness better than light."[83]

As Lynn Marie Getz documents in her skillful study of Hispanos and education, the 1856 vote represented a "myth of Hispano resistance" to public schooling. In subsequent legislation, Hispanos supported public schools, and even the idea of taxing themselves—but they wanted to maintain local funding control.[84]

The Hispano community dominated New Mexico's school leadership and subsequently possessed the type of political power necessary to protect its interests and concerns. For example, in 1875, 27 of the 34 county school commissioners were Hispanos. By 1878, 33 of the 39 elected school commissioners were Hispanos. At the territorial level, Hispanos were also well represented.[85] The office of superintendent of public instruction, created in the 1891 School Law, was held continuously by Hispanos from 1891 to 1905. Amado Chávez (1891–1897), Placido Sandoval (1897–1899), Manuel C. Baca (1899–1901), J. Francisco Cháves (1901–1904), and Amado Cháves (1904–1905) in another term, served as the territory's chief state school officers.[86]

Of all the southwestern states, New Mexico appeared to most fully embrace Spanish/English bilingualism in its public schools. In 1875, for example, two-thirds (86 of 131) of the public schools were conducted exclusively in Spanish, and an additional one-third (38 of 131) were taught in Spanish and English. Thus, only 5 percent of the territory's public schools (7 of 131) were taught in English.

The publication of legislative laws in English and Spanish further highlighted the political strength of Hispanos. Prior to 1886, Hispanos dominated the legislature, a fact that influenced educational matters.[87] As late as 1889 the school laws of New Mexico were printed in both languages, alternating Spanish and English (see document 3.6). Section 1110 of the 1889 school law permitted texts in either language, stating, "It shall be the duty of the School

Directors to adopt text books in either English or Spanish, or both."[88] Thus, the English-only rules that characterized Texas and California by the 1850s had not taken root in New Mexico. However, the territory finally passed a law in 1891 requiring that English be included in all of the public schools. The Hispano superintendent of schools, Amado Cháves, observed of this new ruling, "no opposition has been encountered in any part of the Territory in the matter of introducing English speaking teachers in districts where heretofore Spanish alone had been taught." Furthermore, he was pleased "that the greatest interest is being shown, in the Spanish speaking counties, in behalf of the new law, which requires that the English language shall be taught in all the common schools of the Territory."[89] Although English eventually became the official language of the public schools in New Mexico in 1907, biliteracy was championed through the end of the nineteenth century. Overall, the secure hold on Hispano culture and language persisted in New Mexico far longer than in other parts of the Southwest.[90]

The Catholic Church played an extremely influential role in the educational development of Latinos particularly and in New Mexico in general. A significant irritant to Anglo officials in the territory was the granting of public funds to Catholic schools. Territory secretary William G. Ritch (1873–1884) railed against the "priestly influence" to "gradually subvert the public schools and the school fund."[91] In his 1875 annual report he condemned the fact that "in a majority of the counties, to-day, the school books and church Catechism, published by the Jesuits, and generally in Spanish, constitute the text books in use in the public schools."[92] The separation of church and state that Ritch desired did not materialize in New Mexico during the nineteenth or even early twentieth century. County school commissioners often requested religious orders to organize public schools and paid priests and nuns from the common school fund. The Jesuit press that Secretary Ritch mentioned, Imprenta del Río Grande, published public school textbooks used in the late nineteenth century.[93]

Statistics from the territorial secretaries' reports suggest that Hispanos heavily patronized the Catholic schools, which were also accessible to non-Catholics. St. Michael's College in Santa Fé, for example, was established by the Christian Brothers in 1859. The school offered advanced instruction for young men but also ran a "free department for the poorer class, which has been attended by a yearly average of 180 male [Mexican] pupils." St. Michael's received seven hundred dollars annually from the public school funds for its "charitable services."[94] In addition to academies for boys in Santa Fé and Las Vegas, girls in particular were most likely to attend Catholic schools because of Hispanic cultural practices that kept the sexes apart.

In 1874, Secretary Ritch wrote, "girls are not generally admitted to the public schools. This arises from a belief quite generally prevailing in the territory that there should be separate schools for girls; and not from indisposition to provide for them."[95] Several teaching orders in New Mexico opened their schools to girls. By the early 1890s over one thousand girls in New Mexico were being educated in Catholic schools. The Sisters of Loretto opened the Academy of Our Lady of Light in 1853. The sisters had arrived in

Santa Fé in 1852 but waited one year because of the need "to acquire the Spanish language" first.[96] In 1881 a permanent building costing $24,000 was built and by 1891, 280 girls were enrolled at Our Lady of Light.[97] The Sister of Loretto operated schools in six other New Mexican towns by 1892, totaling almost six hundred girls. The Sisters of Mercy also taught young Hispanic women in New Mexico. In four different locations they enrolled 379 pupils in 1892. Lastly, the Sisters of Charity teachers were actively involved in the educational enterprise. Including 62 girls in the Santa Fé Orphan Industrial school, 18 Sisters of Charity instructed 544 girls by the end of the nineteenth century. A portion of the public school fund was given to the Catholic schools in exchange for their assistance.[98]

Another factor shaping the educational opportunities of New Mexico's Latino population in the late nineteenth century was the presence of Protestant missionary schools. The Presbyterian Church, under the leadership of Sheldon Jackson, viewed New Mexico in particular as a missionary field ripe with possibilities. Jackson wished to seize New Mexico for Protestantism "as the Spanish had, centuries earlier, replaced the Aztec religion in Mexico with Catholicism."[99] Schools run by the Presbyterian Board of Home Missions were determined to save Latino children from Catholicism and ignorance in general.[100]

The missionary enterprise was extensive in New Mexico. In 1891 Superintendent of Public Instruction Amado Cháves noted that over 880 Mexican children were attending missionary "day schools," and 145 were in boarding schools.[101] Presbyterians were not the only denomination in New Mexico. The Methodist Church and the New West Educational Commission opened schools enrolling hundreds of Mexican children.[102] Protestant missionary schools created dilemmas for Hispano parents in New Mexico. The large number of children enrolled suggests that depending upon the availability of other schools, a missionary school was better than no school at all. However, many Hispano families were threatened with excommunication from the Catholic church for sending their children to Protestant schools. For the children themselves, historians point out that there could be long-term benefits to attending missionary institutions. Because most public schools were conducted in Spanish, and Catholic schools in both languages, missionary schools "offered upwardly mobile Hispanos entry to the language, values, and milieu of the Anglo world."[103] Specifically, Susan Yohn found that Hispano youth, especially those who converted, became "recipients of whatever largesse the mission enterprise had to reward" and became well-connected with networks of Anglos who could further their education and assist with employment.[104]

HIGHER EDUCATION AND THE NINETEENTH-CENTURY MEXICAN AMERICAN[105]

The possession of a college degree or even collegiate participation in mid-nineteenth-century America was rare for anyone regardless of race, gender, or ethnicity. College graduates represented only 1 percent of the male workforce before the Civil War.[106] The small numbers of Latinos in nineteenth-century colleges were thus drawn from among the most privileged classes in the new territories as well as families from northern Mexico who sent their sons to receive a bilingual education.[107]

The newly developing state universities in the Southwest provided one pathway for students seeking higher education during this era. The University of California, Berkeley, opened its doors in 1869 with 40 students. The university quickly found itself in a situation familiar to many colleges of that era—few students were adequately prepared for collegiate-level work. As a result, the university opened a preparatory department called the Fifth Class. Standards for admission were lower, and during its brief existence from 1870 to 1872, the Fifth Class enrolled almost two dozen Mexican-born and Californio students; only two Latino students passed the entrance examination and proceeded to the freshman class.[108] The abolition of the preparatory department in order to "raise standards" two years later resulted in the "virtual disappearance of Spanish surnamed students from the University of California."[109] Although some Latino students attended the Berkeley campus at the University of California during the following one hundred years, it was not until after 1970 that the Latino student was more than a rarity.

In Texas, the flagship university campus in Austin was opened in the fall of 1883. Apparently, in 1894 Manuel García was the first Mexican American to graduate from the University of Texas.[110] Little is known about other Latinos in the Texas university system during this early era, although estimates from the 1920s place the Hispanic population at only 1.1 percent of all undergraduates.[111]

Latino participation in late-nineteenth-century higher educational institutions in Arizona, Colorado, and New Mexico has received little scholarly attention, but the sociohistoric context (of a longer, more open attitude toward bilingualism and biculturalism) suggests participation was higher than in Texas or California. The Mexican-born Mariano Samniego arrived in Arizona as a boy in the 1850s. He received his bachelor's degree in 1862 from Saint Louis University and became a prominent businessman. In 1886 he was appointed as a member of the first board of regents for the University of Arizona.[112] His presence may have encouraged Latino access to schools among the student population in a time and place that Latinos still maintained political power. Another Arizonan of Mexican birth, Estévan Ochoa, was elected mayor of Tucson in 1875 and helped found the Arizona public school system.[113]

Parallel to the important role of Catholic parochial schools during the Americanization period, Catholic colleges also provided safe havens for Latinos seeking higher education in the late nineteenth century. Santa Clara College in San José was a popular choice for Latinos during this era. Founded in 1851 as a school only for young men, Santa Clara was enrolling over one hundred students by its third year. Furthermore, "instruction [was] in both English and Spanish."[114] In the 1867–1868 school year, one-quarter of the student population was Spanish-speaking. Between 1851 and 1876 almost four hundred Hispanic-surnamed students attended Santa Clara College. Gerald McKevitt argues that the Jesuits in charge "actively recruited Spanish-speaking students" through the publication of a Spanish-language catalog.[115] Notre Dame College, California, opened as an academy for girls in 1851 and was chartered in 1868 as a college. According to one account from the 1850s, one-half of the pupils were Americans, "many from Protestant families." The school had "one room for the Spanish girls, another for the English, and a third for the smallest children."[116] The school's report cards and bills were printed in Spanish for the Spanish-speaking parents. Other Catholic colleges in the Southwest that enrolled Latino students include St. Mary's Catholic College in San Francisco, the College of San Miguel in Santa Fé, New Mexico, and St. Joseph's College in Texas.[117]

Historians Leonard Pitt and Albert Camarillo suggest that as southwestern Latinos declined in social, economic, and political status during the last half of the nineteenth century they were less likely to have the means to send their sons off to college or their daughters to Catholic academies.[118] The journal of Jesús María del Estudillo at Santa Clara College in the 1850s and 1860s (document 3.7) illuminates one such situation. Although Jesús María led a very typical college life and enjoyed socializing with young women from nearby Notre Dame College, he also worried about his family's ability to pay his tuition. In his private journal Jesús María recounted his family's declining fortunes and legal troubles over land titles and squatters. Furthermore, he disclosed his hurt feelings over one faculty member's harsh judgement of his English skills.

Higher education, a rare commodity for most of the U.S. population at the turn of the century, was pursued among Latinos when finances permitted. The Hispanics of the Southwest heavily patronized Catholic colleges because the Catholic Church represented a smooth continuity and accommodation with the Spanish language and religion.

DOCUMENT 3.1

Language Issues in the Public Schools of Texas, 1871

States in the Southwest and elsewhere compromised on strict English-only rules, particularly when those policies resulted in parents choosing private schools over the newly created public schools. In this document and the one that follows (3.2), Texas officials reveal how the implementation of statewide policies was often difficult at the local levels.

From *First Annual Report of the Superintendent of Public Instruction of the State of Texas, 1871* (Austin, TX: J. G. Tracy, State Printer, 1872), in the Rare Book Collection, Library of Congress, Washington, D.C. (L204 .A19 1871).

INTRODUCTION OF LANGUAGE IN SCHOOLS

The large proportion of citizens of German and Spanish birth and descent in our State, and the large proportion of emigrants of foreign birth that are constantly being added to our population, rendered the introduction of rule 7, rules governing public schools, necessary. Under its provisions teachers are permitted to teach the German, French, and Spanish languages in the public schools of this State, provided the time so occupied shall not exceed two hours each day. This clause has met with much favor throughout the State, as it brings children of scholastic age of foreign birth or descent into the public schools, where otherwise they would be subject to the additional burthens [*sic*] of expense in supporting private schools, where attention would be paid to them familiar with English and German or Spanish and English.

DOCUMENT 3.2

Persistence of Spanish and German Languages in Texas Schools, 1886

From *Fifth Biennial Report of the Superintendent of Public Instruction for the Scholastic Years ending August 31, 1885 and August 31, 1886*, Being the Eleventh Report for the Department of Education, Benjamin M. Baker (Superintendent of Public Instruction, State of Texas) (Austin: State Printing Office, 1886), in State Government Publications, States Other Than Wisconsin, Wisconsin Historical Society, Madison, Wisconsin.

SCHOOLS TAUGHT IN THE GERMAN AND SPANISH LANGUAGES.

The present law declares that the exercises of the schools shall be conducted in the English language. And yet very many complaints have been made each year to the effect that some of the schools are conducted in the German and some of them in the Spanish languages. Investigation disclosed the truth of some of the charges, but the Superintendent in a majority of instances found himself powerless to remedy the evil. I have myself conversed with teachers who could not speak the English sufficiently well to be understood. I am glad to say that as to the German language nearly all the complaints were made by Germans. I do not hold it objectionable to teach other languages than the English in the schools, but they should be pursued as studies, while the language of the school should be the English. This matter should be seriously considered by the Legislature, for if not arranged it promises much harm to public school interests.

DOCUMENT 3.3

Account of a Protestant Missionary in 1850s Brownsville, Texas

Born in Littleton, New Hampshire, on March 21, 1811, Melinda Rankin received a New England education and began teaching in the common schools of that region. Similar to many women who traveled West to spread the gospel through the public schools, Rankin also chose to combine proselytizing and teaching. Melinda's goal was to convert Catholics to the Presbyterian faith in Mexico. However, because the Mexican constitution forbade the teaching of other religions, she chose to open a school and mission in Brownsville, Texas, along the U.S./Mexico border. With the assistance of the American and Foreign Christian Union in New York, Rankin eventually reached her goal of establishing missions inside Mexico. She also taught among the freed people of the American South after the Civil War and lived actively until her death on December 6, 1888.[119]

In this passage from her book, Twenty Years among the Mexicans, the strength of her evangelical spirit is conveyed through her thoughts and actions while teaching Mexicans in the border town of Brownsville. Readers will gain a sense of the intense anti-Catholicism prevalent in the United States of the early 1850s and the means in which native populations such as the Mexicans of Brownsville interacted with missionaries in their midst.

From: Melinda Rankin, *Twenty Years among the Mexicans: A Narrative of Missionary Labor* (Cincinnati: Chase and Hall, Publishers, 1875), pp. 34–45, 57–61, 63–74.

CHAPTER V.

In the spring of 1852 I believed that time had fully come for me to commence my work for the Mexican people. I had gained some very important information in regard to my probable success, through Rev. Daniel Baker, D.D., a Presbyterian minister, who, in 1850, had ascended the Rio Grande River as far as Roma, a distance of two hundred miles, for investigating the condition of the country for evangelical work. He represented the Mexicans as accessible, and many of them manifesting the desire for instruction in the Bible.

I left Jefferson, Eastern Texas, in May, and went to New Orleans to take passage on a vessel for Brazos Santiago, near the mouth of the Rio Grande

River. I purposed going to Brownsville, a place situated about sixty miles up the river, opposite Matamoras, Mexico. The steamship for which I waited brought news of the invasion of Brownsville, by Indians, of a very alarming character. This condition of affairs, prevailing at the place of my destination, presented somewhat of an obstacle in the way of the further prosecution of my journey, as I had hoped that things had become sufficiently settled on the frontier to insure personal safety at least. But could I turn back because of difficulties in the way? I thought of the trials of "Pilgrim," who met lions in the way, and also of the advice given to him "To keep in the middle of the road, and the lions could not harm him." "Duty to God" was my watchword, and on His powerful arm I trusted for protection, and I resolved to go forward. Remaining in New Orleans over the Sabbath, I attended what was then Rev. Dr. Scott's church, where I heard a sermon from a stranger (Dr. S. being absent) which fully established my faith in God's Providential dealings with his people. Although that stranger, who was a foreigner, judging by his dialect, may never know, in this world, the comforting message he brought to me on that occasion, perhaps a future day will reveal that it was a word spoken in season to one soul at least. With renewed courage I took passage, and crossed the Gulf of Mexico, landing at Brazos, and passing over an arm of the sea, arrived at Point Isabel. There I took a stage for Brownsville.

A new sensation seized me when I saw, for the first time, a *Mexican*, a representative of the nation for which I had entertained such profound interest. I did not feel, as many others have expressed, that the *sight* of a Mexican was enough to disgust one with the whole nation. A heartfelt sympathy was revived, not by the prepossessing exterior, surely, but because a priceless soul was incased in it for whom the Savior had died. And a whole nation of souls, shut out from the light of the gospel of salvation, pressed with an increased influence upon my heart. Although I was coming into a land of new and untried scenes, yet I felt God's presence encompassing me, and I repeated the lines of Madame Guyon,

"To me belongs nor time nor space,
My country is in every place;
I can be calm and free from care,
On any shore, *since God is there*."

Just before arriving in Brownsville, the driver of the stage asked me where I wished to be left. I replied, "Take me to the best hotel in town." He answered, "There is no hotel in Brownsville." This intelligence was somewhat of a damper upon my feelings and prospects, and I mused upon the unpleasant condition of a stranger arrived in such a place after nine o'clock at night. After a little time the driver said, "I know a German woman who sometimes takes lady boarders, and I will take you to her house." Accordingly, I was set down at this woman's door, and I found my way inside, and asked for a night's lodging. The woman kindly received me, and I passed the night very comfortably.

At ten o'clock the next day I sallied out in quest of more commodious quarters. I found an American family, with which I was invited to remain a few days, but they could not give me permanent board. After several applications for a boarding-place, I was finally compelled to provide a home for myself, which I did, by renting two rooms, one for a residence, and the other I intended appropriating to school purposes. The day before opening my school, I went to my rooms, but not under very auspicious circumstances. At dark, I had no bed to sleep on, nor did I know how I was to obtain my breakfast, to say nothing of supper. But before the hour of retiring came, a Mexican woman brought me a cot, an American woman sent me a pillow, and a German woman came and said she would cook my meals and bring them to me. Did I not feel rich that night as I retired to my humble cot? Indeed, I never closed my eyes in sleep with more profound feelings of thankfulness to God. I fully believed I was where my Divine Master had called me to go—upon the border of that land where I had so long desired to be—and to whose people I trusted to the Lord would make me eminently useful.

Next morning I opening my school with five pupils, but more promised. The education of the children seemed the most feasible means of benefiting the people at the time, and I opened a school although upon the American side of the Rio Grande. The laws of Mexico, at that time, most positively forbade the introduction of Protestant Christianity in any form, and had I gone into Mexico proper for the purpose of teaching the Bible, I should have been imprisoned.

That portion of Texas between the Rio Grande and Nueces Rivers had been claimed by Mexico previous to the late war, but the United States had conquered, and, consequently, it was under our government. Some thousands of Mexican people preferred remaining in their old homes, which fact gave me an opportunity of laboring among Mexicans under the protection of our own government. I was truly happy in a short time in obtaining some thirty or forty Mexican children, and giving them daily instruction in the Bible, against which the parents manifested no objection. I found some who could read in the Spanish language, and a few who had acquired some knowledge of the English. The parents were greatly desirous their children should learn the English language, and become Americanized, and hence my school received popular favor on that account. To be able to put the Bible into the hands of three or four dozen Mexican children, and give them instruction in its blessed teachings, I felt to be an unspeakable privilege. Although the work might look small to the eye of human reason, yet faith bade me hope it might prove a *beginning*, and I was satisfied to work on, even in this small way. The parables of our Savior afforded me much encouragement, especially those in which He compares the kingdom of heaven to a grain of mustard seed, which, when sown, is the lead of all seeds, yet from it sprang a tree sufficiently large for the fowls of heaven to lodge in its branches; also "to leaven, which a woman took (there was a good deal of significance in the fact that it was a *woman*) and hid in three measures of meal, until the whole was leavened."

. . .

CHAPTER VI.

In the midst of the most sanguine expectations of permanent good upon this frontier, I was surprised, one day, by hearing that several priests and nuns had come from France to establish their head-quarters at Brownsville. They had brought means for erecting a convent, for the evident purpose of educating the youth of the Rio Grande Valley. Suddenly and unexpectedly, all my prospects of usefulness there seemed completely frustrated; for what could I do, with such an array of influences against Protestantism and the Bible? But, could I abandon the field, and leave it in the hands of foreign priests and nuns? Indeed, I could not get my own consent to run before popery, while I held in my hand such a powerful spiritual weapon as the Word of God, and I was enabled to carry the matter to the throne of grace, and wait for Divine direction. I spent whole nights in prayer to God. During one of those seasons in which I was earnestly seeking for guidance, a light suddenly dawned on my mind, from these words in the book of Revelation: "These shall make war with the Lamb, and the Lamb shall overcome them: for he is Lord of lords, and King of kings: and they that are with Him are called, and chosen, and faithful." The impression that these words made upon my heart, immediately settled the question of my remaining. Although single-handed and alone, yet, with the assurance derived from these words of Scripture, I felt stronger than my enemies; and I resolved to stay and maintain my post. In order to make a successful stand, I must have a building which would bear some comparison with the party with which I had to compete. My accommodations hitherto had been exceedingly limited; and, as I could obtain no aid from the inhabitants of the region, I resolved to go to the United States and secure the means for building a Protestant seminary at Brownsville. If France could afford to send four millions of dollars to the United States for educational purposes, (as she did that year) I felt that the Protestant Christians of the United States could afford a few hundred for the Rio Grande; so, I closed my schools, and set my face toward my native land, feeling quite assured of prompt and efficient aid. The scene of my departure was calculated to deepen my interest for these people. As I was about starting on the stage for Brazos, I was surrounded by the Mexican girls and their mothers, each uttering the earnest request, "come back," "*come back very soon*," and they stood and watched me with tearful eyes, until I passed out of their sight. When I arrived at Brazos, I could find no conveyance to New Orleans but a schooner, and that very small and inconvenient.

[*Miss Rankin's ability to obtain financial contributions in New Orleans was limited, but she made many contacts and gained confidence that her work in Brownsville— and efforts to raise money—were appropriate and useful endeavors. She had little luck in Louisville, but convinced members of Philadelphia's Presbyterian Board of Education to help her raise five hundred dollars. In Boston she obtained another five hundred dollars from its churches. These efforts wearied her, but after a few months rest, she returned along the Ohio and Mississippi rivers arguing the importance of her cause at every stop. Occasionally she accepted rides, but the greater part of her journey was made on foot, walking eight to ten miles a day. Many people along the way expressed contempt for Mexicans but sanctioned her work with donations of 10 to 20 dollars. Overall, she described the journey as fatiguing but ultimately rewarding and successful.]*

. . .

In the autumn of 1854 I entered my new seminary. This was an auspicious event. The days of labor and scenes of anxious solicitude were all forgotten on the morning I assembled my pupils for the first time in this Protestant institution. I explained to them that the building had been given by Christian friends abroad for their benefit, and endeavored to impress them with the vast importance of improving the privileges it would afford them to the best advantage possible. With my Mexican girls, I consecrated this new edifice to God by reading a portion of Scripture and by prayer. The American Bible and Tract Societies of New York continued to supply my demands for books; although I often wondered at their liberality, considering the very unpopular work I had in hand. I used often to think, in reference to the indifference which prevailed so extensively towards Mexico and her people, that the Lord had chosen me for the work because I was so very insignificant, and it mattered little if I did spend my poor life and services among the Mexicans.

CHAPTER IX.

In 1855 I felt the need of assistance, and I ventured to write a letter to Rev. Dr. Kirk, of Boston, Mass., asking for a colporteur for the Mexican frontier. The letter, quite unexpectedly to me, was published in the magazine of the American and Foreign Christian Union for August, 1855. I will copy the letter, and also the remarks of the editor. It was headed—

"A Voice From The Rio Grande.

"The following letter from Miss Rankin, one of the worthy daughters of New England, who, by much sacrifice and indomitable perseverance, has succeeded in establishing a seminary for Mexican young ladies, in Brownsville, on the Texas side of the Rio Grande, which separates the United States from Mexico, will be read with much interest."

But to the letter:

"Brownsville, April, *1855.*

"Rev. Dr. Kirk:

"*Dear Sir*—Convinced that you have sympathy with whatever appertains to the interest of Christ's kingdom, I take the liberty of calling your attention to this remote land, where, and on the border of which, are thousands of immortal souls under the influence of Popery, in its most enslaving and debasing forms. You are fully acquainted with Romanism, and, therefore, I need not describe to you the character of this soul-destroying agency of the arch-enemy Satan. I presume also, that I need not describe the painful emotions awakened in the heart by daily witnessing the sad influence of that system, so wisely calculated to lead immortal souls to endless ruin.

"We have in Brownsville some three or four thousand Mexicans, who have escaped the dreaded influence of a corrupt priesthood of their own country, in whose moral condition and wants my sympathies are deeply enlisted, and in whose behalf I now write. The enterprise in which I was engaged when last in your city I have, with the blessing of God, carried out successfully. A Protestant seminary is reared in front of papal Mexico, and within its walls are gathered Mexican girls, whose improvement encourages me to hope that their consciences may become enlightened, and that they will embrace the Gospel, which can save their souls. I trust it may ultimately be seen that this institution is one of the instrumentalities by which God intends to disenthrall benighted Mexico from the dominion of popery . . .

I proposed to the Board of the American and Foreign Christian Union, that if they would furnish me the means for employing an assistant teacher in my school, so I could be, in part, relieved from school duties, I would become their colporteur and Bible reader. The proposal was accepted, and January of 1856 I came under the auspices of that society. Re-enforced by a competent teacher, I was greatly strengthened, and the school and Bible distribution received a new impulse. I visited all the houses of the Mexicans in Brownsville and vicinity, and supplied every family of which any member could read, with a Bible. Only occasionally would I find one who rejected it. It was said by my American friends: "The Mexicans take your Bibles to turn over to the priests to be burned." I would follow up my investigations until I was satisfied that such was not true. Indeed, I never ascertained that a single Bible was destroyed. But *I did* ascertain that the Mexicans concealed them in the most careful manner, taking them out and reading them by night, as they said, "when the priests were not about." I went one day to the house where one of my pupils resided, to inquire after her absence, and also to make inquiry after a Bible I had furnished her. A report had crept into school that she had exchanged it with the nuns for a "saint," and that they (the nuns) had burned it. The mother of the girl met me at the door, and with streaming eyes told me that her daughter had died with yellow fever but a short time before. I asked her, if she had her Bible? She replied, "No, I put her Bible in her coffin, as she loved it so much, and it was buried with her." I found another similar case, where a father had put the Bible by the side of his son in his coffin. Although I could not fully coincide with this use of God's Word, yet there was something pathetic and suggestive in the act of these bereaved Mexican parents.

DOCUMENT 3.4

Transitions in the Public Schools of California, 1852

These selections from the first annual reports of the school superintendent in California illuminate the transitional state of public schooling under the new governance of the United States. Strict separation of church and state was unheard of and school organization reflected the "mixed" population of the children themselves. Mexican/Spanish families intermarried with Anglo settlers pouring in from the east. Some schools were Catholic, and some were conducted in Spanish and others in English. They reflected a bicultural/bilingual society in which one language or culture was not yet subsumed under the other.

In the documents below the county clerks have estimated the number of school-age children in each county and written their observations of the state of schooling.

From *The First Annual Report of the Superintendent of Public Instruction to the Legislature of the State of California. 1852*, Eugene Cassely, State Printer, Appendix F, School Statistics, pp. 48–50. Found in State Government Publications, States Other Than Wisconsin, Wisconsin Historical Society, Madison, Wisconsin.

SAN LUIS OBISPO COUNTY—200 Children

*Pueblo, San Luis Obispo,
January 7, 1852.*

John G. Marvin, Esq.

"Sir:—Yours of December 5, 1851, was duly received with regard to schools in this County; there is not one at present. A school was kept open last summer, at the expense of the County, by a Spanish teacher, but was so grossly neglected that it was abolished. A great part of the children in this County are of American fathers, but none speak the English language; which, of course their parents are most anxious to have them learn; hence the difficulty—that of getting a teacher who understads [*sic*] both English and Spanish—the situation not being sufficiently remunerative for a person having those acquirements. If a school could be established in this Pueblo, I am of opinion, that it would be sufficient for the absolute wants of the County; as in all other parts the population is so sparse, that but a few could attend, on account of the distance from one farm to another. Whereas, in this place, children would be sent to board for the sake of attending school. I have acquainted some of the heads of of [*sic*] the most respectable families with the purport of your letter, and probably you will hear from some more interested, and better able to inform you on this subject than myself.

I remain yours truly,"

JAMES D. HUTTON,
Clerk San Luis Obispo County.

SANTA CRUZ COUNTY—200 Children

In the town of Santa Cruz are two schools—one English, with 40, and one Spanish, with 25 scholars.

They are supported by the tuition money paid by the patrons.

SANTA CLARA COUNTY—300 Children

In San José there are two Select or High Schools. One of them is a young ladies' Seminary, containing about 90 pupils. It is under the management of the Sisters of Charity. The other is the San José Academy. Through the exertion and liberality of the Hon. G. B. Tingley and several other residents of the city of San José, a subscription of $5,000 was raised, in September last, for the purchase of buildings, and endowing, to some extent, the above mentioned Academy. The Rev. E. Bannister is Principal, and his lady and Miss Winlack, are assistant teachers. The school numbers about 60 scholars, each of whom pay from eight to sixteen dollars per quarter for tuition.

At the Mission of San José is a Catholic School, numbering about 20 pupils.

In the village of Santa Clara, there are two Primary Schools, numbering about 64 scholars. "In addition, there have been two other schools taught in the township. The number of scholars will not vary far from 35."

F. COOPER, P. M.

SANTA BARBARA COUNTY—400 Children

"There is one school in the town of Santa Barbara, under the supervision of the City Council. There are two teachers, who receive together $70 per month from the city. Both teachers being Chilanians [Chileans], no English is taught—Geography, History, Writing, and Arithmetic, being the only branches taught. As there is not one Spanish Grammar in the town, that branch is of course entirely neglected; nor do I know that any exertions have been made to get any books. There is a great deficiency here of school books of every description. The teachers, apparently, are excellent men, and understand their profession.

There is also a school in Santa Inez, under the direction of the Catholic Church. English is here taught. There was formerly a school, well attended, in San Buenoventura, but it has been broken up."

S. BARN . . . [illegible]

DOCUMENT 3.5

From the Report of Apache County, Arizona School Superintendent, 1905

Not all Anglo school superintendents viewed English as the only means of instruction for southwestern Latinos. In this passage a local school superintendent is recommending what we call today bilingual instruction for the children of Arizona's rural public schools in the early 1900s. Furthermore, he highlights the need for bilingual teachers. The Spanish language persisted in the schools of Arizona, Colorado, and New Mexico well into the 1900s.

From *Biennial Report of the Superintendent of Public Instruction of the Territory of Arizona, 1905–1906* (Phoenix: The H.H. M'Neil Co. Print, 1906). Found in State Government Publications, States Other Than Wisconsin, Wisconsin Historical Society, Madison, Wisconsin.

The progress made by pupils in Apache county during the school year ending June 30, 1905, has been very satisfactory in the districts where the English is the dominant language spoken. But in districts where the Spanish language dominates and the teachers had little or no knowledge of the Spanish language, the progress made by the pupils has been very discouraging and unsatisfactory.

The average daily attendance of pupils at school in Apache county this year was 523.22 pupils against 553.52 pupils last year, a decrease in the average daily attendance of 30.30 pupils, which is not a very creditable showing.

The number of children of school age in Apache county, who have not attended any school during the last year is 213, an increase of 49 over the previous year, and the delinquents were largely from districts where the pupils were of Spanish parentage, and the teachers had no knowledge of the Spanish language.

The poor progress made by the pupils in Spanish speaking districts in Apache county is, in my opinion, directly attributable to two prime causes, to-wit:

1st. The lack of knowledge of the Spanish language on the part of English speaking teachers and their inability to interest or explain to their pupils.

2d. The absolute failure upon the part of county officials to enforce the law in regard to compulsory education in Spanish speaking districts . . .

It will be observed, by reference to the school statistics, that in school districts where the Spanish language dominates, and the teachers possessed little or no knowledge of the Spanish language, the average daily attendance of pupils at school largely decreased, as compared with the years when the same districts were so fortunate as to secure teachers who had a fair knowledge of the Spanish language. To illustrate: Take District No. 6 for the school year ending June 30, 1905; this district had a teacher who was a good Spanish as well as English scholar, the average daily attendance of pupils was 70 scholars. For the last year, ending June 30, 1906, this district, No. 6, employed a teacher who possessed no knowledge of the Spanish language, and the average daily attendance of pupils at school was 51.90, a decrease of 18.10 pupils in the average daily attendance, and [the school] made poor progress, and this is the result always in Spanish speaking districts where the teachers employed [possessed no] knowledge of [the] Spanish language.

In all districts where the Spanish language prevails I respectfully suggest that the teachers employed should have a practical knowledge of that language, otherwise, they will be unable to do either themselves or the pupils justice, or render value received for the salary they receive.

DOCUMENT 3.6

Bilingualism in the Territory of New Mexico, 1889

A strong Hispano concentration in the territory of New Mexico contributed to the persistence of Spanish side by side with English as the language of instruction in New Mexican public schools. After 1912, when New Mexico (and Colorado and Arizona) became states, bilingualism was slowly squeezed out of the schools and as a feature of state legal documents.

From *Compilation of the School Laws of New Mexico*, containing "All Laws and Parts of Laws Relating to Public Schools of the Territory of New Mexico" (East Las Vegas: J. A. Carruth, Printer, Binder and Blank Book Manufacturer, 1889), p. 14. Rare Book Room, Library of Congress, Washington D.C. (LB2529. N6 1889)

Sec. 1110. It shall be the duty of the School Directors to adopt text books in either English or Spanish, or both, and when adopted they shall not be changed for a period of five years. Any School District Treasurer who shall pay to any teacher any money until the requirements of this section are complied with, shall be deemed guilty of a misdemeanor and subject to a fine in double the amount of the money so paid, or to imprisonment for a period not exceeding thirty days, and such Treasurer and the sureties on his official bond shall be liable for the payment of such fine.

Sec. 1110. Será el deber de los directores de escuelas adoptar libros de texto en Inglés y Español ó en ambos idiomas y cuando adoptados no serán cambiados por el período de cinco años. Cualquier Tesorero de distrito de escuela que pagare á algun dinero hasta que los requisitos de esta seccion sean cumplidos será considerado culpable de un mal proceder y sujeto á una multa en doble la suma de dinero pagada así, ó á encarcelamiento por un término que no exceda trienta dias, y tal Tesorero y los fiadoresen su fianza oficial serán responsables al pago de tal multa

DOCUMENT 3.7

Higher Education for a Californio, Santa Clara College

In these 1862 Diaries of Jesús María Estudillo, we are privy to the thoughts and feelings of a young man from one of the older Californio families. Like many Mexican families of the nineteenth century, their fortunes were eroded by the legal costs of fighting land grant title adjudications and threats from squatters on their land. Furthermore, the young man's sensitivity toward criticism of his English skills is revealed as he partakes in college life in San José during the 1860s.

Courtesy of the Bancroft Library, University of California, Berkeley, and Santa Clara College Archives, San José, California.

January 28.

"First day I went to class. I made a composition in the English class. Commenced today in chemistry on Metallic elements. The last study hours I spent putting down some scraps in my scrap book. This day has been extremely cold. The snow still appears on the mountains. The ice was very thick this morning. After dinner (noon) I went to Father Mengarini's room and gave him one hundred and fifty dollars on account for my tuition. Molina received Solari's letter and wished me to pay him eight dollars for the desk; but I would not give him but six."

February 8.

"This morning I was examined in grammar and arithmetic, as for the former, I stood a middling examination, but for the latter, I passed much better than English. We had Philosophy and some experiments performed on Liquids or Non-Elastic Fluids in motion. Today we had a short drill. I was nominated and elected 2nd Corporal by acclamation, at first I did not wish to accept the office but afterwards I consented, not that I cared for the office. Palmer went to the Mission [San Jose], promised to come back tomorrow."

February 28.

"When I came to the study for Chemistry class, I found two letters in my desk, one from my mother and another from Delores. Lola (Delores) tells me that she has been very sick, that they did not expect her to live. Of course these sad news pierced my heart at the first sight . . . We had speaking class. I spoke, 'Unfurl the banner of freedom.' We had in chemistry class a lecture by Roche and soda water was made. I took a glass of it, it was very good. I got this afternoon the 'Spectator' from the library, I commenced reading it this evening. Lola tells me that Dona Estudillo [is] not pleased with my epistolary style, that I have not improved my English. Indeed, I am very sorry that she has such a bad opinion of me, also says that I spent too much money on my way to the college."

April 21.
" . . . Father Caredda this morning reminded us of a few points in the rules of the College and during the forenoon the principal rules of the College were hung in the windows of the study room . . . After he had read the written rules, he said that there were four or five different spirits in the College but how they came in, he did not know—viz, Spirit of Novel Reading, 2nd, Spirit of Gambling, 3rd, Spirit of Laziness, 4th, Spirit of Destruction, by which different kinds of furniture had been spoiled, such as doors, and desks disfigured in the like manner."

May 1.
"Splendid time! Picnic, picnic!!! I got permission to go out, and soon did I put my desires into execution. I got to San Jose about ten o'clock A.M. and then I met Palmer, Keating and Murphy. Palmer and myself got ready to go to the Burnett's picnic. I was invited by Prevost. Palmer drove the horse and buggy, after an hour and a half of traveling we came to the desired spot; but nevertheless, had to make a small mistake, we expected the picnic was at the house of Mr. Burnett, not so, it was two miles further on Steven's Creek. After a hard travel we got to the desired spot. For a picturesque scene, this spot can hardly be surpassed. Here indeed the work of Nature has displayed its wondrous hand in the landscape of the country around the spot of the pleasure enjoyment. Flowery green meadows with a beautiful running stream was a sight for a poet to contemplate upon. The fair sex, among whom there were many handsome ones, seemed to enjoy themselves under the shade of an alder tree whose branches covered us from the sun. I did not partake in the dance. My acquaintances were, I mean in the female line, were Miss Sunol, Miss Bascom, to the latter I did not speak. Of course, Lizzie Miller was there. We had plenty of champagne, cakes and everything we could wish. Soon after we left on our way home, but we could not find the right road; but after Bernard Murphy took the lead and we found ourselves in Santa Clara. I saw at Cameron's Hotel some of the Hoges family. The afternoon after we got to Santa Clara was very windy and during the whole of the night the wind blew extremely hard. It was a little too warm when we left San Jose, but as soon as we got to the mountains a few drops of rain fell but did not continue."

May 18.
"I went out walking with some boys and Father Nattini, walked as far as the first bridge and then came home . . . After supper I spoke to Breen to write me

a speech for the exhibition and he very kindly proposed to do it. Palmer this evening was caught making a cigarrito in the Refectory and Father Caredda has given him punishment, but Palmer would not submit and proposes to leave the College tomorrow."

May 19.

"It is sad to record that a friend, whose friendship I always keep sealed with the truest bond of affection, was turned out of the College last night, he was not even permitted to pass the night at the College. Father Caredda spoke to us in the Refectory and for many other reasons, better known to them (the faculty), Edward Palmer was expelled after dinner. I heard Palmer had been in the College last night but did not see him . . ."

June 9.

"The private examinations began today, they commenced with all the small boys in the preparatory Dept. Received this afternoon a letter from Dolores and the papers; I see by Dolores' letter that Magdalena [his sister, Mrs. John Nugent] has brought forth a child, a girl named Sybil, and she is well. This evening during the first hour of study I and Marks and Duffy were in Father Caredda's room to write the invitations for the exhibition . . . I gave Father Caredda the name of my speech of the exhibition, to be put in the programme, he told me that he would not promise to call me as there were so many speeches already . . ."

June 10.

"Today I was very much displeased in the way Father Young treated me concerning my speech after having given it to him two weeks ago, he comes now, that there is no time to speak it in the exhibition, yes, for me there is no time, but for the rest, there is plenty. In chemistry class the compositions were brought in, but I promised Father Messea to give him mine this evening, but I did not on account of having gone to the exhibition of the other college and coming home so late. Seven of us boys went with Father Young. If at our exhibitions, we could not do better I would duck my head. The most ridiculous speaking I have ever heard, for example; 'Policy of the Administration' and two others. I saw Miss Hall and five girls of the Seminary sat right behind us and with their fan, they gave me and James Hughes more than the air needed."

October 13.

"Of all my College days, this has been the wretchedest, no peace has dwelled within this troubled bosom in the whole day since grammar class to the hour of writing these few lines, seven o'clock in the evening. I have wished that I would not have had to come back this session and I declare that if I am kept in the same English class after Christmas, I will not come back, at least if the same teacher teaches the class. After the class was over this morning, I took out my grammar to the study room to have Father Young explain something I did not understand, when some three or four boys called him and he commenced to speak to them. They told him some trifles, the fact was, that he left me standing with my book in my hand and did not finish his explanation, the small boys were more important to him; this I considered the worst kind of insult and hope I shall see the time when I can have an explanation of this act of my teacher. This evening the boys went to the Circus, I did not go because I had too many duties to perform and expected to write to San Diego, this last thing I did not do."

October 20.

"After supper I went down and sat by myself on the last bench by the corridor of the dormitory and contemplated for a good while what course in life I should follow out of College. Sometimes I thought or [*sic*; of] remaining till I would graduate; at others, I thought of not coming any more after this session and if circumstances would not permit, I would not come back after Christmas. For a long while these thoughts were in my mind, they were expelled when Townsend came up to where I was and soon after the bell rang for beads [rosary] and we had to repair thither."

October 24.

"I received this morning a letter from Mr. Ward, brought to me from San Jose by Crandell. I learnt that he had sent two hundred dollars by the stage. I asked Father Caredda's permission to go and bring money. I got the money from Mr. Crandell. I bought several articles of stationery, a Spanish Diction-ary, English and Spanish, paid $6.75, Lord Chesterfield's Letters to his son, sundries, $4.00. I saw in San Jose, James Breen, H. Farley, B. Murphy, we had a game of billiards. Farley was my partner, we beat them. There was great excitement this morning when I came to San Jose, on account of the scape [*sic*] of Felipe, the prisoner who was to have been hung this morning. He killed the jailer and four others made the scape [*sic*]. I came back to Santa Clara about half past seven . . ."

"I asked him [Father Neri] when he was going to give me my lecture back and to my great astonishment, he said he had already given it to Valenzuela because he found many things to insert in mine. I was raging mad to hear this, after he promised it to me, and gave it to somebody else. I thought there must have been something else besides connected with it, not that something was to be inserted in my lecture, why did he not tell it to me before, he had it already when he told me I was to deliver it. I shall see that I have this explained to me. I wrote this evening about a page letter to Miss Mary Gray of El Rancho Puente. I heard that Ed Palmer was going to the Oakland College in a few days."

November 14.

"I heard from Faure & Bowie that a paper of candies with a china image of something like it, but anyhow, that they were introduced into the Convent and directed to Miss Grace Riddle and signed by Bowie. Father Caredda had this morning spoken with Bowie, Faure and I believe Hastings, to find out who was the person that sent this article into the Convent. Of course they directed themselves to Bowie as it was signed by him, but he denied having done such a thing. Whoever did this, indeed, proves to be a black-guided person, for no gentleman would have done such a thing. The picture that was painted was the doing of no gentleman. Father Caredda says that if it is found out that such a thing was done by anybody from this college, that he will be expelled immediately and so he deserves."

CHAPTER FOUR

Education and Imperialism at the Turn of the Century

Puerto Rico and Cuba, 1898–1930

Cuba and Porto Rico, were intrusted [sic] to our hands by the war, and to that great trust, under the providence of God and in the name of human progress and civilization, we are committed. It is not a trust we sought; it is a trust from which we will not flinch.

—*President William McKinley*[1]

INTRODUCTION

American victory in the Spanish-American War of 1898 launched the United States into the role of imperial power. American support for Cuban independence from Spain initially brought the U.S. military to the Caribbean in 1898. When the *U.S. Maine* battleship blew up in the Havana harbor, killing over two hundred U.S. servicemen, Congress authorized war against Spain. The intervention to free Cuba from the "tyranny" of Spanish colonialism expanded to include Spain's Pacific colonies. The Treaty of Paris in December 1899 concluded the brief war and established the United States as an imperial world power.[2] According to the terms of the Treaty of Paris, the United States acquired the Philippine Islands, Guam, Puerto Rico, and the naval base at Guantánamo Bay, Cuba. Cuba itself was protected from outright acquisition in the 1898 Teller Amendment, which authorized the president to intervene in Cuba but did not grant him the power to establish rule over the island.[3] In return, Spain received 20 million dollars. The treaty also guaranteed religious freedom in the new territories, but the U.S. Congress held the power to determine the "legal, civil and political status" of the newly acquired peoples.[4] The former Spanish subjects of Puerto Rico and Cuba thus found themselves transferred from one imperial power to another.

Public education as a vehicle of Americanization had been utilized in the southwestern U.S. territories acquired in the 1848 Treaty of Guadalupe Hidalgo. Fifty years later, the American government maintained the faith that public schools and the English language would also convince Puerto Ricans and Cubans of the superiority of American culture and democracy. However, different social and political contexts resulted in distinct education systems on the two islands. The wealth of resources and political power among Cubans, backed by the Teller Amendment, resulted in less direct intervention in Cuban affairs. Nevertheless, Cuba was governed from 1901 to 1934, during which time the Platt Amendment established Cuba as an American protectorate.[5]

In contrast, Puerto Rico was declared a U.S. territory. Congressional passage of the Foraker Act in 1900 determined Puerto Ricans' constitutional, civil, and legal rights. Under the Foraker Act the U.S. president was given authority to appoint the governor of Puerto Rico and to remove him at will. The president also appointed the heads of all six administrative departments, including education. Universal male suffrage was lost in addition to a series of political liberties, resulting in historian José Trías Monge's conclusion that as a result of the Foraker Act, Puerto Rico "lost the right to government by consent of the governed."[6] In 1917 the Foraker Act was dismissed when Congress passed the Jones Act, granting U.S. citizenship rights to Puerto Ricans, many of whom desired independence, not American citizenship. In 1947 Congress finally permitted Puerto Ricans to elect their own governor. In 1952 Puerto Rico was established as a commonwealth with its own constitution, a political status that remains nebulous and unchanged to this date. Puerto Rico's colonial legacy, first under Spain and then the United States, has incalculably shaped its history. The island of Puerto Rico was acquired during an era of U.S. history when scientific racism was gaining credence and conquered peoples were viewed as genetically inferior. Although some American officials in Cuba and Puerto Rico eventually treated the island cultures with respect, most officials aspired to bring the unenlightened and ignorant out of the darkness of four hundred years of Spanish empire and into the light of U.S. governance, culture, and the English language.[7] The public schools were thus viewed as the perfect vehicle for Americanization by inculcating the English language and American values upon the youth.

AN EDUCATIONAL EXPERIMENT WITH CUBAN TEACHERS

> *The Cubans are excitable and naturally cannot yet fully comprehend the benefits which will come to them through a safe and beneficent government . . . It will take time to accomplish what is yet to be desired. It can be achieved only slowly and may best be attained through a system of education.*
>
> —*Governor General Leonard Wood, 1899*[8]

Although the Teller Amendment placed restrictions on U.S. domination of Cuba, historians have argued that the eventual annexation of Cuba was a goal of U.S. politicians at the turn of the twentieth century.[9] Americanization in Cuba followed a different path than in Puerto Rico. As Louis A. Peréz points out, instead of imposing English and American ways on the children, U.S. officials in Cuba "decided to Americanize the teachers."[10] Through exchange programs and scholarships an estimated total of 3,000 Cuban teachers traveled to the United States for study and exposure to the English language, American ideas, and customs. In most cases, English was valued among Cubans, particularly among the elite, who saw its utility in future advancement and trade with the United States. Through the training of teachers in U.S. ideals, adoption of American textbooks translated in Spanish, and other Americanization measures, the brief Cuban contact with U.S. education most likely contributed to positive portrayals of the United States in the classrooms of Cuba.

One example of this unique educational experiment occurred at Harvard. Harvard alumni—in cooperation with the U.S. government, philanthropists, and the university—conducted summer sessions for Cuban teachers in 1900 and 1901. More than 1,300 teachers arrived in Cambridge in 1900 and spent several weeks learning English, studying pedagogical methods, and taking excursions to nearby sites. (See document 4.1 for a thorough description.) The program was so popular and successful that it was repeated in 1901, although on a smaller scale.[11] Other universities in the United States also became involved with Cuban teacher training. For example, the New York State Normal School at New Paltz annually invited 30 Cuban teachers to participate in an educational session.[12]

In addition to the Harvard and New Paltz programs, a Cuban Educational Association was formed with the idea of underwriting the expense of sending Cuban youth to college in the United States. Through this program over 2,500 young men and women, mostly from the elite class, were sent to U.S. colleges. Students receiving scholarships were required to return to the island after completion of their studies.[13] In both Cuba and Puerto Rico, the large mixed-race population posed dilemmas for host U.S. institutions during an era of racial segregation. Pérez found that Afro-Cubans had difficulty securing scholarships through the Cuban Educational Association. The president of the University of Missouri, for example, told the organization that while there was "no law forbidding Negroes to enter our university, if one should attempt to do so, he would surround himself with no end of trouble owing to the status of public opinion in the state. You understand therefore that the offer [to come study] is not extended to the Cubans that may be Negroes."[14] In contrast, Booker T. Washington opened the doors of his African American institution to Afro-Cubans. Washington wrote in 1898, "it is our duty to follow the work of destruction in Cuba with that of construction. One half the population of Cuba is composed of mulattos or Negroes. All who have visited Cuba agree that they need to put them on their feet, the strength that they can get by thorough intellectual, religious, and industrial training, such as is given at Hampton and Tuskegee."[15] As will be seen in the case of Puerto Rico, darker-skinned and mixed race Cubans and Puerto Ricans were purposefully channeled to the U.S. industrial education schools designed for African Americans.

EDUCATION IN PUERTO RICO, 1898–1930

The Federal Government of the United States has been generous in all its dealings with Porto Rico, and more than just, but a wise and farseeing statesmanship will point out to the people of the United States that colonization carried forward by the armies of war is vastly more costly than that carried forward by the armies of peace, whose outposts and garrisons are the public schools of the advancing nation. Five hundred thousand dollars for one year . . . would not support a very extensive military campaign; but that sum spent on education would work such a change in Porto Rico as to put beyond the question of a doubt the ultimate and splendid success of the ingrafting [sic] of American institutions in Spanish America.

—Dr. Samuel McCune Lindsay,
Commissioner of Education, Porto Rico, 1902[16]

The educational strategies attempted during Puerto Rico's first decade as a U.S. territory were varied and idiosyncratic, reflecting a revolving door of appointed military and civil governors and commissioners of education. The initial strategy pursued under the second military governor, Guy V. Henry, reflected a zeal for Americanization with little realistic understanding of implementation. Henry established a Code of School Laws in 1899 urging districts to organize public schools. In his circular dated January 19, 1899, he required that teachers "shall be expected to learn English," and in new appointments, English-speaking teachers "shall be preferred." Furthermore, all candidates for teacher diplomas were required to pass examinations in English. Community criticism of the new laws arose immediately and even U.S. policymakers acknowledged that Henry's mandate was more adaptable to the "school system of Massachusetts."[17] Even upon his departure in May of 1899 Henry advised Puerto Ricans that the school children were the "seeds of the earth, and their proper culture depends upon us. You people have to change and adopt the American customs."[18] (See document 4.2.)

More realistic measures for public education were pursued under the leadership of General John Eaton upon his arrival in January 1899. Eaton had extensive experience both as superintendent of schools under the Freedmen's Bureau during the Reconstruction of the American South and as U.S. commissioner of education. Governor Henry reorganized the administrative departments in February 1899 and created a formal Bureau of Education under the U.S. Department of Interior. Eaton was placed in charge of educational matters with the assistance of Victor S. Clark, his eventual successor. Under Eaton's brief leadership, (January–May 1899), steps were taken to centralize educational matters in the colony. Then, in May of 1899,

Guy V. Henry was relieved of his command and General George W. Davis became military governor.

Under General Davis the Bureau of Education continued, but he also appointed a Board of Education, which allowed for more Puerto Rican participation. The board originally consisted of five members, three of whom were Puerto Rican natives. By December 1899 the board was increased to nine members, seven of whom were from Puerto Rico. General Davis granted broad power to the board for the "general and superintending capacity over the educational interests of Puerto Rico." He believed that Puerto Rican representation ensured that "On the one hand, the modern and progressive methods would be introduced, and on the other, that any proposed measure in opposition to local sentiment or custom would be promptly and fully brought to his attention."[19] Davis was not the only U.S. official to incorporate Puerto Rican opinions in his decision making. Historians have acknowledged the role of key Americans in demonstrating sensitivity to Puerto Rican culture and opinions during military occupation.[20]

Under John Eaton's brief leadership in 1899, attitudes toward the acquisition of the English language were softened from the administration's initial stance. Eaton asked how English could be "introduced with the least friction." The 1899 reorganization of the schools called for the country to be divided into 16 districts and an American supervisor of English to be hired for each district. Eaton directed the English supervisors not to "remove or disparage Spanish, for four hundred years the language of this people, in the use of which their ancestors lived and died and recorded their history." (See document 4.3.) Instead, Eaton cautioned that English was to be introduced as a way of becoming "acquainted with the principles of American liberty, with American affairs, American commerce and trade, and thereby share in their benefits . . . familiarize yourself with the principles of the American Constitution, and be sure that your official conduct is guided by them. Do not indulge in fault-finding."[21]

According to Eaton, the fears that had been raised among the schoolteachers of Puerto Rico when General Henry had imposed the English language were slowly ebbing. By late 1899 Eaton believed that "the public saw how the attention to the two languages would harmonize." Most importantly, the Puerto Rican teachers who had naturally been threatened began to feel that the "fear of losing the Spanish was groundless and more and more the advantage of knowing English became manifest. Moreover the regular teachers saw that its introduction, instead of displacing them or in any way operating against their interest, was altogether to their advantage."[22] Eaton's work in Puerto Rico was cut short by illness and Victor S. Clark led the bureau through the end of military rule (April 30, 1899) and through the first year of civil government, May 1, 1899 until May 1, 1900.

Another means of Americanizing the youth of Puerto Rico during the early years of U.S. rule and occupation was through college programs. In 1903 the University of Puerto Rico was created with a Normal (teacher training) Department. The university's emphasis upon teacher training for American assimilation as well as agricultural and mechanical arts resembled that of many post–Civil War institutions on the mainland for African Americans. Prior to the expansion of the university in 1913, when the liberal arts college opened, Puerto Ricans were sent to the United States mainland. As early as 1901 the school commissioner in Puerto Rico reported "provision has been made by the legislature of the island to send a limited number of young men and women to complete their education in the United States."[23] The Puerto Rican colonial government sponsored approximately 45 "poor young men of robust constitution and good conduct" per year between 1901 and at least 1907 to attend colleges in the United States. Section 73 of the colonial legislation specified the institutions available: "The colleges or institutions designated to which the said students shall attend are Hampton Institute, Hampton, VA. and Tuskegee Institute, Tuskegee, Ala., and such other similar educational institutions as the commissioner of education may from time to time specify."[24] Tuskegee and Hampton, African American industrial education colleges, were also the recipients of advanced students from Native American reservations.[25]

A loophole in the colonial legislation enabled the commissioner to permit some Puerto Rican scholarship students to attend institutions with broader academic missions than Tuskegee and Hampton. In fact, in 1903–04, the commissioner of education in Puerto Rico reported 18 students in their third academic year at institutions as varied as Haverford, Rutgers, Cornell, Wesleyan, MIT, University of Michigan, University of Maryland Medical School, and Lehigh University.[26] By 1905, almost five hundred Puerto Ricans were attending American institutions as a means of building pride in the United States and educating officials to staff the colonial government.[27] Despite the exception of some students who attended more academically prestigious institutions, part of the colonial legacy was thus an attempt to narrow educational opportunities for Puerto Rican youth into government-approved institutions for industrial training. The college program disbanded after 1907 as more funds were channeled into developing normal schools and the fledgling University of Puerto Rico.

Americanization occurred in the island's primary and secondary schools throughout the first half of the twentieth century. Governance of schools was changed in Puerto Rico with the Foraker Act of 1900. The act required the president to appoint a commissioner of education who would "superintend public instruction throughout Puerto Rico, and all disbursement . . . [and] make such reports through the governor as may be required by the Commissioner of Education of the U.S., which shall annually be transmitted to Congress."[28] During the period of the Foraker Act, five different commissioners held office, each introducing different policies concerning the scope of English language instruction in the schools, the required level of English proficiency of teachers, and other curricular and administrative matters.[29]

The shifting rules on English language instruction in the primary grades provide an excellent example of the influence of political forces on educational policy and the disorder caused by constant change. As historian Juan

José Osuna observed of these years, "the school system of Puerto Rico, like the political status, has been like a ship without a haven to anchor in, roaming the seas with no definite home port view."[30] After more than 16 years of controversy and experimentation with bilingual education in the urban schools (rural schools were largely exempt from English language instruction), English was suspended as a form of instruction in the first four grades. Experts brought to study the Puerto Rican system concluded that most Puerto Ricans left school after three or four years and thus it was wiser to focus on basic Spanish language instruction. Furthermore, in the U.S. Department of Education's 1915 report, *The Problem of Teaching English to the People of Puerto Rico*, investigator José Padín identified the severe limitations of English instruction: "the probable cause of this failure lies in the misconception of the method and material best suited to teach English to non–English speaking children who are studying at the same time their mother tongue." He concluded that this was particularly difficult because of the lack of "regard to the fact that they live in a non-English environment, and utilizing the advantages which accrue to the children from linguistic training in their native language."[31]

Under the administration of Commissioner Paul G. Miller, Spanish thus became the language of instruction in the first four grades of elementary school and English was taught as a separate subject between 1915 and 1928. This decision was validated by expert advice. For example, in 1926 a panel of experts from Columbia University's International Institute of Education was called in to study Puerto Rico's schools. The survey commission recommended, as had earlier experts, that "English be not taught in any schools below the fourth grade . . . that English as a subject be taught intensively in the fourth, fifth, and sixth grades and that it be used as the language of instruction beyond the sixth grade."[32] Despite an educational climate in the 1920s that generally favored the opinions of educational experts and surveys as scientifically based, the Columbia University recommendations were ignored.[33] Miller's policies were not continued. Instead the political forces of Americanization pressured the administration to initiate English-language instruction from the earliest grades. Not surprisingly, the Department of Education valued the advice of a 1928 report commissioned to the Brookings Institute of Washington, D.C. In its report, *Porto Rico and Its Problems*, the authors asserted that "English is the chief source, practically, the only source, of democratic ideas in Puerto Rico. There may be little that they remember, but the English school reader itself provides a body of ideas and concepts which are not to be had in any other way. It is the only means which these people have of communication with and understanding of the country which they are now a part." The Brookings Institute thus came out directly in opposition to the experts at Columbia, stating, "it is our conclusion that the Puerto Rican Department of Education, in refusing to follow the recommendations of the Educational Survey Commission, acted with a true instinct." This conclusion was reached despite the report's own admission of the ineffectiveness of English language instruction, stating, "in spite of the fact that the mass of the people can make little use of the very poor English they acquire in the present-day schools of the country."[34]

The prolonged colonial status of Puerto Rico profoundly influenced its educational development. The U.S. Congress controlled internal affairs in Puerto Rico until 1952, when a constitution was approved; Puerto Rico became a commonwealth and elected its own governor. Prior to 1952 educational matters were tied with political currents from the United States. As presidential appointees, the actions of commissioners of education were closely scrutinized on the mainland. Thus, Americanization measures, particularly English language instruction, were imposed. These policies were followed even when educational experts ruled that the type of English instruction utilized was impairing—rather than assisting—the intellectual development of Puerto Rican students. As the Puerto Rican migration to the U.S. began in the 1920s, 1930s, and 1940s, these young American citizens from the island brought with them far fewer English skills than expected and encountered serious difficulties in the U.S. schools.

DOCUMENT 4.1

Expedition of Cuban Teachers to Cambridge, Massachusetts

After the U.S. assisted Cuba with independence from Spain in the Spanish-American War (1898–99), Cuba was established as an American protectorate from 1901 to 1934. U.S. officials in the occupied territories of Cuba, Puerto Rico, and the Philippines viewed education as a key vehicle for imposing on residents of these islands the superiority of American culture and governance. The teacher training program described below sheds light on the attitudes and beliefs that Americans held toward Cubans at the turn of the century and on some of the broader intellectual and academic practices and ideologies of the era.

From *Report of the U.S. Commissioner of Education for the Year 1899–1900*, vol. 2 (Washington: Government Printing Office, 1901), pp. 1643–1647.

The expedition of Cuban teachers to Cambridge in the summer of 1900 originated in the following letter dated February 6, written in Habana, and signed by Ernest Lee Conant (A.B. Harv. 1884, LL. B. and A. M. Harv. 1889), who had been practicing law in Habana since the end of the war with Spain, and Alexis E. Frye (LL. B. Harv. 1890, A. M. Harv 1897), who had been for a few weeks superintendent of schools for Cuba by military appointment:

Headquarters Division of Cuba, Habana,
February 6, 1900.

President Charles W. Eliot,
Cambridge, Mass.

Dear President Eliot: We are planning to carry as many Cuban teachers as possible (perhaps 1,000 or more) to the United States next summer, and as alumni of old Harvard and with the firm belief that our alma mater offers the best facilities, we naturally turn to her for help.

These teachers will have for their object hard study as well as a tour of observation through our country. The general plan will be as follows: The party will leave Cuba on government transports or on chartered steamers about the last of June. It is our wish that the steamers may land us directly in Boston, and that the teachers may attend the Harvard summer school for six weeks. The next four weeks will then be given to travel and visits to the great cities, perhaps crossing the continent to San Francisco. We are sure that this brief outline will tell you the whole story. You can readily see what tremendous results would follow with 1,000 intelligent men and women (after such a broadening experience) scattered over the island.

Of course the one great item is expense. Can it not be arranged so that the instruction for six weeks at Harvard shall be free? With this as a starting point, we shall organize a committee in Cambridge and Boston with a view to securing free accommodation in homes during the six weeks. We shall ask various cities to plan temporary entertainment. If we can not secure Government transports, it may be possible to secure some appropriation in the island to pay the cost of steamer travel. The teachers are poor; they need this summer's outing and work. They need it for themselves and they need it for the sake of our own country.

The school laws of Cuba (see article 23 of decree sent you) require courses of summer study from the teachers. This will be one of the great means of educating teachers now in the schoolroom and who can not attend normal schools. Many of these teachers lack even the elements of education; many of them have hardly been beyond the city limits of their own towns. We can not carry normal schools to every town and city; but we can carry the teachers to educational institutions, and we want the best, namely, Harvard. We want the teachers to breathe the atmosphere of the greatest school in America. We want them to feel the history and associations, to enjoy the facilities of libraries and laboratories. We want them to come in contact, not only with the strong minds of the professors, but also with hundreds of the brightest and best teachers in America who will this summer be in Cambridge. We want these teachers to have the culture that comes from travel; we want them to carry this culture back into the Cuban homes and the Cuban schools. We want these teachers to know our country, to know our people. We want the ties between the two countries drawn closer, so that all feeling of antagonism may melt away, in order that our country may do a higher and better work for Cuba.

Of course we know that the work ordinarily done in the Harvard summer school would need to be adapted to the teachers of Cuba. The work is of too high a grade in general, and the subjects as a whole are such as are not taught in the public schools of Cuba. Without interfering in the slightest degree with the summer school, could you not plan a parallel school with a course specially fitted to the needs of the Cuban teachers? More than nine-tenths of these teachers can neither speak nor understand English. There are enough, however, with a knowledge of English to form a medium for transmitting the work of the summer school to the others.

As soon as we know whether Harvard University will extend this invitation and will do this grand work we will bend every energy to complete the plans, and we shall succeed. We have submitted the proposition to General Wood, and it goes almost without saying that he will give his powerful support to the movement.

Sincerely yours,
Ernest L. Conant.
Alexis E. Frye.

Approval of the Plan.

This letter [preceding document], which was received in Cambridge on the 12th of February, was considered on the 13th at a special meeting of the president and fellows; and the president was then authorized to reply in the affirmative, if General Wood favored the plan. A few days afterwards a telegram was received from General Wood strongly endorsing the project, whereupon the following telegram was sent to Superintendent Frye: "Frye, Habana. Yes. Eliot." Notices of the project and of the affirmative answer of Harvard were thereupon published in the Cuban newspapers, and an active discussion immediately arose as to the feasibility of the plan. It was contended that it would be impossible for young women to go on such an expedition, in violation of the social habits of the Cuban people; the Catholic Church in some places manifested opposition to the project; and at first the general sentiment of the people seemed to be adverse. Superintendent Frye was at some disadvantage, because he had not traveled over the island, and was personally known in Habana and the immediate neighborhood only. Nevertheless, in the course of a month it became evident that there was so much interest in the project that it was expedient to devise the arrangements for the expedition in detail, and to announce them as soon as possible. Thereupon, Mr. Frye visited Washington and Cambridge about the 1st of April. In Washington he secured the cordial cooperation of Secretary Root, who subsequently expressed his approval in a cordial letter to President Eliot, dated May 8.

Subscriptions for the Cuban Summer School.

When Mr. Frye began to discuss the details of the expedition with the Harvard authorities, it soon appeared that the university would really become responsible for the health and safety of the members of the expedition while in Cambridge, and that it would, therefore, be expedient for the university to supervise the lodging, feeding, and protecting of the members of the expedition during the six weeks of their stay there. It also appeared that the regular summer school would not be suitable for the Cuban teachers, and that special courses of instruction would be needed. Thereupon, a public meeting was held in Boston to describe the objects of the proposed expedition and call attention to them; and a circular was issued by the president and fellows of Harvard College asking the community for the means of paying all the expenses of the expedition during its six weeks in Cambridge, including board, lodging, instruction, excursions, and entertainments. Subscriptions began to come in before the end of April, and continued to flow in until the middle of August. The sum asked for was $70,000; and that sum was ultimately provided, and a little more, the total subscribed being $71,145.33.

The subscription list is an interesting one because of the large number and the variety of persons who took part in it. It was emphatically a popular subscription, and represented all classes of the community. Very little personal solicitation was necessary. The circular was disturbed widely, and the newspapers from time to time called attention to the state of the subscription. One large contribution came by order of the court from the unused balance of the fund raised near the outbreak of the war with Spain to provide means of caring for the sick and wounded among the troops in Cuba (the volunteer aid fund). When this fund was distributed in accordance with the order of the court $20,000 of it came to the subscription for the Cuban teachers.

Plan of Instruction.

The plan for the instruction comprehended (1) two lessons a day in English; (2) a course of eighteen lectures in Spanish on physiography, illustrated by as many excursions to different points of geographical interest in the neighborhood of Boston; (3) two courses of lectures in Spanish on historical subjects—one on the history of the United States, the other on the history of the Spanish colonies in North and South America; and (4) lectures on free libraries, on the organization of the American schools, and on imitation and allied faculties in children. Through special gifts received from Mrs. Quincy A. Shaw, a course of illustrated lectures on the kindergarten was provided for the Cuban women teachers, and a workshop course on American sloid for a selected number of Cuban men. Laboratory instruction in physiography being out of the question for so large a number of persons, field study was adopted as the best substitute. The instruction in English was to be given in 40 sections—20 for men, and 20 for women. The teachers selected for these sections were in general young graduates and undergraduates of Harvard College and Radcliffe College. Each teacher of English was to give two lessons a day to his or her section—one from 8 o'clock till a quarter before 9, and the other from half past 11 till 12. The lectures were all to come between these two English lessons, and no lesson or lecture was to be more than three-quarters of an hour long. Sanders Theater was to be used for all the lectures; and the English lessons were to be given in 40 rooms, all of which were in the college yard. The afternoons were to be devoted to excursions, each Cuban teacher being provided with at least three excursions each week. Sundays and evenings were to be left free.

Arrangements in Cuba and Cambridge.

On the 16th of May a circular was issued by Superintendent Frye in Habana, setting forth the project as fully as was then possible, giving all details concerning the transportation of the teachers to Boston on Government steamers, describing the arrangements made in Cambridge for the accommodation of the visiting teachers and the provable advantages of the trip. The circular also gave instructions concerning clothing, baggage, medical attendance, health certificates, vaccination, and other details. The university had limited the number of Cuban teachers to 1,450, which is the capacity of its largest lecture room, Sanders Theater. Moreover, the two dining halls would not accommodate well more than 1,450 persons in addition to the regular summer school. Superintendent Frye was therefore obliged to provide means of selecting these 1,450 persons from the 3,500 teachers who were already at work in the public schools of Cuba. The selections were made by Cuban authorities exclusively—in general by the school boards already established all over the island. As soon as Superintendent Frye's circular had been distributed through the Cuban towns and villages, the work of selection began.

In the meantime, the following arrangements had been made in Cambridge: Students occupying rooms in college dormitories offered their rooms in sufficient number to accommodate all the Cuban men teachers. Rooms enough were then engaged in houses within half a mile of University Hall to accommodate all the women teachers in groups of from 8 to 16 in a house. Each householder undertook, for a price agreed upon, to receive a certain number of teachers, provide them with furnished rooms, and give them a simple breakfast. The use of three houses was given without rent; and several others were offered but not accepted because they were far too far from the yard. It was necessary to engage a business agent who should have charge of all the arrangements for the accommodation of the visitors in Cambridge; and his first task was to provide rooms for the women teachers. Since many of the students who offered their rooms in college dormitories were unwilling that their beds, linen, and blankets should be used, it was necessary to hire these articles in large quantity for six weeks' use. It was decided that the Cuban women should eat their luncheons and dinners in Memorial Hall, the capacity of which is 756 seats; and that the men teachers should eat all their meals in Randall Hall, a portion of that hall however, being reserved for the regular summer school, which consists of both men and women, the women being in the majority. In both halls the Cuban teachers were to be provided with a bill of fare for each meal arranged by the steward, and every teacher was to take whatever he or she wanted from that bill of fare. In Randall Hall, the members of the regular summer school followed the ordinary rule of that hall, which is to order by the plate and pay for exactly what is ordered. Two methods were in use, therefore, at every meal in Randall Hall—one for the Cubans, the other for the American summer school.

By the end of June the business manager, Mr. Clarence C. Mann (A.B., Harv., 1899) had completed his arrangements, and had opened an office in Holden Chapel as headquarters for information—in fact, for all the business of the expedition. He had also engaged about twenty chaperons to live in or near the houses in which the women were lodged, and a large number of clerks and guides, most of whom were Harvard students in the law school, the college, and the scientific school. All the chaperons, and most of the guides, spoke some Spanish. In addition, a few interpreters were employed. Subsequently it became necessary to engage an additional number of chaperons. These ladies lived in the houses with the Cuban women teachers, ate with them at Memorial Hall, helped them with their English lessons, went shopping with them, adjusted their difficulties, attended to their ailments, tried to prevent overwork and overexcitement, directed them gently, and befriended them heartily. The success of the expedition, so far as the women teachers were concerned, was largely due to these ladies . . .

The first positive statement of the number of persons to be entertained at the university came by telegraph from General Wood as follows:

Habana, June 29, 1900—2.19 P.M.

President Eliot, *Harvard, Boston:*

Transports left Cuba as follows * * * June 25, *McPherson* from Gibara, 110 males, 96 females; total 206 * * * June 26, *Crook* from Matanzas, 295 males * * * June 26, *Buford* from Cienfuegos, 51 males, 67 females; total 118 * * * June 28, *Sedgwick* from Sagua la Grande, 428 females. Total 1,047 so far. *McClellan* leaves from Nuevitas. As soon as her departure is reported will wire you.

Wood.

Habana, June 30,1900–11.56 A.M.

President Eliot, *Harvard, Boston:*

In addition to my telegram of yesterday, *McClellan* left from Nuevitas 29th with 156 males, 70 females: total 226 * * * Total teachers sailed to date, 612 males, 661 females; total 1,273.

Wood.

The expedition was, then, 177 persons short of the maximum number named by the university; but in a country where the means of communication are few and difficult it was a remarkable feat to get 1,273 teachers on board the transports within six weeks of the issuing of the first circular letter of instructions from Superintendent Frye's office.

Hospitalities and Excursions.

The first lesson was given on the morning of Thursday, July 5, when the division of the whole body into 40 sections was made at Memorial Hall, and each section was guided from the hall to the recitation room which that section was to occupy throughout the six weeks. The first excursion, which started on Thursday afternoon, labored of course under some difficulties, because the meeting places were unfamiliar and most of the teachers knew nothing about electric cars, but in two days the whole machinery of the Cuban school was in operation, and thereafter it ran with remarkable smoothness. The excursions were of three kinds: The geographical excursions, which formed a portion of the instruction in geography; the excursions to several characteristic manufacturing establishments, and the excursions of social nature. Only one of these last was provided by the university, but there were many others that were arranged by private persons.

The Catholic societies of Boston and Cambridge had made arrangements, with the cooperation of the university, to offer to the Cuban teachers facilities for reading and writing in rooms provided by the university within the college yard. For the men, Harvard 1 was devoted to this purpose; for the women, rooms in Phillips Brooks House. In both places the Catholic societies kept their representatives throughout the day and evening, and were enabled to show the Cubans very acceptable hospitality. The Catholic societies also gave two concert dances each week for the Cuban teachers in the Hemenway Gymnasium and took all the responsibility for the management of these entertainments. Three concerts, which were very largely attended and were much enjoyed, were given in Sanders Theater—one by the Baptist societies of Cambridge, one by the Catholic societies, and one by the Cubans themselves. Each week a programme in Spanish was issued, in which all the lessons or lectures and all the excursions were carefully described, and the numbers assigned to each excursion were given . . .

At the Catholic Church on Holyoke Street, St. Paul's, special services were held for the benefit of the Cuban visitors throughout their stay, and these services were well attended. Through the good offices of Archbishop Williams, Father Fidelis, a graduate of Harvard College in 1861, who had become familiar with the Spanish language through long residence in South America, was brought to Cambridge for the express purpose of attending to the religious wants of the visiting Catholics.

The attendance at the English lessons was excellent, hundreds of the teachers being very regular in their attendance. At the lectures in Spanish in Sanders Theater the attendance was not so good, and yet it was creditable, particularly at the lectures on physiography, which were handsomely illustrated by means of lantern slides. The lessons in sloyd were followed eagerly; and the kindergarten lessons were well attended, considering that hours could not be found for all of them which were altogether free from other appointments. The attendance at the excursions was about 60 per cent of the whole number of teachers. The weather was hot much of the time, and the Cubans were not accustomed to walking any distance. Those excursions which demanded much walking were not pleasurable for them, and were attended as a matter of duty.

Physique of the Visitors.

The physique of the visitors necessarily attracted the immediate attention of those who were responsible for their welfare. The ages of the Cuban teachers ranged from 16 to 60, but the extremes were not numerously represented. The selecting bodies in Cuba had selected too many elderly people, who were, of course, incapable of learning English, or indeed of absorbing readily new ideas. About 10 per cent of the men were over 44 years of age, and about 10 per cent of the women were over 38. To the Cuban authorities, however, it may have seemed expedient to select for the excursion some persons of influence or high standing in their several communities, whose presence would be a safeguard for the younger members, and who would be able to impress their views on their own people after the return of the expedition. There at first seemed to be too large a proportion of delicate and feeble persons, but the very favorable physical experience of the expedition shows that this feebleness was more apparent than real. It was obvious at first sight that the Cuban men were decidedly shorter than the American men, and Dr. Sargent subsequently confirmed this general observation by the measuring of 479 of the Cuban men. He found that the medium height of the Cuban male teachers was 64.3 inches—a height surpassed by over 90 per cent of American male students. The Cuban women were also decidedly shorter than American women; thus, only 20 per cent of the Cuban women attained a stature of 62.2 inches—a stature which is surpassed by 50 per cent of American women students. As to weight, although the Cuban teachers were older than American students, more than 90 per cent of American male students surpass in weight the 114 pounds attained by only 50 per cent of the Cuban teachers. The medium weight of the American female student is 114.6 pounds, and the medium weight of the Cuban female teacher was 102 pounds. Eighty per cent of American female students surpass the medium weight of the Cuban female teachers. Physically the Cuban women seemed decidedly superior as women to the Cuban men as men; and this appearance was borne out by the measurements taken by Dr. Sargent, the Cuban women comparing more favorably with the American women than the Cuban men with the American men. Most of the Cuban teachers gained steadily in weight while they were in Cambridge, and many returned to Cuba in a better condition of health than when they came thence. This gain of weight may have been due to the fact that they were much more active while in Cambridge than they are habitually in Cuba. The men had to walk to and from all their meals and to their language lessons and their lectures, and there was some walking on the excursions. The women walked from their rooms to luncheon and dinner and to their daily lessons and lectures, and many of them went on from two to three excursions per week. Going up and down stairs was also an unwonted exercise for most of the visiting teachers, rural Cuban houses being in general only one story in height.

What the Cubans Learned.

The chief result of the expedition was the opening of the minds of these 1,300 intelligent people to a flood of new observations and new ideas. There was a great diversity among them as regards education and capacity. As

General Wood said in a letter written from Habana on the 24th of February to Maj. Henry L. Higginson, "You will find all classes among them, from the highly educated to those of very limited education, but they are all enthusiastically interested in educational matters, and to these people and to the children they are teaching we must look for the Cuba we hope to build up. These men and women will come back to Cuba with very many new ideas and very much better fitted to teach." A fair proportion of them learned much English and got a new conception of science teaching and history teaching, but many of them were too old to learn a new language, or, indeed, to acquire much intellectual training of any sort, yet all saw with their eyes the American ways of living and the outside, at least, of many American institutions, such as schools, hospitals, asylums, libraries, churches, and theaters. They made two voyages on the ocean; they had a hasty view of New York, Philadelphia, and Washington; they caught a glimpse of the country on their rides through New Jersey, Pennsylvania, and Maryland, and they became well acquainted with Cambridge and the neighborhood of Boston, from Marblehead on the one side to Point Allerton and Nantasket on the other. They came in contact with a considerable number of American educated young people and found them serviceable, cordial, and friendly. When the expedition was about to leave Cambridge for the fortnight's journey, the Cubans wished to have the young men who had worked for them and with them in Cambridge accompany them on their journey, and Superintendent Frye so arranged it; and it was with real regret that the guides and guided parted at Philadelphia, whence the transports sailed for Cuba.

DOCUMENT 4.2

Farewell Address to General Guy V. Henry, Former Military Governor of Puerto Rico

General Guy V. Henry, second military governor of Puerto Rico (December 1898–May, 1899) established an unpopular Code of School Laws in January 1899 requiring rapid Americanization of Puerto Rico's public schools. Requirements for the teaching of English at all levels and for English-speaking teachers proved unrealistic for the first year of U.S. occupation. Subsequent governors and education commissioners relaxed some of the most extreme requirements, particularly regarding English-language instruction. In this selection Puerto Rican schoolchildren reveal the trappings of U.S. culture that have been absorbed into their schools—the American flag, patriotic songs, and a display of thankfulness to the beneficent American liberators.

From "Education in Porto Rico," by Hon. John Eaton, formerly U.S. Commissioner of Education, and director of public instruction for Puerto Rico, chapter 4 in *Report of the Commissioner of Education for the Year 1899–1900*, vol. 1 (Washington: Government Printing Office, 1901), pp. 252–254.

What occurred is here taken from the San Juan News, the only paper published in English in the island:

Promptly at 2 o'clock yesterday afternoon the school teachers, in honor of Generals Henry and Davis, assembled at the theater on the Colon Plaza. Two thousand children were present from the schools of San Juan and Santurce, and twenty-four schools were represented.

The exercises were under the control of the English supervisor of San Juan. The time for preparation was short, but the exercises were nevertheless wonderfully successful. The following was the programme rendered:

Salute.
Star-Spangled Banner, band of Beneficencia.
Farewell address in English to General Henry.
Same in Spanish.
Response of General Henry.
Song, Columbia, by the girls of the Normal College.
Address of welcome to General Davis, in English.

Same in Spanish.
Response.
National anthem, by all, with band.

The singing of Columbia by the young ladies of the Normal college was nicely rendered, the words being very clearly pronounced. The band of the Beneficencia assisted by playing the "Star-Spangled Banner" and "America." The speeches in English were by two pupils of Mr. Timothee, 22 Sol, and the same in Spanish by two pupils of Mr. Saavedra, of Santurce.

The singing of the national anthem was accomplished with precision and enthusiasm.

Address to the retiring governor-general:

"General Henry: We, who are about to live, salute you. When we heard the thunder of American ships bombarding our homes our hearts were filled with fear, and when we saw the proud flag go down which had floated so long over our land we were told that the end of things was near; but the stars did not fall. We soon saw that the Yankees who walked our streets were not pigs. Slowly we forgot our fears. The Lafayette Post brought to each school a Star Spangled Banner, hereafter to be the flag of our island. Now, every morning we salute it, and pledge ourselves to defend it. Since you have been our commander we have learned many things. By your order our schools are becoming better. A map of your great country, and a little book in your language has been put in our hands, to teach us to speak and write it and read its stories. It opens to us a new door of opportunity. When our teachers tell us of the boyhood of Washington, Lincoln, and Grant, and how they became great and good men under the Star Spangled Banner, and now that we have the same liberty which they had to learn and to do, our childish fears are changed to hopes.

"We rejoice that ours has become the land of the free and the brave, and that we can be what we make ourselves. You have removed the burden of tax from the daily pursuits by which our fathers supported our homes, and from the bread which we eat. You have been thoughtful of our child life, and ordered the clothing of the naked and promised us a park for sports. You have abolished the payment of the school fee, which drew the line between the rich and the poor, led teachers to be partial, and caused us to spend our school days with a growing sense of class differences, cultivating on the one hand offensive pride and on the other disturbing jealousies. We expect that no more this dislike for each other will grow with our growth. We shall receive equal attention from our teachers, sit side by side on the same bench, and learn to think that our merits and our honors depend on the good things we do. We hear of your interest in our having a house specially fitted for a school, apart from all bad conditions, as have the children in the States, and we thank you. We part from you with regret, and trust that we shall, by becoming worthy American citizens, show our appreciation of what you have done for the many children of Porto Rico. Young as we are, we are not without some thought of the cares which have burdened you and injured your health. We trust that rest will make you well again . . . In your departure you have our best wishes for your welfare, long life, and happiness."

General Henry's response was as follows:

"This is a very beautiful sight before me. These children are the seeds of the earth, and their proper culture depends upon us. You people have to change and adopt the American customs.

I do not think there are any brighter children in the world than the Porto Rican . . ."

[Shortly after his departure from Puerto Rico, General Henry died October 27, 1899 in New York City.]

DOCUMENT 4.3

Teaching English in Public Schools, Puerto Rico

In this excerpt from John Eaton, director of public education in Puerto Rico, he reveals the dilemma thoughtful educators faced in their political positions as they introduced the American language and government to newly colonized peoples. Unlike his military governor predecessor, Guy V. Henry, Eaton focused on teaching English in addition to Spanish rather than replacing the Spanish language. Contemporary educators have called this type of approach "additive" because it does not involve the destruction of one culture or language over the other.

From "Education in Porto Rico," by Hon. John Eaton, formerly U.S. commissioner of education and director of public instruction for Puerto Rico, in *Report of the Commissioner of Education for the Year 1899–1900*, vol. 1 (Washington: Government Printing Office, 1901), pp. 236–238.

The absence of the English language furnished the greatest difficulty in the way of those who wished to become American in thought, belief, and loyalty. How should English be introduced with the least friction? The imagination is likely to magnify greatly the difficulty of a child's learning any other than its native tongue. The way children learn their own language may be called the natural method. A simple text-book with appropriate lessons, pictures, words, and sentences is selected, and the child goes forward from lesson to lesson by easy stages. In Porto Rico it was important that the learning of English should as far as possible be divested of imaginary objections in the minds of the teachers of the public schools. The 10,000 copies of Appleton's First Reader ordered were considerably delayed in coming. Every Saturday I had been meeting the teachers in San Juan an hour, enforcing a few important simple methods in regulating the schools. The Saturday after the arrival of the First Readers I took with me a sufficient number to supply a copy to each teacher. And beginning with the first lessons, which is wholly in English—a picture of a cat and a few simple words—I began to drill them as beginners. They at once mastered the lesson, and we went on from lesson to lesson through the hour, when they found they had begun to learn English, and no bugbear had turned up as expected. From that time they went forward giving the lessons to their respective schools. In some cases the pupils learned faster than their teachers. Some teachers employed private instructors and specially prepared themselves in each lesson before giving it to their schools. The people of wealth could employ instructors in English at will. The question of its introduction into the public schools was mainly of importance to those who had not the means of doing as they chose in providing for the instruction of their own families, or, in other words, those who were entirely dependent on the public schools for education. It was clearly their right to demand that these schools should furnish them the privilege of learning English, upon which so much of their future depended. It was remembered that German had been extensively introduced into the city schools in the States by putting a German text-book in the hands of the pupils who should give it the prescribed attention under the direction of the regular teacher, who spoke only English, and an expert in German should visit the schools once or twice a week, as prescribed, and see that the German was correctly pronounced and written. The instructive expedient of Jacotot was recalled. Native teachers and pupils used Spanish. They all, teachers and pupils, were together beginners in English. What could be better than the natural method of acquiring the new language? Objections to a new language were expected. One might say: "I heard the first words from my mother's lips in Spanish. I breathed the first words of tenderness to my love in Spanish. Would you have me forget my mother tongue?" The answer was: "By no means. The American plan would teach a better Castilian than you were taught, but it would also teach English as opening the greatest door of opportunity to the rising generation."

Supervisors of English.

When the commanding general, therefore, had so far advanced in the knowledge and control of affairs as to find the general or insular treasury receiving sufficient money to warrant expenditure, the island was divided into sixteen parts, and a supervisor of English, native to the language, was employed at $50 a month for twelve months of the year to visit each school as frequently as the number would permit and to see that English was correctly pronounced and written. This was an aid both to the pupils and the regular teachers. No native teachers were removed; their fear of loss of place disappeared; they foresaw that in the nature of things they would be expected to teach English themselves some time in the future and many of them specially appreciated this opportunity to gain a knowledge of it themselves.

United States Flags and Patriotic Songs

Col. A.C. Blakewell, of the Lafayette Post of the Grand Army of the Republic of New York, had visited the island and had commenced to supply the United States flags to every school, together with directions in English for saluting the flag. The exercise interested both teachers and scholars. Later a flag for every school was supplied.

No singing was found in the schools, but patriotic songs in English were introduced as far as possible, the teachers in many instances copying the words and the music for their schools, as no appropriate text-books were to be had. The interest was so great that shortly a visitor from the States calling at a school would be surprised by a reasonably good recitation in English reading and the salutation of the flag in English, together with the singing of "America."

Meantime Spanish was taught as efficiently as in the past, and teachers, pupils, and the public saw how the attention to the two languages would harmonize, and it was apparent that the fear of losing the Spanish was groundless and more and more the advantage of knowing English became manifest. Moreover the regular teachers saw that its introduction, instead of displacing them or in any way operating against their interest, was altogether to their advantage. And thus the question was settled in a way to win popular approval instead of disapproval and resistance . . .

The following letter was addressed to the [English] supervisors:

San Juan, May 13, 1899.

My Dear Sir: Supervising the teaching of English in Porto Rico, which you now undertake, is one of the most important duties of the hour. Porto Ricans manifest a great interest in the subject, and the eye of the United States is upon it. You start with a small book in your hand, but in it are the greatest possibilities, warranting the concentration of all your powers and the utmost singleness of purpose. You should spare no thought, you should omit no preparation, in order to do it well. You are not to remove or disparage Spanish, for four hundred years the language of this people, in the use of which their ancestors lived and died and recorded their history. But you are to teach the reading and writing of English, and you are selected because it is believed you can do it correctly. You do not displace other teachers. They are paid from the municipal treasuries, and you are paid from the State treasury. You do not come to disparage or antagonize their work; but you go to them as their special friend, to aid them in making the changes which the new conditions require. In the ocean currents which swell around Porto Rico the island has always been continually nearer American than Europe. Now the currents of history have swept around the people and separated them from Europe and united them to the United States, the most powerful nation on the American continent, to a people enjoying the largest liberty of any in the world—not license to do evil, to harm others, but liberty regulated by law, in which there is found the largest good for the greatest number. Porto Ricans are Americans. The rich can buy what they desire, but those without money or wealth are dependent upon the opportunities which the law assures them . . . The school system, common and equal for all, is their only hope for gaining the training by which they can fully enter into the enjoyment of their new privileges. To them, the gaining of the knowledge of English is the medium through which they will become acquainted with the principles of American liberty, with American affairs, American commerce and trade, and thereby share in their benefits. All that they expect from their new relations must come to them through the English language, in which are to be found American history, literature, art, science, statesmanship, and in the use of which they are to enter into industrial and commercial relations with the business of the States and take a share in their civil administration. This great boon you carry to them in the little reader in your hands . . . Familiarize yourself with the principles of the American Constitution, and be sure that your official conduct is guided by them. Do not indulge in fault-finding.

Remember that "the joys of victory are the joys of man," and so try to assure the success of teachers and pupils that they may share in these joys. Be sure that your example is worthy to be followed in all things. "Peace hath her victories no less renowned than war." May you add a conspicuous illustration to the truth of this saying by fidelity and success in the discharge of your important duties.

Sincerely yours,
John Eaton,
Director, etc.

CHAPTER FIVE

Segregation and New Arrivals, 1898–1960

I recently saw that in Donna, Texas, the Mexican children who went to school there were bathed with gasoline, especially their heads. The teachers of the school did that and they not only bathed those who went more or less dirty but also those who were clean. One of my countrymen who was indignant because of this action tried to get the Mexican parents to get together and make a protest before the school board but the other Mexicans told him, "What is it for us to protest when they won't pay any attention to us?"

—*Alonso M. Galván, 1931*[1]

INTRODUCTION

After 1900, while students in newly colonized Puerto Rico were being Americanized on their island or sent to the United States for advanced training, children of Mexican descent in the Southwest United States experienced increasing segregation. Between 1898 and 1960, economic, political, and social turmoil in Latinos' home countries, along with the demand for labor in the United States, contributed to increasing immigration to the United States, particularly in the urban areas of the Northeast, midwestern cities such as Chicago, and the Southwest. Specifically, factors such as the 1910 Mexican Revolution, the displacement of thousands of Puerto Rican agricultural workers from their farms as a result of the U.S. government and industry's influence in narrowing the island to a one-crop economy, and the demand for railroad and seasonal agricultural workers contributed to a continuing flow of Mexicans and Puerto Ricans to the United States.[2] The Great Depression of the 1930s curbed immigration, particularly when the U.S. government began a campaign to repatriate Mexicans in order to permit more jobs for Americans. Historians have estimated that between one-third and one-half of the Mexican population in the United States left during the depression, many involuntarily.[3]

The mid-twentieth century, however, was also a time of recognition among scholars and government workers of the hopes and needs of Mexican Americans and Puerto Ricans. During the 1930s, 1940s, and 1950s Latino and

American academics and writers began to educate the public about Latino concerns. George I. Sánchez's, *Forgotten People* (1940) and Carey McWilliams's lucid *North from Mexico: The Spanish-Speaking People of the United States* (1950) had a broad impact in bringing the Hispanic story to the nation's attention.[4] The first generation of Latino scholars trained in the United States, such as George I. Sánchez and Carlos E. Castañeda at the University of Texas, Austin, also shaped the study and publication of Latino issues.

Latinos were well capable of organizing themselves on behalf of social organizations. In the mid-twentieth century, Latinos used this ability to organize politically. From Harlem to the smaller towns in Texas and California, Latinos formed associations to protect their rights as residents and citizens.[5] Many of these organizations provided a training ground for the leaders who emerged during the 1960s and 1970s Civil Rights era. In short, the years from the 1920s through the 1950s witnessed not only the creation of second- and sometimes third-generation Mexicans moving into the middle class, but also the infusion of newly arrived immigrants, fleeing from political and social turmoil.[6]

SEGREGATION IN THE SOUTHWEST

Historians have found that prior to 1900, Mexican Americans were often integrated in the public schools and Mexican Americans were hired as public school teachers (see document 5.1). After 1900, newly implemented linguistic and cultural policies increasingly segregated Mexican American children from Anglos and deprived the former of equal educational opportunities. According to historian Gilbert Gonzalez, several factors contributed to this increased segregation: Anglo fear of the rapid influx of Mexican Americans into Southwestern communities (particularly after the 1910 Mexican Revolution), residential segregation, racism, and a political economy unwilling to provide more than a rudimentary level of schooling for the agricultural workforce.[7]

Unlike the rigid, de jure segregation of African Americans from whites in Southern public classrooms, statutes for Southwestern school districts rarely included segregation clauses. Rather, Anglo school administrators utilized vague and often unwritten justifications to place Mexican children into separate classrooms or entirely separate schools from their Anglo peers. Administrators justified segregation based on the perception that the children possessed deficient English language skills, scored low on intelligence tests, and/or practiced poor personal hygiene.[8] Although many school districts claimed that Mexican children were only segregated in the early grades, they were rarely transferred to the upper grades in Anglo schools.

Economic reliance upon migrant agricultural workers in the Southwest also resulted in nonenforcement of compulsory school attendance laws for Mexican-origin children. One study of selected Southwestern counties in the 1930s revealed that attendance among enrolled white children ranged from

71 to 96 percent, while school attendance for enrolled Mexican children only ranged from 39 to 89 percent.[9] Furthermore, among Mexican Americans who were in school, 85 percent attended segregated schools in the 1930s; expenditures and supplies overwhelmingly favored the white schools.[10] Among other devices used to maintain an inequitable system, Mexican American children were given used textbooks from the white schools, and Mexican American athletic teams were not allowed to compete in Anglo sports leagues. (See document 5.2.)

Historians such as Guadalupe San Miguel, Jr., have documented how the Mexican American community in Texas reacted with agency, not passive acceptance, to the increasingly inequitable educational opportunities of the twentieth century. Through the creation of grassroots organizations such as the League of United Latin American Citizens (LULAC) in 1929 and a shifting coalition of community, state, and regional organizations, Mexican Americans responded proactively to protect the future of their children in a rapidly changing U.S. economy and society.[11] (See document 5.3.)

For example, Mexican Americans in the Southwest initiated school desegregation cases decades earlier than the landmark case of *Brown v. Board of Education* (1954). In 1930, with the help of LULAC, parents in the Del Rio Independent School District of Texas sued, citing that they had been denied use of facilities used by "other white races."[12] The plaintiff, Jesús Salvatierra, lost because the court found that Mexican children were separated as a result of "special language needs." The next year, parents in California were more successful. In *Alvarez v. Lemon Grove* (1931) Mexican parents argued successfully that the school board had no right to segregate children based on Hispanic surname or Mexican "look."[13]

Two more court cases in the 1940s provided broader jurisdiction. In *Méndez et al. v. Westminster School District of Orange County* (1947), Mexican California parents argued that their children were unconstitutionally segregated. (See documents 5.4 and 5.5.) Since the 1860s, California school law provided for separate schools for "Negro, Mongolian and Indian children." Thus, the plaintiffs in *Méndez* had to demonstrate that Mexicans were not Indians. They won their case on several grounds, and the judge stated that "evidence clearly shows that Spanish-speaking children are retarded in learning English by lack of exposure to its use by segregation." The National Association for the Advancement of Colored People (NAACP) also joined this suit, testing sociological arguments against segregation for the first time.[14]

LULAC and the post–World War II advocacy group, the American G.I. Forum, provided the resources to launch the next desegregation case in Texas: *Delgado v. Bastrop Independent School District* (1948). Once again, Mexicans argued that they were Caucasian, not black, and thus were illegally being segregated. Furthermore, in *Delgado*, attorney for the plaintiffs, Gus García, argued that the schools were depriving children of Mexican descent of equal facilities, services, and education. (See documents 5.6, 5.7, and 5.8.) Judge Ben H. Rice agreed and ordered the end of segregation by September 1949. The court did allow, however for separate classes—only in the first grade and on the same school grounds. African Americans in Texas were left in segregated schools.[15]

Not all advocacy programs in the mid-twentieth century focused on school desegregation. San Miguel documented the creation of an innovative program called the Campaign of Little Schools of 400 in Texas, a preschool program designed to prepare Latino children for the English-speaking public schools. This LULAC-sponsored activity was a forerunner to the U.S. government's Head Start program.[16]

Historians of the Mexican American experience, similar to historians of African American history, have emphasized agency as well as the positive aspects of segregated schools. For instance, Mexicans in Houston during the segregation era did not necessarily view the creation of a "Mexican" school in their community as a form of oppression. The establishment of the Lorenzo de Zavala School in 1920, named after a Mexican patriot, was "heartily supported" by the Mexican population and witnessed high enrollments. Likewise, in El Paso, the Mexican Aoy Preparatory School boasted the best attendance of any of the city's public schools in the early 1900s. (See document 5.9.) San Miguel also documents the presence of successful Mexican American youth in Houston's secondary schools and colleges, thus raising questions about "the popular and historical interpretation of the Mexican experience in education" in which "all Mexican origin children were non-achievers."[17]

In New Mexico and Colorado, during the years of segregation, Hispanos emphasized their distinct heritage as something to be celebrated. Political power within this group helped secure and finance educational facilities. As a result, Hispano public school officials determined who taught and administered, dictated the nature of social and academic environments, and determined which students should prepare for college.[18]

One example of Hispano political clout in New Mexico was the establishment of a bilingual teacher training school. In 1909, the state legislature founded the Spanish- American Normal School at El Rito. The legislature charged the institution to educate "Spanish-speaking natives of New Mexico for the vocation of teachers in the public schools of the counties and districts where the Spanish language is prevalent."[19] The school enrolled over one hundred future teachers by 1918. In the 1930s the Normal School was still open. Eventually it was absorbed into the New Mexico higher education system.[20]

The 1920s through 1950s witnessed an increasing number of Mexicans entering college following two decades of minimal participation.[21] Philanthropy, increasing numbers of middle-class Latinos, and the G.I. Bill were major contributors to this shift. These Latino college students were pioneers. Often the only Latinos in their classes, they provided leadership and talent to the formation of the Chicano/Puerto Rican civil rights movement of the 1960s and 1970s.

The college-enrolled pioneers of the 1920s-1950s were clearly exceptional. Unlike the late-nineteenth-century participation of Latinos from older, elite Hispano families (as described in chapter two), students from middle- and working-class Latino families were finally entering higher education. Still, the barriers to high school graduation were formidable. Lack of enforcement of school attendance laws, language difficulties, classroom harassment, and racism resulted in scarce numbers of Mexican American children reaching eighth grade.[22] The pipeline to higher education was thus choked off early in most Latino children's lives. Despite these obstacles, which impeded most Mexican Americans from collegiate participation prior to the 1960s, at least four factors contributed to the success of the few who broke through the barriers.

First, community and charitable organizations became involved. During the Great Depression of the 1930s, the Protestant Young Men's Christian Association (YMCA) in Los Angeles committed $30,000 to work with Mexican American youth. The YMCA hired role model and social worker Tom García to head this project. García created boys' clubs, organized the first Mexican Youth Conference, and provided training and leadership to adolescent boys.[23] Significantly, YMCA officials provided contacts with higher education leaders. Scholarships, admissions information, and important networks were made available to Latino male youth. As an offshoot of the YMCA club, Mexican American students at UCLA created the first Latino student organization, called the Mexican-American Movement (MAM). Under the direction of student Felix Gutierrez from 1938–1944, the first Latino college student newspaper, *The Mexican Voice*, was in print at UCLA. After 1944 the title was changed to *The Forward* and the tone of the paper changed as it focused on war-time activities of members of MAM.[24]

A second factor opening access to higher education for nonelite Latino families involved what historian Muñoz described as the "active support of individual teachers, clergy, or social workers that were sympathetic and in a position to identify youth with exceptional intelligence."[25] For example, Frances Esquivel secured a U.C. Berkeley alumni scholarship through the efforts of her high school history teacher, Miss Helen Grant, a U.C. Berkeley alumna.[26] Similarly, the writer and scholar Ernesto Galarza entered Occidental College in 1923 and then became the first Mexican American to enter Stanford through the active assistance of interested teachers.[27]

Third, the passage of the Servicemen's Readjustment Act of 1944, or G.I. Bill, also assisted in expanding higher education access in the mid-twentieth century. Muñoz argued that "among the thousands of returning Mexican American veterans who took advantage of this opportunity to pursue a higher education were Americo Paredes, Octavio Romano V., and Ralph Guzman. They were destined to become . . . significant contributors to Mexican American intellectual life."[28] The American G.I. Forum, created in 1948, was composed of Latino World War II veterans who actively worked to ensure that the G.I. Bill and other benefits were extended to veterans.[29]

For example, Donato demonstrated how Hispano veterans in Colorado demanded local access to higher education and were responsible for creating the San Luis Institute, a public two-year college. One San Luis veteran recalled, "I remember that almost all of us who discharged from the military went to college." According to Donato "the sense of camaraderie among San Luis students who went on to Adams State" aided their access and retention at a four-year institution.[30]

Fourth, the Latino community contributed to the increased college participation during the 1920s through 1950s. Previously mentioned for its work in desegregation cases, LULAC and numerous other Latino-based organizations provided college scholarships. The 1920s-1950s also witnessed the entrance of Latino faculty into higher education, further providing role models and encouragement for higher learning. Anecdotal evidence suggests that prior to the 1940s, Hispanic-surnamed faculty at white colleges and universities were generally from Spain and clustered in the romance language and literature departments.[31] Key role models and intellectuals who trained the leaders of the Chicano generation include George I. Sánchez, first at the University of New Mexico in the 1930s and then from 1940 until his death at the University of Texas at Austin. Historian Carlos Castañeda was also a significant figure at the University of Texas at Austin. He devoted his life's work to documenting and correcting Latino history as a professor in the Department of History.[32]

MAINLAND PUERTO RICAN EDUCATION IN THE POST-DEPRESSION ERA

While school conditions for Southwestern Mexicans remained fairly static during the 1930s, 1940s, and 1950s, this was a time when Puerto Ricans began arriving to other areas of the United States, transforming the Hispanic population into a national, and not just a Southwestern, phenomena. Small numbers of Puerto Ricans had lived in the United States in the nineteenth and early twentieth centuries. The period of greatest immigration, however, began after World War II. In 1940, almost 70,000 Puerto Ricans lived on the mainland. By 1950 that number had increased to 300,000 and in 1960 was 887,661. New York City and surrounding areas absorbed most of this immigration.[33] Most historians agree that the following factors spurred migration to the mainland: the Jones Act of 1917, which granted U.S. citizenship to Puerto Ricans; the Johnson Acts of 1921 and 1924, which curtailed European immigration; labor shortages in the United States during World Wars I and II; and relatively inexpensive transportation costs to the United States.[34]

During the years prior to World War II, Puerto Ricans in the United States established strong networks, created communities, and formed mutual aid associations. Concerns regarding the public schools were explored through avenues such as the Puerto Rican association Madres Y Padres Pro Niños Hispanos (Mothers and Fathers in Support of Hispanic Children) during the 1930s and early 1940s. This organization, for example, questioned the school officials' use of intelligence testing, which channeled Puerto Rican children into classrooms for "backward" children rather than recognizing the inherent language bias in such testing.[35]

The 1940s and 1950s witnessed a sharp jump in the number of Puerto Rican children in the New York City schools. In 1949, there were 29,000 Puerto Rican children in the schools; four years later they numbered about 54,000. By 1968 almost 300,000 Puertorriqueños attended New York City schools. Ten years earlier, in response to rapid increases, the city had commissioned an intensive, multi-year investigation, *The Puerto Rican Study, 1953–1957.* In this study researchers recommended extensive bilingual preparation of teachers and support staff, but teachers and administrators were overwhelmed in the late 1950s and 1960s by new student arrivals—it was estimated that less than one-quarter of students could speak English.[36] (See document 5.10.)

Puerto Rican women hired as substitute auxiliary teachers (SATs) in New York City during the 1940s and 1950s were able to assist new pupils. There were too few in number compared to the demand, and these teachers found themselves struggling to "provide alternative modes of instruction for the increasing numbers of Spanish-speaking youngsters arriving to this city."[37] Furthermore, Sánchez Korrol persuasively demonstrates how the SATs lent "recognition" and "legitimacy" to the introduction of alternative methods of teaching English as a second language. Her analysis of the struggles of the post–World War II era documents both the problems of the times and the involvement of Puerto Ricans in the search for solutions.

The experience of Puerto Rican children arriving into the New York City schools before 1965 has been captured in memoirs, such as that of Esmeralda Santiago. She came as a child on the cusp of adolescence to the unknown urban world of English language and the hidden rules of the public school realm. (See document 5.11.) In this environment, Spanish-speaking children were viewed as defective, in need of remedial instruction, and of lesser intelligence than white counterparts. Other Latino children came to the rural U.S. as migrant agricultural workers. Most of their experiences consisted of poor working conditions and low pay.[38] In some instances, innovative programs were developed to broaden, rather than narrow, their life chances. The Civil Rights movement of the 1960s and 1970s would wipe away some of the most egregious practices of school segregation based upon language skills and color, but it would not entirely erase deep-seated prejudices.

DOCUMENT 5.1

Teacher's Certificate, Zapata County, Texas, 1898

Mercurio Martinez was born in San Ygnacio, Zapata County, Texas, on October 27, 1876. He attended the local public schools and went to Austin, Texas to study at St. Edward's College from 1895 to 1898. After graduation from St. Edwards he began his teaching career in Zapata County and was employed in that county from 1898 to 1920 in various teaching and administrative capacities. Mr. Martinez's biography is illustrative of the experiences of many of the early and unknown Latino teachers employed in the public schools of the American Southwest in the early twentieth century.

From Mercurio Martinez Collection, Cushing Memorial Library and Archives, Texas A and M University, College Station, Texas.

DOCUMENT 5.2

Mexican American Schooling in the Southwest circa 1930s

During the 1930s and 1940s several influential studies pertaining to the educational conditions of Mexican American children were published. The photographs and captions below are from Professor Herschel T. Manuel's The Education of Mexican and Spanish-Speaking Children in Texas *(Austin: University of Texas, 1930). Conditions for Mexican American children were often substandard and enforcement of school attendance lax. Reports such as Annie Reynold's* The Education of Spanish-Speaking Children in Five Southwestern States *(1933); Wilson Little's* Spanish-Speaking Children in Texas *(1944) and two works by George I. Sánchez,* The Education of Bilinguals in a State School System *(1934) and* Concerning the Segregation of Spanish-Speaking Children in the Public Schools *(1951) highlighted the educational needs of Latino children in the Southwest.*

From H. T. Manuel, *The Education of Mexican and Spanish-Speaking Children in Texas* (Austin: The University of Texas, 1930) p. 71.

These Mexican children are holding state textbooks discarded by the school for other whites and then issued to them. When this picture was taken they had received no new books, nor enough of the worn ones.

Drinking facilities such as these are common. Incidentally, the children pictured are themselves an interesting study.

DOCUMENT 5.3

Founding Principles of LULAC

The League of United Latin American Citizens (LULAC) was founded in Corpus Christi, Texas, on February 17, 1929, as a result of the merger of four small organizations. It was founded by middle-class Mexican Americans frustrated with continuing discrimination and violation of the political and civil rights of Latinos in the Southwest. In 1933 women were encouraged to join LULAC Ladies Councils that focused on civic activities, including the improvement of educational facilities and bilingual training. LULAC became involved in several key desegregation cases involving Mexican American children, in addition to securing broader political rights for Latinos in the Southwest. The following document is from Article 2 of the original 1929 constitution.

From *LULAC News*, August 1931, in the LULAC Archives, Benson Latin American Collection, University of Texas, Austin.

Article 2. The Aims and Purposes of This Organization Shall Be:

1. To develop within the members of our race the best, purest, and most perfect type of a true and loyal citizen of the United States of America.
2. To eradicate from our body politic all intents and tendencies to establish discriminations among our fellow citizens on account of race, religion or social position as being contrary to the true spirit of Democracy, our Constitution, and our Laws.
3. To use all legal means at our command to the end that all citizens in our country may enjoy equal rights, the equal protection of the laws of the land, and equal opportunities and privileges.
4. The acquisition of the English language, which is the official language of our country, being necessary for the enjoyment of our rights and privileges, we declare it to be the official language of this organization and we pledge ourselves to learn and speak the same to our children.
5. To define with absolute and unmistakable clearness our unquestionable loyalty to the principles, ideals, and citizenship of the United States of America.
6. To assume complete responsibility for the education of our children as to their rights and duties and the language and customs of this country; the latter, in so far as they may be good customs.

7. We solemnly declare once and for all to maintain a sincere and respectful reverence for our racial origin, of which we are proud.

8. Secretly and openly, by all lawful means at our command, we shall assist in the education and guidance of Latin-Americans and we shall protect and defend their lives and interests whenever necessary.

9. We shall destroy any attempt to create racial prejudices against our people, and any infamous stigma which may be cast upon them, and we shall demand for them the respect and prerogatives which the Constitution grants to us all.

10. Each of us considers himself with equal responsibilities in our organization, to which we voluntarily swear subordination and obedience.

11. We shall create a fund for our mutual protection, for the defense of those of us who may be unjustly prosecuted, and for the education and culture of our people.

12. This organization is not a political club, but as citizens we shall participate in all local, state, and national political contests. However, in doing so we shall ever bear in mind the general welfare of our people; and we disregard and abjure for all in any personal obligation which is not in harmony with these principles.

13. With our vote influence we shall endeavor to place in public office men who show by their conduct respect and consideration for our people.

14. We shall select as our leaders those among us who demonstrate, by their integrity and culture, that they are capable of guiding and directing us properly.

15. We shall maintain public means for the diffusion of these principles and the expansion and consolidation of this organization.

16. We shall pay our poll tax, and that of the members of our families, in order that we may enjoy our rights fully.

17. We shall diffuse our ideas by means of the press, lectures, and pamphlets.

DOCUMENT 5.4

Méndez v. Westminster School District, Orange County, California, 1946

In 1931 a lower jurisdictional court in California had ruled against the school segregation of Mexican American children in Lemon Grove v. Alvarez, CA *(1931). Méndez and the accompanying appeal provided broader jurisdiction, making it against the law in Orange County, California, to segregate children of "Mexican or Latin descent." The court ruled that segregating children by their ethnicity denied them equal protection of the law, "notwithstanding English language deficiencies of some of the children." For the first time, the NAACP joined in the school district's unsuccessful appeal, as an amicus curiae (friend of the court). (See document 5.5).*

From: *Méndez et al. v. Westminster School District of Orange County et al.* Civil Action No. 4292. 64 F. Supp. 544 District court, S. D. California, Central Division, February 18, 1946 (excerpts).

McCormick, District Judge.

Gonzalo Méndez, William Guzman, Frank Palomino, Thomas Estrada and Lorenzo Ramirez, as citizens of the United States, and on behalf of their minor children, and as they allege in the petition, on behalf of "some 5000" persons similarly affected, all of Mexican or Latin descent, have filed a class suit pursuant to Rule 23 of Federal Rules of Civil Procedure, 28 U.S.C.A. following section 723c, against the Westminister, Garden Grove and El Modeno School Districts, and the Santa Ana City Schools, all of Orange County, California, and the respective trustees and superintendents of said school districts.

The complaint, grounded upon the Fourteenth Amendment to the Constitution of the United States[1] and Subdivision 14 of Section 24 of the Judicial Code, Title 28, Section 41, subdivision 14, U.S.C.A.,[2] alleges a concerted policy and design of class discrimination against "persons of Mexican or Latin descent or extraction" of elementary school age by the defendant school agencies in the conduct and operation of public schools of said districts, resulting in the denial of the equal protection of the laws to such class of persons among which are the petitioning school children.

Specifically, plaintiffs allege:

"That for several years last past respondents have and do now in further-ance and in execution of their common plan, design and purpose within their respective Systems and Districts, have by their regulation, custom and usage and in execution thereof adopted and declared: That all children or persons of Mexican or Latin descent or extraction, though Citizens of the United States of America, shall be, have been and are now excluded from attending, using, enjoying and receiving the benefits of the education, health and recreation facilities of certain schools within their respective Districts and Systems but that said children are now and have been segregated and required to and must attend and use certain schools in said Districts and Systems reserved for and attended solely and exclusively by children and persons of Mexican and Latin descent, while such other schools are maintained, attended and used exclu-sively by and for persons and children purportedly known as White or Anglo-Saxon children.

"That in execution of said rules and regulations, each, every and all the foregoing children are compelled and required to and must attend and use the schools in said respective Districts reserved for and attended solely and exclusively by children of Mexican and Latin descent and are forbidden, barred and excluded from attending any other school in said District or System solely for the reason that said children or child are of Mexican or Latin descent."

The petitioners demand that the alleged rules, regulations, customs and usages be adjudged void and unconstitutional and that an injunction issue restraining further application by defendant school authorities of such rules, regulations, customs, and usages.

It is conceded by all parties that there is no question of race discrimination in this action. It is, however, admitted that segregation per se is practiced in the abovementioned school districts as the Spanish-speaking children enter school life and as they advance through the grades in the respective school districts. It is also admitted by the defendants that the petitioning children are qualified to attend the public schools in the respective districts of their residences.

In the Westminister Garden Grove and El Modeno school districts the respective boards of trustees had taken official action, declaring that there be no segregation of pupils on a racial basis but that non-English-speaking children (which group, excepting as to a small number of pupils, was made up entirely of children of Mexican ancestry or descent), be required to attend schools designated by the boards separate and apart from English-speaking pupils; that such group should attend such schools until they had acquired some proficiency in the English language.

The petitioners contend that such official action evinces a covert attempt by the school authorities in such school districts to produce an arbitrary discrimination against school children of Mexican extraction or descent and that such illegal result has been established in such school districts respec-tively. The school authorities of the City of Santa Ana have not memorialized any such official action, but petitioners assert that the same custom and usage exists in the schools of the City of Santa Ana under the authority of appropriate school agencies of such city.

The concrete acts complained of are those of the various school district officials in directing which schools the petitioning children and others of the same class or group must attend. The segregation exists in the elementary schools to and including the sixth grade in two of the defendant districts, and in the two other defendant districts through the eighth grade. The record before us shows without conflict that the technical facilities and physical conveniences offered in the schools housing entirely the segregated pupils, the efficiency of the teachers therein and the curricula are identical and in some respects superior to those in the other schools in the respective districts.

The ultimate question for decision may be thus stated: Does such official action of defendant district school agencies and the usages and practices pursued by the respective school authorities as shown by the evidence operate to deny or deprive the so called non-English-speaking school children of Mexican ancestry or descent within such school districts of the equal protection of the laws?

The defendants at the outset challenge the jurisdiction of this court under the record as it exists at this time. We have already denied the defendants' motion to dismiss the action upon the "face" of the complaint. No reason has been shown which warrants reconsideration of such decision.

1. and 2. While education is a State matter, it is not so absolutely or exclusively. Cumming v. Board of Education of Richmond County, 175 U.S. 528, 20 S.Ct. 197, 201, 44 L.Ed. 262. In the Cumming decision the Supreme Court said: "That education of the people in schools maintained by state taxation is a matter belonging to the respective states and *any interference on the part of Federal authority with the management of such schools cannot be justified except in the case of a clear and unmistakable disregard of rights secured by the supreme law of the land.*" (Emphasis supplied.) . . .

Obviously, then, a violation by a State of a personal right or privilege protected by the Fourteenth Amendment in the exercise of the State's duty to provide for the education of its citizens and inhabitants would justify the Federal Court to intervene. State of Missouri ex rel. Gaines v. Canada, 305 U.S. 337, 59 S.Ct. 232, 83 L.Ed. 208. The complaint before us in this action, having alleged an invasion by the common school authorities of the defendant districts of the equal opportunity of pupils to acquire knowledge, confers jurisdiction on this court if the actions complained of are deemed those of the State. Hamilton v. Regents of University of California, 293 U.S. 245, 55 S.Ct. 197, 79 L.Ed. 343; cf. Meyer v. Nebraska, 262 U.S. 390, 43 S.Ct. 625, 67 L.Ed. 1042, 29 A.L.R. 1446.

Are the actions of public school authorities of a rural or city school in the State of California, as alleged and established in this case, to be considered actions of the State within the meaning of the Fourteenth Amendment so as to confer jurisdiction on this court to hear and decide this case under the authority of Section 24, Subdivision 14 of the Judicial Code, supra? We think they are.

3. In the public school system of the State of California the various local school districts enjoy a considerable degree of autonomy. Fundamentally, however, the people of the State have made the public school system a matter of State supervision. Such system is not committed to the exclusive control of local governments . . .

4. The Education Code of California provides for the requirements of teachers' qualifications, the admission and exclusion of pupils, the courses of study and the enforcement of them, the duties of superintendents of schools and of the school trustees of elementary schools in the State of California. The appropriate agencies of the State of California allocate to counties all the State school money exclusively for the payment of teachers' salaries in the public schools and such funds are apportioned to the respective school districts within the counties. While, as previously observed, local school boards and trustees are vested by State legislation with considerable latitude in the administration of their districts, nevertheless, despite the decentralization of the educational system in California, the rules of the local school district are required to follow the general pattern laid down by the legislature, and their practices must be consistent with law and and [*sic*] with the rules prescribed by the State Board of Education. See Section 2204, Education Code of California.

When the basis and composition of the public school system is considered, there can be no doubt of the oneness of the system in the State of California, or of the restricted powers of the elementary school authorities in the political subdivisions of the State . . .

5. We therefore turn to consider whether under the record before us the school boards and administrative authorities in the respective defendant districts have by their segregation policies and practices transgressed applicable law and Constitutional safeguards and limitations and thus have invaded the personal right which every public school pupil has to the equal protection provision of the Fourteenth Amendment to obtain the means of education

We think the pattern of public education promulgated in the Constitution of California and effectuated by provisions of the Education Code of the State prohibits segregation of the pupils of Mexican ancestry in the elementary schools from the rest of the school children.

Section 1 of Article IX of the Constitution of California directs the legislature to "encourage by all suitable means the promotion of intellectual, scientific, moral, and agricultural improvement" of the people. Pursuant to this basic directive by the People of the State many laws stem authorizing special instruction in the public schools for handicapped children. See Division 8 of the Education Code. Such legislation, however, is general in its aspects. It includes all those who fall within the described classification requiring the special consideration provided by the statutes regardless of their ancestry or extraction. The common segregation attitudes and practices of the school authorities in the defendant school districts in Orange County pertain solely to children of Mexican ancestry and parentage. They are singled out as a class for segregation. Not only is such method of public school administration contrary to the general requirements of the school laws of the State, but we think it indicates an official school policy that is antagonistic in principle to Sections 16004 and 16005 of the Education Code of the State.[3]

Obviously, the children referred to in these laws are those of Mexican ancestry. And it is noteworthy that the educational advantages of their commingling with other pupils is regarded as being so important to the school system of the State that it is provided for even regardless of the citizenship of the parents. We perceive in the laws relating to the public educational system in the State of California a clear purpose to avoid and forbid distinctions among pupils based upon race or ancestry[4] except in specific situations[5] not pertinent to this action. Distinctions of that kind have recently been declared by the highest judicial authority of the United States "by their very nature odious to a free people whose institutions are founded upon the doctrine of equality." They are said to be "utterly inconsistent with American traditions and ideals." Kiyoshi Hirabayashi v. United States, 320 U.S. 81, 63 S.Ct. 1375, 1385, 87 L.Ed. 1774.

Our conclusions in this action, however, do not rest solely upon what we conceive to be the utter irreconcilability of the segregation practices in the defendant school districts with the public educational system authorized and sanctioned by the laws of the State of California. We think such practices clearly and unmistakably disregard rights secured by the supreme law of the land. Cumming v. Board of Education of Richmond County, supra.

6. and 7. "The equal protection of the laws" pertaining to the public school system in California is not provided by furnishing in separate schools the same technical facilities, text books and courses of instruction to children of Mexican ancestry that are available to the other public school children regardless of their ancestry. A paramount requisite in the American system of public education is social equality. It must be open to all children by unified school association regardless of lineage.

8. We think that under the record before us the only tenable ground upon which segregation practices in the defendant school districts can be defended lies in the English language deficiencies of some of the children of Mexican ancestry as they enter elementary public school life as beginners. But even such situations do not justify the general and continuous segregation in separate schools of the children of Mexican ancestry from the rest of the elementary school population as has been shown to be the practice in the defendant school districts—in all of them to the sixth grade, and in two of them through the eighth grade.

The evidence clearly shows that Spanish-speaking children are retarded in learning English by lack of exposure to its use because of segregation, and that commingling of the entire student body instills and develops a common cultural attitude among the school children which is imperative for the perpetuation of American institutions and ideals.[6] It is also established by the record that the methods of segregation prevalent in the defendant school districts foster antagonisms in the children and suggest inferiority among them where none exists. One of the flagrant examples of the discriminatory results of segregation in two of the schools involved in this case is shown by the record. In the district under consideration there are two schools, the Lincoln and the Roosevelt, located approximately 120 yards apart on the same school grounds, hours of opening and closing, as well as recess periods, are not uniform. No credible language test is given to the children of Mexican ancestry upon entering the first grade in Lincoln School. This school has an enrollment of 249 so-called Spanish-speaking pupils, and no so-called English-speaking pupils; while the Roosevelt, (the other) school, has 83 so-called English-speaking pupils and 25 so-called Spanish-speaking pupils. Standardized tests as to mental ability are given to the respective classes in the two schools and the same curricula are pursued in both schools and, of course, in the English language as required by State law. Section 8251, Education Code. In the last school year the students in the seventh grade of the Lincoln were superior scholarly to the same grade in the Roosevelt School and to any group in the seventh grade in either of the schools in the past. It further appears that not only did the class as a group have such mental superiority but that certain pupils in the group were also outstanding in the class itself. Notwithstanding this showing, the pupils of such excellence were kept in the Lincoln School. It is true that there is no evidence in the record before us that shows that any of the members of this exemplary class requested transfer to the other so-called intermingled school, but the record does show without contradiction that another class had protested against the segregation policies and practices in the schools of this El Modeno district without avail.

While the pattern or ideal of segregating the school children of Mexican ancestry from the rest of the school attendance permeates and is practiced in all of the four defendant districts, there are procedural deviations among the school administrative agencies in effectuating the general plan.

In Garden Grove Elementary School District the segregation extends only through the fifth grade. Beyond, all pupils in such district, regardless of their ancestry or linguistic proficiency, are housed, instructed and associate in the same school facility.

This arrangement conclusively refutes the reasonableness or advisability of any segregation of children of Mexican ancestry beyond the fifth grade in any of the defendant school districts in view of the standardized and uniform curricular requirements in the elementary schools of Orange County.

But the admitted practice and long established custom in this school district whereby all elementary public school children of Mexican descent are required to attend one specified school (the Hoover) until they attain the sixth grade, while other pupils of the same grade are permitted to and do attend two other elementary schools of this district, notwithstanding that some of those pupils live within the Hoover School division of the district, clearly establishes an unfair and arbitrary class distinction in the system of public education operative in the Garden Grove Elementary School District.

The long standing discriminatory custom prevalent in this district is aggravated by the fact shown by the record that although there are approximately 25 children of Mexican descent living in the vicinity of Lincoln School, none of them attend that school, but all are peremptorily assigned by the school authorities to the Hoover School, although the evidence shows that there are no school zones territorially established in the district.

The record before us shows a paradoxical situation concerning the segregation attitude of the school authorities in the Westminister School District. There are two elementary schools in the undivided area. Instruction is given pupils in each school from kindergarten to the eighth grade, inclusive. Westminister School has 642 pupils, of whom 628 are so-called English-speaking children, and 14 so-called Spanish-speaking pupils. The Hoover School is attended solely by 152 children of Mexican descent. Segregation of these from the rest of the school population precipitated such vigorous protests by residents of the district that the school board in January, 1944, recognizing the discriminatory results of segregation, resolved to unite the two schools and thus abolish the objectionable practices which had been operative in the school of the district for a considerable period. A bond issue was submitted to the electors to raise funds to defray the cost of contemplated expenditures in the school consolidation. The bonds were not voted and the record before us in this action reflects no execution or carrying out of the official action of the board of trustees taken on or about the 16th of January, 1944. It thus appears that there has been no abolishment of the traditional segregation practices in this district pertaining to pupils of Mexican ancestry through the gamut of elementary school life. We have adverted to the unfair consequences of such practices in the similarly situated El Modeno School District.

Before considering the specific factual situation in the Santa Ana City Schools it should be noted that the omnibus segregation of children of Mexican ancestry from the rest of the student body in the elementary grades in the schools involved in this case because of language handicaps is not warranted by the record before us. The test applied to the beginners are shown to have been generally hasty, superficial and not reliable. In some instances separate classification was determined largely by the Latinized or Mexican name of the child. Such methods of evaluating language knowledge are illusory and are not conducive to the inculcation and enjoyment of civil rights which are of primary importance in the public school system of education in the United States.

It has been held that public school authorities may differentiate in the exercise of the reasonable discretion as to the pedagogical methods of instruction to be pursued with different pupils.[7] And foreign language handicaps may be to such a degree in the pupils in elementary schools as to require special treatment in separate classrooms. Such separate allocations, however, can be lawfully made only after credible examination by the appropriate school authority of each child whose capacity to learn is under consideration and the determination of such segregation must be based wholly upon indiscriminating foreign language impediments in the individual child, regardless of his ethnic traits or ancestry.

9.–11. The defendant Santa Ana School District maintains fourteen elementary schools which furnish instruction from kindergarten to the sixth grade, inclusive.

About the year 1920 the Board of Education, for the purpose of allocating pupils to the several schools of the district in proportion to the facilities available at such schools, divided the district into fourteen zones and assigned to the school established in each zone all pupils residing within such zone.

There is no evidence that any discriminatory or other objectionable motive or purpose actuated the School Board in location or defining such zones.

Subsequently the influx of people of Mexican ancestry in large numbers and their voluntary settlement in certain of the fourteen zones resulted in three of the zones becoming occupied almost entirely by such group of people.

Two zones, that in which the Fremont School is located, and another contiguous area in which the Franklin School is situated, present the only flagrant discriminatory situation shown by the evidence in this case in the Santa Ana City Schools. The Fremont School has 325 so-called Spanish-speaking pupils and no so-called English-speaking pupils. The Franklin School has 237 pupils of which 161 are so-called English-speaking children, and 76 so-called Spanish-speaking children.

The evidence shows that approximately 26 pupils of Mexican descent who reside within the Fremont zone are permitted by the School Board to attend the Franklin School because their families had always gone there. It also appears that there are approximately 35 other pupils not of Mexican descent who live within the Fremont zone who are not required to attend the Fremont School but who are also permitted by the Board of Education to attend the Franklin School.

Sometime in the fall of the year 1944 there arose dissatisfaction by the parents of some of the so-called Spanish-speaking pupils in the Fremont School zone who were not granted the privilege that approximately 26 children also of Mexican descent, enjoyed in attending the Franklin School. Protest was made en masse by such dissatisfied group of parents, which resulted in the Board of Education directing its secretary to send a letter to the parents of all of the so-called Spanish-speaking pupils living in the Fremont zone and attending the Franklin School that beginning September, 1945, the permit to attend Franklin School would be withdrawn and the children would be required to attend the school of the zone in which they were living, viz., the Fremont School.

There could have been no arbitrary discrimination claimed by plaintiffs by the action of the school authorities if the same official course had been applied to the 35 other so-called English-speaking pupils exactly situated as were the approximate 26 children of Mexican lineage, but the record is clear that the requirement of the Board of Education was intended for and directed exclusively to the specified pupils of Mexican ancestry and if carried out becomes operative solely against such group of children.

It should be stated in fairness to the Superintendent of the Santa Ana City Schools that he testified he would recommend to the Board of Education that the children of those who protested the action requiring transfer from the Franklin School be allowed to remain there because of long attendance and family tradition. However, there was no official recantation shown of the action of the Board of Education reflected by the letters of the Secretary and sent only to the parents of the children of Mexican ancestry.

The natural operation and effect of the Board's official action manifests a clean purpose to arbitrarily discriminate against the pupils of Mexican ancestry and to deny to them the equal protection of the laws.

The court may not exercise legislative or administrative functions in this case to save such discriminatory act from inoperativeness. Cf. Yu Cong Eng v. Trinidad, 271 U.S. 500, 46 S.Ct. 619, 70 L.Ed. 1059.

There are other discriminatory customs, shown by the evidence, existing in the defendant school districts as to pupils of Mexican descent and extraction, but we deem it unnecessary to discuss them in this memorandum.

We conclude by holding that the allegations of the complaint (petition) have been established sufficiently to justify injunctive relief against all defendants, restraining further discriminatory practices against the pupils of Mexican descent in the public schools of defendant school districts. See Morris v. Williams, 8 Cir., 149 F.2d 703.

Findings of fact, conclusions of law, and decree of injunction are accordingly ordered pursuant to Rule 52, F.R.C.P.

Attorney for plaintiffs will within ten days from date hereof prepare and present same under local Rule 7 of this court.

1. "Section 1. All persons born or naturalized in the United States, and subject to the jurisdiction thereof, are citizens of the United States and of the State wherein they reside. No State shall make or enforce any law which shall abridge the privileges or immunities of citizens of the United States; nor shall any State deprive any person of life, liberty, or property, without due process of law; nor deny to any person within its jurisdiction the equal protection of the laws."

2. "The district courts shall have original jurisdiction as follows: * * *"
   64 F.SUPP.–35
   Sec. 41, subd. (14) "Suits to redress deprivation of civil rights. Fourteenth. Of all suits at law or in equity authorized by law to be brought by any person to redress the deprivation, under color of any law, statute, ordinance, regulation, custom, or usage, of any State, of any right, privilege, or immunity, secured by the Constitution of the United States, or of any right secured by any law of the United States providing for equal rights of citizens of the United States, or of all persons within the jurisdiction of the United States."

3. "Sec. 16004. Any person, otherwise eligible for admission to any class or school of a school district of this State, whose parents are or are not citizens of the United States and whose actual and legal residence is in a foreign country adjacent to this State may be admitted to the class or school of the district by the governing board of the district."

   "Sec. 16005. The governing board of the district may, as a condition precedent to the admission of any person, under Section 16004, require the parent or guardian of such person to pay to the district an amount not more than sufficient to reimburse the district for the total cost, exclusive of capital outlays, of educating the person and providing him with transportation to and from school. The cost of transportation shall not exceed ten dollars ($10) per month. Tuition payments shall be made in advance for each month or semester during the period of attendance. If the amount paid is more or less than the total cost of education and transportation, adjustment shall be made for the following semester or school year. The attendance of the pupils shall not be included in computing the average daily attendance of the class or school for the purpose of obtaining apportionment of State funds."

4. Sec. 8501, Education Code. "Children between six and 21 years of age. The day elementary school of each school district shall be open for the admission of all children between six and 21 years of age residing within the boundaries of the district."

  Sec. 8002. "Maintenance of elementary day schools and day high schools with equal rights and privileges. The governing board of any school district shall maintain all of the elementary day schools established by it, and all of the day high schools established by it with equal rights and privileges as far as possible."

5. Sec. 8003. "Schools for Indian children, and children of Chinese, Japanese, or Mongolian parentage: Establishment. The governing board of any school district may establish separate schools for Indian children, excepting children of Indians who are wards of the United States Government and children of all other Indians who are descendants of the original American Indians of the United States, and for children of Chinese, Japanese, or Mongolian parentage."

  Sec 8004. "Same: Admission of children into other schools. When separate schools are established for Indian children or children of Chinese, Japanese, or Mongolian parentage, the Indian children or children of Chinese, Japanese, or Mongolian parentage shall not be admitted into any other school."

6. The study of American institutions and ideals in all schools located within the State of California is required by Section 10051, Education Code.

7. See Plessy v. Ferguson, 163 U.S. 537, 16 S.Ct. 1138, 41 L.Ed. 256.

DOCUMENT 5.5

Appeal from Westminster School District, California, 1947

The defendant in Méndez *(Westminster School District) appealed Judge McCormick's finding in favor of the plaintiffs. As a result, the school district appealed the decision in the Ninth Circuit but the finding was upheld. Note the support (amicus curiae) filed on behalf of the Mexican Americans from a broad range of organizations including the American Jewish Congress, the NAACP, and the American Civil Liberties Union.*

From: *Westminster School District of Orange County et al. v. Méndez et al.* No. 11310 161 F.2d 774 Circuit Court of Appeals, Ninth Circuit, April 14, 1947, as Corrected Aug. 1, 1947 (excerpts).

Appeal from the District Court of the United States for the Southern District of California, Central Division; Paul J. McCormick, Judge.

Action by Gonzalo Mendez and others, by their father and next of friend, Gonzalo Mendez, and others, against Westminster School District of Orange County, and others, to redress alleged violations of civil rights. Judgment for plaintiffs, 64 F.Supp. 544, and defendants appeal.

Affirmed.

Joel E. Ogle, County Counsel, George F. Holden and Royal E. Hubbard, Deputies County Counsel, all of Santa Ana, Cal., for appellant.

David C. Marcus, Los Angeles, Cal. (William Strong, of Los Angeles. Cal., of counsel), for appellees.

Thurgood Marshall, and Robert L. Carter, both of New York City, and Loren Miller, of Los Angeles, Cal., for Nat. Ass'n Advancement of Colored People, amicus curiae.

Will Maslow and Pauli Murray, both of New York City, Anne H. Pollock, of Los Angeles, Cal. (Alexander H. Pekelis, of New York City, Spe. Advisor), for American Jewish Congress, amicus curiae.

Julien Cornell, Arthur Garfield Hays and Osmond K. Fraenkel, all of New York City, A. L. Wirin and Fred Okrand, both of Los Angeles, Cal., for American Civil Liberties Union, amicus curiae.

Charles F. Christopher, of Los Angeles, Cal., for Nat. Lawyers Guild, Los Angeles Chapter, amicus curiae.

A. L. Wirin and Saburo Kido, both of Los Angeles, Cal., for Japanese-American Citizens League.

Robert W. Kenney, Atty. Gen., of Cal., and T. A. Westphal, Jr., Deputy Atty. Gen., for Atty. Gen., for Atty. Gen of Cal., amicus curiae.

Before GARRECHT, DENMAN, MATHEWS, STEPHENS, HEALY, BONE, and ORR, Circuit Judges.

STEPHENS, Circuit Judge.

The petition herein which prays for present and future relief and costs is filed under authority of section 24, subdivision 14, of the Judicial Code, 28 U.S.C.A. § 41 (14),[1] and section 43 of 8 U.S.C.A.,[2] and is based upon alleged violations of petitioners' civil rights as guaranteed by the 5th and 14th amendments to the Constitution of the United States. No argument as to the application of the 5th amendment is made in this appeal and it need not be considered.

The petition contains allegations to the following effect. A number of minors (at least one each from each school division herein mentioned) for themselves and for some 5000 others as to whom the allegations of the complaint apply,[3] citizens of the United States of Mexican descent, who attend the public schools of the State of California in Orange County, filed a petition by their fathers, as next friends, for relief against trustees and superintendents of several school districts and against the superintendent and secretary and members of a city board of education. Unless we shall indicate otherwise, our use of the terms "school districts," "districts" or "schools" will be understood as inclusive of both district and city school territories or schools. The term "school officials" includes all respondents.

All petitioners are taxpayers of good moral habits, not suffering from disability, infectious disease, and are qualified to be admitted to the use of the schools and facilities within their respective districts and systems.

A common plan of the school officials has been adopted and practiced, and common rules and regulations have been adopted and put into effect, whereby (using the words of the petition) "petitioners and all others of Mexican and Latin descent" are "barred, precluded and denied," "attending and using and receiving the benefits and education furnished to other children," and are segregated in schools "attended solely by children of Mexican and Latin descent." To such treatment, petitioners and others in the same situation have objected, and they have demanded and have been refused admission to schools within their respective districts which they would attend but for the practice of segregation. "That by this suit and proceedings, petitioners seek to redress the deprivation by respondents herein [school officials] under color of regulation, custom and usage of petitioners' civil rights, privileges and/or immunities secured to them by the Laws of the United States, and guaranteed to each of them by the Laws and Constitution of the United States of America."

To the petition, the school officials respond by a motion to dismiss for lack of federal court jurisdiction, because (to use the words of the motion) "this is not a suit at law or in equity authorized by law to be brought by any person to redress the deprivation, under color of any law, statutes, ordinances, regulation, custom, or usage, of any state, of any right, privilege, or immunity, secured by any law of the United States providing for equal rights of citizens of the United States or of all persons within the jurisdiction of the United States," and because the "petition fails to state a claim upon which relief can be granted." The motion was denied without prejudice to the assertion of any available legal defenses by way of answers to the petition. Respondents in their answer reassert their position as to the law in the motion to dismiss, and put in issue all of the allegations relating to the subject of segregation.[4]

. . .

Summed up in a few words it is the burden of the petition that the State of California has denied, and is denying, the school children of Mexican descent, residing in the school districts described, the equal protection of the laws of the State of California and thereby have deprived, and are depriving, them of their liberty and property without due process of the law, as guaranteed by the Fourteenth Amendment of the Constitution of the United States.

Respondents are officers of the State of California in the Department of Education of that state, and as it will hereinafter be shown their action under the intendment of the Fourteenth Amendment is the action of the state in all cases where such action is taken under color of state law. We must, therefore, consider the questions: Are the alleged acts done under color of state law, and do they deprive petitioners of any constitutional right?

. . .

We hold that the respondents acting to segregate the school children as alleged in the petition were performing under color of California State law.

The court found that the segregation as alleged in the petition has been for several years past and is practiced under regulations, customs and usages adopted more or less as a common plan and enforced by respondent-appellants throughout the mentioned school districts; that petitioners are citizens of the United States of Mexican ancestry of good moral habits, free from infectious disease or any other disability, and are fully qualified to attend and use the public school facilities; that respondents occupy official positions as alleged in the petition.

In both written and oral argument our attention has been directed to the cases in which the highest court of the land has upheld state laws providing for limited segregation of the great races of mankind. In Roberts v. City of Boston, 5 Cush. Mass., 198,[6] a law providing for the segregation of colored school children was held valid in an opinion by Chief Justice Shaw of the Supreme Judicial Court of Massachusetts, but that equal facilities must be provided for the use of the colored children. Chief Justice Wallace of the Supreme Court of California in Ward v. Flood, 48 Cal. 36, 17 Am.Rep. 405, followed with approval. Cumming v. Board of Education, 175 U.S. 528, 20 S.Ct. 197, 44 L.Ed. 262, reaffirmed the principle. In Gong Lum v. Rice, 275 U.S. 78, 48 S.Ct. 91, 72 L.Ed. 172, the principle of the Roberts case, supra, was followed in the opinion written by Chief Justice Taft and affirmed the State Supreme Court of Mississippi in its application of the "colored" school segregation statute to an American citizen of pure Chinese blood. Plessy v. Ferguson, 163 U.S. 537, 16 S.Ct. 1138, 41 L.Ed. 256, was upon the right of the state to require segregation of colored and white persons in public conveyances, and the act so providing was sustained again upon the principles expressed by Chief Justice Shaw. This list of cases is by no means complete.

It is argued by appellants that we should reverse the judgment in this case upon the authority of the segregation cases just cited because the Supreme Court has upheld the right of the states to provide for segregation upon the requirement that equal facilities be furnished each segregated group. Appellees argue that the segregation cases do not rule the instant case. There is argument in two of the amicus curiae briefs that we should strike out independently on the whole question of segregation, on the ground that recent world stirring events have set men to the reexamination of concepts considered fixed. Of course, judges as well as all others must keep abreast of the times but judges must ever be on their guard lest they rationalize outright legislation under the too free use of the power to interpret. We are not tempted by the siren who calls to us that the sometimes slow and tedious ways of democratic legislation is no longer respected in a progressive society. For reasons presently to be stated, we are of the opinion that the segregation cases do not rule the instant ease and that is reason enough for not responding to the argument that we should consider them in the light of the amicus curiae briefs. In the first place we are aware of no authority justifying any segregation fiat by an administrative or executive decree as every case cited to us is based upon a legislative act. The segregation in this case is without legislative support and comes into fatal collision with the legislation of the state.

3. and 4. The State of California has a state-wide free school system governed by general law, the local application of which by necessity is to a considerable extent, under the direction of district and city school boards or trustees, superintendents and teachers. Section 16601 of the California Educational Code requires the parent of any child between the ages of eight and sixteen years to send him to the full time day school. There are some few exceptions, but none of them are pertinent here. There are no exceptions based upon the ancestry of the child other than those contained in §§ 8003, 8004, Calif.Ed.C. (Both repealed as of 90 days after June 14, 1947.), which includes Indians under certain conditions and children of Chinese, Japanese or Mongolian parentage. As to these, there are laws requiring them in certain cases to attend separate schools. Expressio Unius Est Exclusio Alterius. It may appropriately be noted that the segregation so provided for and the segregation referred to in the cited cases includes only children of parents belonging to one or another of the great races of mankind.[7] It is interesting to note at this juncture of the case that the parties stipulated that there is no question as to

race segregation in the case. Amicus curiae brief writers, however, do not agree that this is so. Nowhere in any California law is there a suggestion that any segregation can be made of children within one of the great races. Thus it is seen that there is a substantial difference in our case from those which have been decided by the Supreme Court, a difference which possibly could be held as placing our case outside the scope of such decisions. However, we are not put to this choice as the state law permits of segregation only as we have stated, that is, it is definitely confined to Indians and certain named Asiatics. That the California law does not include the segregation of school children because of their Mexican blood, is definitely and affirmatively indicated as the trial judge pointed out, by the fact that legislative action has been taken by the State of California to admit to her schools, children citizens of a foreign country, living across the border. Calif.Ed.C. §§ 16004, 16005. Mexico is the only foreign country on any California boundary.[8]

It follows that the acts of respondents were and are entirely without authority of California law, notwithstanding their performance has been and is under color or pretense of California law. Therefore, conceding for the argument that California could legally enact a law authorizing the segregation as practiced, the fact stands out unchallengeable that California has not done so but to the contrary has enacted laws wholly inconsistent with such practice. By enforcing the segregation of school children of Mexican descent against their will and contrary to the laws of California, respondents have violated the federal law as provided in the Fourteenth Amendment to the Federal Constitution by depriving them of liberty and property without due process of law and by denying to them the equal protection of the laws.

5. It may be said at this point that the practice could be stopped through the application of California law in California State Courts, and this may be so but the idea is of no relevancy. Mr. Justice Douglas made this point clear in the case of Screws v. United States, supra, when he said that the Fourteenth Amendment does not come into play merely because the federal law or the state law under which the officer purports to act is violated. *"It is applicable when and only when some one is deprived of a federal right by that action."* (Emphasis ours.) And it is as appropriate for us to say here, what Mr. Justice Douglas said in a like situation in the cited case, "We agree that when this statute is applied [in our case when § 41(14) of 28 U.S.C.A. is applied] it should be construed so as to respect the proper balance between the states and the federal government in law enforcement." Punishment for the act would be legal under either or both federal and state governments. United States v. Lanza, 260 U.S. 377, 382, 43 S.Ct. 141, 67 L.Ed. 314; Hebert v. Louisiana, 272 U.S. 312, 47 S.Ct. 103, 71 L.Ed. 270, 48 A.L.R. 1102. However, since the practice complained of has continued for several consecutive years, apparent to California executive and peace officers, and continues, it cannot be said that petitioners violated Mr. Justice Douglas' admonition in taking their action in a federal court.

In the view of the case we have herein taken the contention that the Findings of Fact do not support the Conclusions of Law and the Judgment is wholly unmeritorious. The pleadings, findings and judgment in this case refer to children of "Mexican and Latin descent and extraction," but it does not appear that any segregation of school children other than those of Mexican descent was practiced. Therefore, we have confined our comment thereto. If the segregation of all children of Latin descent and extraction in addition to those of Mexican descent were included in the practice and the plan, its illegality would, of course, be upon the same basis as that herein found. In addition, however, the impossibility of there being any reason for the inclusion in the segregation plan of all children of Latin descent and extraction and the palpable impossibility of its enforcement would brand any such plan void on its face.[9]

Affirmed.

1. Section 41. (Judicial Code, section 24, amended.) Original jurisdiction. The district courts shall have original jurisdiction as follows: * * * (14) Suits to redress deprivation of civil rights. Fourteenth. Of all suits at law or in equity authorized by law to be brought by any person to redress the deprivation, under color of any law, statute, ordinance, regulation, custom, or usage, of any State, of any right, privilege, or immunity, secured by the Constitution of the United States, or of any right secured by any law of the United States providing for equal rights of citizens of the United States, or of all persons within the jurisdiction of the United States."

2. "§ 43. Civil action for deprivation of rights
"Every person who, under color of any statute, ordinance, regulation, custom, or usage, of any State or Territory, subjects, or causes to be subjected, any citizen of the United States or other person within the jurisdiction thereof to the deprivation of any rights, privileges, or immunities secured by the Constitution and laws, shall he liable to the party injured in an action at law, suit in equity, or other proper proceeding for redress."

3. Rule 23 of Federal Rules of Civil Procedure, 28 U.S.C.A. following section 723c as to class suits.

4. It is alleged in the answer that a large number of school children concerned are unfamiliar with and unable to speak the English language. Other affirmative defenses are alleged but they need not be mentioned for the reason that the findings of fact are not attacked and the appeal is based upon the question as to whether or not petitioners' civil rights under the Fourteenth Amendment to the Constitution of the United States have been violated.

6. The decision in the case of Roberts v. City of Boston, 5 Cush. 198, cited in the majority opinion in the above entitled case (April 14, 1947), was not founded directly upon a state statute. A state statute granted certain discretionary powers to an elected School Committee, but these powers did not specifically provide for any segregation of school children on the basis of race or color. However, Boston had long conducted separate schools for colored school children. Shortly before institution of the case (the case antedated the Civil War), which was for damages allegedly suffered by the plaintiff, a colored child, for being excluded from the school nearest her residence, the School Committee had adopted a resolution approving the policy of continuing the separate schools. The decision in the case upheld the acts of the Committee. (Stephens, C.J.)

7. Somewhat empirically, it used to be taught that mankind was made up of white, brown, yellow, black and red men. Such divisional designation has little or no adherents among anthropologists or ethnic scientists. A more scholarly nomenclature is Caucasoid, Mongoloid and Negroid, yet this is unsatisfactory, as an attempt to collectively sort all mankind into distinct groups.

8. The right of children to attend schools organized under laws of the state has been termed a fundamental right. See Wysinger v. Crookshank, 82 Cal. 588, 23 P. 54. Education "is a privilege granted by the state constitution, and is a legal right as much as is a vested right in property." 23 Cal.Jur. pp. 141, 142. In the same volume, p. 161: "It is now settled that it is not in violation of the organic law of the state or of the nation to require children in whom racial differences exist to attend separate schools, provided the schools are equal in every substantial respect. But only in the event such schools are established may children be separated in respect of race. And no separation may be had, in the absence of statutory or constitutional authority therefor."

9. The case of Lopez v. Seccombe, D.C.S.D.Cal., 71 F.Supp. 769, cited and commented upon in the concurring opinion, went to uncontested judgment upon stipulation, and is supported alone by formal findings of facts and conclusions of law. No discussion of principles appears in the record, no opinion or memorandum was filed, and no counsel in the instant case mentioned it in his brief, notwithstanding the same lawyer was chief counsel in both cases. (Stephens, C.J.)

DOCUMENT 5.6

Delgado vs. Bastrop Independent School District, Bastrop County, Texas, 1948

In 1947 six-year-old Minerva Delgado was denied entrance to a so-called white school because she was of Mexican American descent. On her behalf, Delgado's grandfather, Samuel Garcia, sued "in behalf of all school children of Mexican descent within the School District . . . and this suit is filed as a Class Suit . . . in that they are all required to attend segregated schools and classes solely because they are of Mexican descent." Although the defendants attempted to demonstrate (see transcripts below) that lack of English facility required the segregation of Mexican and Latin-origin children, the judge ruled on the children's behalf. One of the key points that Delgado's lawyer, Gus Garcia, had to demonstrate, was that segregating Latin children, although not a law in Texas, was the custom and thus could be tried in a court of law. Both Delgado v. Bastrop *and* Méndez v. Westminster *were county or district level desegregation cases that preceded* Brown v. Board of Education *(1954).*

> From: *Minerva Delgado, et al. v. Bastrop Independent School District of Bastrop County, Texas, et al.* Civil Action no. 388 United States District Court Western District of Texas, Austin Division, June 15, 1948. In National Archives and Records Administration, Southwest Region, Fort Worth, Texas.

The Plaintiffs allege:

This Court has jurisdiction under the provisions of 28 U.S. Code, Sec. 41 (14), in that this is a suit, in equity and at law, to redress the deprivation of civil rights; said deprivation being by officials of the State of Texas, under color of custom and/or usage by said officials, acting for and in behalf of said State of Texas, of rights, privileges, and immunities secured by the Constitution of the United States under the Fourteenth Amendment, and rights secured by the laws of the United States, including particularly 8 U.S. Code, Sec. 43 . . .

IV.

1. The defendants have exceeded the authority vested in them by the Constitution and Laws of the State of Texas in carrying out a policy of segregating children of Mexican descent from other children as hereinafter

set forth. For several years last past, the Defendants adopted a common custom, plan, or usage and/or practice as follows: That children of Mexican descent be barred, prohibited and excluded, solely because of said Mexican descent, from attending those certain public schools and classes under Defendants' charge, which said schools and classes (hereinafter referred to as regular schools and classes) are exclusively established or maintained by said Defendants for attendance by school children of so-called white or Anglo-American parents (these latter children being hereinafter referred to as other white children); and that said children of Mexican descent be segregated into public schools and classes, commonly referred to as "Mexican" schools (hereinafter referred to as segregated schools and classes), established or maintained exclusively for the attendance of school children of Mexican descent.

2. Pursuant to said custom, usage and/or common plan, the above-named Defendants have, for several years last past, prohibited, barred and excluded, and do now prohibit, bar and exclude, the Plaintiffs, and all such school children of Mexican descent, from attending the certain regular schools and classes, within their charge and under their control, reserved by said Defendants for the exclusive attendance of the other white school children; and said Defendants have thus prevented said Plaintiffs, and said school children of Mexican descent, from receiving the educational, health and recreational benefits which such other white children receive in said regular schools and classes; and the said Plaintiffs, and said children of Mexican descent have, for several years last past, been generally and continuously assigned to certain segregated schools and classes intended exclusively for said children of said Mexican descent.

3. There is no provision in the Constitution of the State of Texas or in any Statute of said State authorizing or permitting the segregation, into segregated schools and classes, by Officers of the State of Texas engaged in the administration of the Public School Laws of the State of Texas, of school children of Mexican descent.

4. The exclusion of the Plaintiffs and said children of Mexican descent from said regular schools and classes, and the segregation of said children of Mexican descent in said segregated schools and classes as aforesaid, is solely because said children are of Mexican descent; and said exclusion and segregation are intended to, and have the effect of, discriminating against the Plaintiffs and the said children of Mexican descent solely because of their Mexican ancestry.

5. Said segregation aforesaid is unjust, capricious and arbitrary and in violation of the Constitution of the United States in that it deprives the plaintiffs and said children of Mexican descent of liberty and property without due process of law, and denies them the equal protection of laws, and of privileges and immunities as citizens of the United States, as guaranteed by the Fourteenth Amendment to the Constitution of the United States; and said segregation further deprives said Plaintiffs, and said children of Mexican descent, of rights under the 8 U.S. Code Sec. 43.

V.

The Attorney General of the State of Texas has rendered an opinion that it is illegal for School Officials of the State of Texas to segregate children of Mexican descent into separate schools and classes solely because of such descent. Despite said opinion, said Defendants above-named have continued the segregation aforesaid.

VI.

1. The practice, custom and/or usage of segregating school children of Mexican descent, as aforesaid, is general in the State of Texas, and obtains in many school districts of the State of Texas in addition to the school districts named as defendants herein; and said existence of practice is of general and common knowledge in the State of Texas, and of general and common knowledge to the educational officials of the State of Texas. Said practice has been expressly called to the attention of the defendant L.A. Woods, as State Superintendent of Public Instruction. Said defendant, as said State Superintendent, has issued no orders, instructions or regulations, pursuant to his duty so to do [sic], to the defendant School Districts and their officers, nor to any school district in the State of Texas, directing that said practice be discontinued; on the contrary, he has participated in said practice by allocating certain school funds, which said school funds were, to the knowledge of said Defendant, to be used for the maintenance of segregated schools and classes, and to be used for the maintenance of the practice of segregation hereinabove complained of.

2. The existence of said practice is known to Defendant members of the State Board of Education, and said Board and the members thereof have issued no orders, instructions, or regulations to the State Superintendent of Public Instruction nor to the Defendant School Districts and their officers and members named herein through said Superintendent of Public Instruction, nor to any School District in the State of Texas, directing that such practice be discontinued; on the contrary they have participated in said practice by allocating certain school funds and textbooks which said schools funds and textbooks were, to the knowledge of said Defendants, to be used for the general maintenance of the practice of segregation hereinbefore complained of.

VII.

This suit is brought by the Plaintiffs for and in behalf of themselves and for and in behalf of all school children of Mexican descent within the School District within which said Plaintiffs are respectively residenced, and this suit is filed as a Class Suit for and in behalf of said children of Mexican descent so residenced in that, as heretofore set forth in this complaint, all of said children of Mexican descent are in the same class as said Plaintiffs in that they all are required to attend segregated schools and classes solely because they are of Mexican descent; said Class is so numerous that it is impracticable to bring all of its members before the Court; and the character of the right sought to be enforced herein is several and there are common questions of fact and of law affecting the several rights of the school children of Mexican descent constituting said class; and a common relief is sought by said school children of Mexican descent against all the Defendants herein.

VIII.

Unless enjoined by order of this Court both by permanent injunction, and by injunction pendente lite, the Defendants intend to continue to practice the custom and/or usage aforesaid, and to continue the general practice of segregation aforesaid. The Plaintiffs and the Class in whose behalf this proceeding is filed, have no plain, speedy or adequate remedy at law, and will suffer great and irreparable injury unless an injunction pendente lite and a permanent injunction are issued by this Court enjoining said practice, custom and/or usage.

xxxxxxxxxxxxxxxxxx

For a second cause of action, being a cause of action for damages against the Defendants Bastrop Independent School District of Bastrop County [et. al.] . . .

I.

The Plaintiffs incorporate herein all of the allegations set forth in the Plaintiffs' first cause of action . . .

II.

The Plaintiffs, as aforesaid, by the acts of the Defendants complained of, were deprived of their rights under the Constitution of the United States and the Laws of the United States to be free from discrimination solely because of their ancestry; and were thus denied the right, by the said Defendants, to receive an education in the regular schools of Texas, free from such discrimination; and were further deprived by said Defendants from securing the educational, recreational and health benefits accorded by said Defendants to other white children, to the damage of the Plaintiffs, and of each of them, in the sum of Five Thousand ($5,000) Dollars.

III.

The acts of the Defendants Bastrop Independent School District of Bastrop County, [et al], were wantom [*sic*], reckless and with a complete disregard of the rights of the Plaintiffs by virtue whereof the Plaintiffs, and each of them, are additionally entitled to punitive or exemplary damages in the sum of Five Thousand ($5000) Dollars.

xxxxxxxxxxxxxx

Wherefore the Plaintiffs pray for relief as follows:

Under the first cause of action:

For a judgment and decree granting a permanent injuction, and for an order granting an injunction pendente lite, against all the Defendants in behalf of the Plaintiffs, and the school children of Mexican descent represented by them, enjoining said Defendants and their agents from in any manner assigning into segregated schools and classes school children of Mexican descent under their control; and from in any manner, directly or indirectly, participating in said practice of segregation, including the proration and/or payment of any funds and/or instructional materials of the State of Texas, used, or to be used, for the purpose of maintaining segregated schools and classes for school children of Mexican descent.

Under the second cause of action:

For damages in behalf of the Plaintiffs, and each of them, against the . . . Bastrop Independent School District of Bastrop County, [et al.], in the sum of Five Thousand ($5000) Dollars actual damages, and, additionally, the sum of five thousand dollars as punitive and/or exemplary damages.

And Plaintiffs further pray for such other relief as may be proper.

_____

Gus C. Garcia
Attorney for Plaintiff

Robert C. Eckardt
A. L. Wirin
State of Texas
County of Travis

Before me, the undersigned authority, on this day personally appeared Gus C. Garcia, known to me to be the person whose name is subscribed below, who states upon oath that he had read the foregoing complaint, designed to be used in the case of *Minerva Delgado, By Her Grandfather and Next of Friend, Samuel Garcia, et al vs. Bastrop Independent School District of Bastrop County, et al,* that he knows the contents of said complaint, and that the facts and statements therein contained are true and correct.

(signed Gus. C. Garcia)
Affiant.

SUBSCRIBED and sworn to before me, the undersigned authority, on this 17th day of November, A.D. 1947.

GIVEN under my hand and seal of office.

(signed Louise Vine)
Louise Vine
Notary Public in and for Travis
County, Texas.

DOCUMENT 5.7

Desegregation Orders, Bastrop County, Texas, 1948

Honorable Ben H. Rice Jr. ruled in favor of Minerva Delgado, declaring that the segregation of Latino pupils was "arbitrary and discriminatory" and violated the equal rights clause of the Fourteenth Amendment. The judge did permit schools to clearly demonstrate through "scientific and standardized tests" separation of children who lacked English proficiency during their first scholastic year, but only for one year.

From: *Minerva Delgado, et al. v. Bastrop Independent School District of Bastrop County, Texas, et al.* Civil Action no. 388 United States District Court Western District of Texas Austin Division, filed on June 15, 1948. In National Archives and Records Administration, Southwest Region, Fort Worth, Texas.

FINAL JUDGMENT

Abstract of Principal Features

This action came on for trial on the 15th day of June, 1948, before the Honorable Ben H. Rice, Jr., Judge Presiding, the plaintiffs . . . and the defendants . . . being represented by their attorneys. . . . The plaintiffs having heretofore voluntarily dismissed their second cause of action in their complaint; and the evidence having been introduced both oral and documentary and said action having been submitted for decision on the merits, the court being fully advised in the premises, and findings of fact and conclusions of law having been waived by stipulation of the parties,

It Is Therefore Ordered, Adjudged, and Decreed that:

1. This action by plaintiffs is a representative class action on behalf of themselves and of all pupils of Mexican or other Latin-American descent, and the action has been properly brought as such class action pursuant to law.

2. The regulations, customs, usages, and practices of the defendants, Bastrop Independent School District of Bastrop County, et al, and each of them in so far as they or any of them have segregated pupils of Mexican or other Latin-American descent in separate classes and schools within the respective school districts of the defendant school districts heretofore set

forth are, and each of them is, arbitrary and discriminatory and in violation of plaintiff's constitutional rights as guaranteed by the Fourteenth Amendment to the Constitution of the United States, and are illegal.

[3.] The [defendant school districts and their officers] are hereby permanently restrained and enjoined from segregating pupils of Mexican or other Latin American descent in separate schools or classes within the respective school districts of said defendants and each of them, and from denying said pupils use of the same facilities and services enjoyed by other children of the same ages or grades; provided, however, that this injunction shall not prevent said defendant school districts or their trustees, officers, and agents from providing for, and maintaining, separate classes on the same campus in the first grade only, and solely for instructional purposes, for pupils in their initial scholastic year who, at the beginning of their initial scholastic year in the first grade, clearly demonstrate, as a result of scientific and standardized tests, equally given and applied to all pupils, that they do not possess a sufficient familiarity with the English language to understand substantially classroom instruction in first-grade subject matters.

4. If in any school district obedience to this decree renders it practically necessary, in the discretion of the school district, that additional school buildings be provided or moved from one campus to another, then a reasonable time is hereby allowed for compliance, but in no event beyond September 1949.

5. The defendant, L. A. Woods, as State Superintendent of Public Instruction, is hereby permanently restrained and enjoined from in any manner, directly or indirectly, participating in the custom, usage or practice of segregating pupils of Mexican or other Latin-American descent in separate schools or classes.

6. The motion of the State Board of Education and the members thereof to be dropped as parties hereto is sustained, and they are hereby dropped and dismissed from this suit with their costs.

> *Dated at Austin, Texas, this 15th day of*
> *June, 1948.*
> (Ben H. Rice, Jr.)
> United States District Judge

DOCUMENT 5.8

Deposition for Delgado v. Bastrop, 1948

As part of the depositions for Delgado v. Bastrop, lawyer Gus Garcia questioned numerous school officials regarding the custom and practice of segregating Mexican and Latin children into different schools. As the following testimonial reveals, the number of grades available to Mexican children, the condition of the schools, and quality of teachers was often greatly inferior to that available to Anglo children.

> Source: *Minerva Delgado, et al. v. Bastrop Independent School District of Bastrop County, Texas, et al.* Civil Action no. 388 United States District Court Western District of Texas Austin Division. In National Archives and Records Administration, Southwest Region, Fort Worth, Texas.

Appearances:

A. L. Wirin
Gus Garcia

Attorneys for Plaintiffs

Ireland Graves
J. Chrys Dougherty

Attorneys for defendants, Bastrop Independent School district and others.

Joe R. Greenhill, Asst. Atty. General
Martin Harris, Asst. Atty. General

Attorneys for L. A. Woods, State Supt. of Schools

- - - - - - - - -

Depositions of Gus F. Urbantke, I. W. Popham, P. J. Dodson, Claud E. Brown, C. R. Akin, and F. Kenneth Wise, defendants, taken on the 30th day of April, 1948, before me, Mrs. Opal Looke, a Notary Public in and for the County of Travis, State of Texas, in the Capital National Bank Building, Austin, Texas, between the hours of 9:30 o'clock A.M. and 4:30 o'clock P. M. of said day. And the said witnesses personally appeared before me to depose in a civil case pending in the District Court of the United States for the Western District of the State of Texas, wherein Minerva Delgado et al. are plaintiffs and the Bastrop Independent School District of Bastrop et al. are defendants; and that I was then and there attended by counsel, as above set out; and that said Gus F. Urbantke, I. W. Popham, P. J. Dodson, Claud E. Brown, C. R. Akin, and F. Kenneth Wise, each being of lawful age and sound mind, and each being by me first duly examined, cautioned and sworn to tell the truth, the whole truth, and nothing but the truth, touching his knowledge of the matters and things in controversy in said civil cause, depose and say as hereinafter set out.

––––––––

. . . agreeing that same [deposition] may be reduced to typewriting and thereafter used in evidence by either party upon the trial of said cause, subject to all legal objections and exceptions, except as to the manner and form of taking. It is further expressly agreed that the depositions may be signed by the witnesses at any time before the trial of said cause.

. . .

P. J. Dodson, having been duly sworn, testified as follows:

Direct Examination

Questions by Mr. Wirin:

Q   Are you the Superintendent of the Bastrop Independent School District?
A   Yes, sir.
Q   How long have you been superintendent for that district?
A   This is my 17th year.
Q   Generally, as superintendent, what are your duties?
A   Well, supervising the school system entirely, all sections of it.
Q   And as superintendent, you carry out the policies of the school district?
A   Yes, sir.
Q   How many grade schools are there in the district?
A   We have, I guess you would call it three. We have the—the colored school has a grade school; then we have a grade school incorporated with the White school, in with the high school, a so-called White building; and then we have the Manor Ward School.
Q   The Manor Ward School ?
A   Yes, sir.
Q   What is that school?
A   That is a school for Latin-American children primarily.
Q   Where is the White school located?
A   It is on the corner of Farm and Hill Street.
Q   With respect to that location, where is the school for Latin-American children?

A I imagine it is about six blocks away; it is out at North and Main.
Q How long has the "White" school been in existence, to your knowledge?
A I don't know.
Q Many years?
A The building got torn down and replaced with another building and was built in '92; there was a school there before that, so I don't know how long.
Q Anyway, it was there before you became superintendent?
A That is right.
Q How long has the Latin-American school been in existence?
A The Manor Ward School was built the first year after I went there. We did have a one-teacher school with about 15 Latin-American students in it when I went there.
Q Well, in any event, after the first year you went there, there have been these two schools?
A Well, this one-teacher Latin-American school was a separate school when I went there; you see, we didn't have many Latin-Americans in town, only 15 or 16, but there were Latin-American schools all around us that had a large enrollment. One had fifty some-odd; one had 90 some-odd; with one teacher and very few pupils going, I made an arrangement with the County Superintendent, if he would transfer them to me, I would take the State money and buy a bus, give them a building and give them a good school; so we took our 16 out of that tumble-down building, and about 125 from the rural districts surrounding us, and created a good school for the Latin-Americans.
Q And since then that school has been used for the Latin-Americans?
A Yes, sir.
Q And the other school has been used by "White" children?
A Not entirely; we have had some Latin-Americans, of course, in the other school. There have been no American children in the Latin-American school, but there have been Latin- Americans in all the grades in the so-called "White" school.
Q As of what grade is the—what grades are taught in the Latin-American school?
A At the present time we have four grades .
Q And has there been any change in the number or amount of grades?
A Yes, sir; when we first established it, we had eight grades.
Q When was the four-grade curriculum put into effect?
A Four-grade? This past September.
Q And prior to this past September, what grades were in the Latin-American school?
A Last year I believe we had fifth and sixth over there.
Q And before last year?
A And the year before, I think we had the seventh over there; about two years before that, we had the eighth over there.
Q Now, during the period when you had up to the eighth grade, was it the regular practice that children attending the Latin-American school,

when they reached the eighth grade, were then all transferred to the "White" school?

A   Well, about the second year after I got there, it was. Up until that time we didn't teach any Latin-Americans in the "White" school at all.

Q   I see.

A   In other words, I have been trying gradually to bring about a social revolution; you know what is necessary to do that.

Q   When you got there or after you got there, you arranged for the Latin-American children who got through, say, the seventh grade to be transferred to the "White" school?

A   That is right; at that time it was the seventh grade was the top grade.

Q   Seventh grade?

A   The eighth grade was the top grade later on, but I don't know just how long it was before I got the Latin-Americans started; but I know for the last 10 or 12 years we have had Latin-Americans in high school, at least.

Q   All right. Now, let's see. In other words, you arranged for the entire class, for all the pupils in the entire class to be transferred when they reached the seventh grade?

A   That is right; when they finished the grade that we taught over there.

Q   Finished the grade that you taught over there; and that is the practice now, except the grade has been lowered to the fourth grade rather than originally the seventh?

A   No, not necessarily; we have some Latin-American students in the first, second, third, and fourth in the "White" school, and since—

Q   But is it the general practice for the transfer to be made when the group of pupils, as a group, reach a certain grade?

A   As a general practice, unless they come and ask to go to the other school and speak English; there are some that start in the first grade in the other school.

Q   Now, you say there are some children of all grades of Latin-American descent who are in the "White" school'?

A   Yes, sir.

Q   How many such children are there?

A   Not very many in the first four grades.

Q   Well—

A   I have that information in my car down there, if you want me to go down and get that envelope. I forgot to bring it up.
    (Discussion omitted. The envelope referred to above was obtained.)

Q   All right; now, you said if a child can speak English, then he is allowed to go to the "White" school?

A   No, you understand I don't go over there and ask them if they can speak English and bring them in; but if the kids want to go to the school and can speak English, we bring them in. We haven't made a practice of giving them an examination; I recognize that we should have, but I have never thought of it; but all who want to come, come over.

Q   You say you have not had the practice of giving any tests in English?

A   No, sir.

Q   How, then, is a determination made as to whether a child of Latin-American descent can speak sufficient English to be allowed to go to the "White" school?

A   Well, merely by questioning. As a matter of fact, there has never been but one child that anybody asked to go to the "White" school that we didn't take into the "White" school.

Q   Do you know—will your records show something about that child?

A   I can tell you.

Q   Suppose you do.

A   This past fall, the first day of school, one of our Latin-American students in the freshman class in high school called me that afternoon and said, "Mr. Dodson," says, "Mama wants Minerva Delgado to go to the 'White' school." I said, "Why" He says, "She is too far from the Latin-American school." I said, "Does she speak English?" He says, "No, sir." "Does your Mother speak English?" He says, "No, sir." I says, "She will have to go up there until she can speak English well enough to do the work."

That is the only child I know that has ever been turned down; that was a six-year-old beginner who didn't speak English by their own admission.

Q   To your knowledge are there any children of Anglo-American descent who attend the Latin-American school?

A   No, sir.

Q   Has there ever been a case—

A   Not that I know of.

Q   —when a child of Anglo-American descent attended ?

A   There have been some half Anglo-American.

Q   But a full Anglo-American, so far as you know, has never attended the Latin-American school?

A   Has never, so far as I know.

Q   What would you say are the reasons for the maintenance of the different or separate school buildings?

A   I don't know how it became started; just custom, I suppose . That is what started it before I got there.

Q   In other words, when you got there, it was just a custom?

A   That is right.

Q   Have you just continued the custom, except the grades have been lowered?

A   The grades have been lowered, and I more or less continued the custom except for the cases I have talked of where they came and asked, and could speak English.

Q   And the records which are coming up will have those cases?

A   Yes; they will have the names of the children, some of them. Now, you understand I just made a hurried check over the old registers for about the last five or six years.

Q   Now, in addition to the custom which was responsible for this practice before you became superintendent, are there any other reasons for this practice?

A   Well, common sense, in my viewpoint, says if a child can't speak English, I don't know what I would do with him if I put ten of them that can't speak English in with twenty that can speak English. He might be able to learn to speak English under those conditions, but if I did, I know I would be slowing down those that do speak English.

Q   So that in addition to custom, the other reason for the maintenance of separate schools is the matter of English or the teaching of English?

A   That is the primary interest now; that is the primary reason.

Q   Is there any reason besides that reason for your maintenance of separate school buildings now?

A   No, sir.

Q   Now, are there some children, let us say for the purpose of this question, comparatively few, of Latin-American descent who attend the Latin-American school who, let's say, in the third grade show some proficiency in the acquirement of English; are there?

A   If they could speak English before they started.

Q   There would be some such children?

A   Yes.

Q   As a matter of regular practice, are those Latin-American children who show proficiency in the acquirement of English in the third grade transferred to the "White" school?

A   I don't make any investigation and bring them over.

Q   Are there some children, perhaps, a few in the Anglo-American school who somehow or other don't seem to acquire a proficiency in English even up to the third grade?

A   No, sir.

Q   There aren't?

A   No, sir, I wouldn't think so; you mean in the American—

Q   American school.

A   Anglo-American school that don't speak English?

Q   Wouldn't there be some children who come from under-privileged or poor parents who for one reason or another don't speak English well or don't speak the kind of English which it is necessary to know in order to progress in other studies in the school?

A   Oh, there is some of them that will fail English in school, yes, in high school.

Q   Well, let's talk about an Anglo-American child who fails English with a very low mark in the Anglo-American school, is he transferred to the Mexican—

A   No, sir; he doesn't fail English because he doesn't understand it. He fails English because he doesn't know grammar and doesn't know the technical part in English; but as far as the teacher telling him to do something, he is able to understand and follow directions. This big trouble with Latin-Americans, if you tell them to do something and they don't understand enough English to follow directions and do what the teacher tells them to, then we have got to teach them basic English. When you talk about deficiency in English, we have plenty students [*sic*], of course, that are not finished English students; they don't make "A's" in their English subjects; but when the teacher tells them to do such and such, they understand what she says. They might not be able to do it, but they understand what she says.

Q   Well, is the standard which is used in determining whether or not a child speaks English, the kind of English which the child speaks in order to be able to get promoted from one class to another because the child is able to understand the matters which are taught in the class, or is the kind of English which the child needs to know in order to follow the general instructions of the teacher?

A   Well, for putting them into your Anglo-American school, I would think the English we would want them to know—we can teach them grammar and things like that, but we would want them to know enough so they understand what we are saying when we try to tell them to do something.

Q   Suppose that instead of having separate schools, children of Mexican descent with a deficiency in the knowledge of English for the purpose of following the instructions of the teacher were allowed to go to the "White" school and were given special instruction in the "White" school in English so that they could understand and follow the teacher, wouldn't that adequately dispose of the problem?

A   Well, that is what we are doing; it doesn't make any difference to me; I am perfectly satisfied to put them over in the other school. I don't have room for them; I don't have trained teachers that have taught Spanish-speaking children enough to know how to teach them English; but as far as association is concerned, we have no resentment at all to putting Latin-American students in our school, the "White" school.

Q   In other words, then, if there were the facilities from — so far as teaching the children of Mexican descent proficient English, that can be done just as well when children of Mexican descent are in the same building as when the children of Mexican descent are in a different building?

A   In the same building and separate classes, yes.

Q   Yes.

A   My only reason has always been the fact that if I put them in the same room, those who speak English are going to be suffering while they wait for those to be taught English.

Q   If they were put in separate rooms, that problem would be disposed of?

A   That is right, yes, sir.

Q   As a matter of fact, also, as an experienced educator of young school children, wouldn't you say that children who do not have a proficiency in the English language might derive some benefit from rather continuous association with children who speak the English language well—I mean on the playground?

A   I wouldn't doubt that a bit in the world. As a matter of fact, when they get out of school up there at the Latin-American school, about 90 per cent of them, you see, these children don't belong to us very much; most of them belong to the rural districts around there; they ride in on buses, all right on the same bus. They get out up at the other school at 2:30 or 2:20; the bus leaves at 4:00. They leave the Latin-American school and come down on the campus and play with the "White" children waiting for buses; they get a lot of association that way; I am sure that helps.

Q   So that as an educator, you would say that except for the problems in physical arrangement and perhaps, even, financial problems, there would be an advantage in the acquirement of English—

A   That is right.

Q   If there were no separate building?

A   If they were on the same grounds in separate classes, yes, sir.

Q   Yes, sir.

A   That would be ideal from my mind, as a matter of fact.

Q   Because then the child would have the advantage of acquiring English by speaking it with his friends and associates?

A   That is right.

Q   As well as merely listening to English from the teacher?

A   That is right.

Q   Which is all that a child gets when it attends a separate school building; you agree to that?

A   Yes. Because of that, we insist that those children speak English on the school ground up there at the Latin-American school.

Q   And of course, if they played with Anglo-American children and associated with them more, they would more naturally and readily—

A   That is right.

Q   —just naturally speak better English?

A   That is right.

Q   Wouldn't you say, then, as an educator, that from your experience the maintenance of separate school buildings tends to retard the acquirement of proficient English by those who are not already proficient, rather than improve it, so far as the separation is concerned?

A   Well, now, I don't know.

Q   I think you have already said that the association which comes by children not proficient in English with children who are, through continuous association and exchange, is an important factor in the acquirement of proficiency in English?

A   I think it would help, yes.

Q   Well, maintenance of separate schools don't permit that [*sic*], in so far as they are separate?

A   Well, I don't know. We have, of course, we have children in the third and fourth grade who have learned a lot of English.

Q   You mean—

A   In the Latin-American school; and by requiring them to play on the ground together—to speak English on the playground and to associate with those who speak English—I am not exactly certain that they don't learn just as much English that way as they would with the American children.

Q   In any event, I take it as your opinion that when children who are not proficient in English associate and talk with children who are proficient, the more of that that is done, the more readily the children who are not proficient acquire proficiency in English?

A   I would say so.

Q   And that is particularly true with children who are not proficient, when they are just naturally associated with in friendly relations on the playgrounds and buses with the Anglo-American children?

A   Of course, I don't know if we would have a great deal more of that if we had them in the same building than we do having them on the same grounds from 2:30 to 4:00 o'clock; there wouldn't be a great deal more of that. They play on the playground together.

Q   Is there a cafeteria in the "White" school?

A   Yes, sir.

Q   Is there a cafeteria in the Mexican school?

A   Not at present, no, sir. We had one there before we had one in the "White" school, but limited space has forced us to close that.

.   .   .

Q   Generally, is the curriculum in the Mexican school the same as in the "White" school?

A   Yes, sir; put a little more stress on certain things in the Latin-American school.

Q   What stress is—put on what things in the Latin-American school?

A   Health, English, and citizenship.

Q   With respect to English, just tell us specifically, if you can, in what particular respects the instruction or the treatment of the children in the Mexican school is different from the Anglo-American school.

A   Well, of course, in the first grade at the Mexican school we have practically none of them that can speak a word of English when they start. The teacher has to use signs and everything else, just like we would have to in our high school Spanish class, teaching our students Spanish. We have to hold up a lemon and tell them "limon, lapel, ventana, and puerto."

Q   That, of course, could be done just as satisfactorily and efficiently if the Mexican-American children were in a separate class in the same building as the Anglo-American students?

A   Yes.

.   .   .

Q   With respect to college training which the teachers have before they teach school, do you have an estimate as to the amount of college training which your teachers on an average have for the Anglo-American school?

A   We have—most of them have degrees; we have probably four or five in the Anglo school that haven't completed their school; we have one that has only six hours of formal college work. We have four or five that have three years or less, and the other 18 all have degrees, Bachelor's and some Master's.

Q   What about the Latin-American school, so far as college training is concerned?

A   We have one in the Latin-American school has only six hours of formal college work, and we have two, I believe, that have three and three and a half.

.   .   .

Q   Have some papers come up that might give answers as to the number of Mexican children who have been attending the "White" school?

A   (Refers to records.) According to this hurried check that I made, in 1941–42 we had Dorothy Gonzales in the first grade.

Q   All right; in the first grade.

A   Clarence Gonzales in the fourth grade.

Q   Are they related, do you know?

A   I suppose they are brother and sister, but I don't know.

Q   All right.

A   And Richard Alvarado in the second grade.

Q   That is the entire—

A   That was in 1941–42.

Q   All right.

A   Now, I might have overlooked some. I just leafed through, looking for Latin-American names, is the principal thing about this.

Q   All right.

A   1943–44 I found two in the second grade: Dorothy Rial and Carl Rial.

Q   All right.

A   1944–45 I found Viola Martinez in the fifth grade.

Q   All right.

A   1945–46 I found Joseana Rogers; that wasn't her true name. She was a step-child of a man named Rogers; she was a full blood Latin-American. In the sixth grade a Donald Pacheco; in the fifth—

Q   Rogers was Anglo-American, the parent?

A   Yes.

Q   All right.

A   1944–45 I had Ralph Barrera, in the first grade; 1945–46 I had—

Q   You have given us that, I think.

A   I had Ralph and Janet Barrera at the same time. Ralph was still in the first grade, and Janet was starting the first grade.

Q   All right.

A   1946–47 Ralph was in the second, and Janet in the second. 1947–48— let's see here; 1946–47 too there was Charles Schotz and Helen Schotz, second and fourth grade. I understand probably their father was German, but they moved to us from Mexico City, where they had been attending the American school down there. Their mother was Mexican. Then 1947- 48 I have got three Barrera children: Ralph in the third and Janet in the third and Shirley Ann in the first.

DOCUMENT 5.9

Aoy (Mexican Preparatory) School, El Paso, Texas, 1905

In 1887 Olives Villanueva Aoy (1823–1895) opened a bilingual private school in El Paso, Texas, for Spanish-speaking pupils. In 1888 the El Paso School Board incorporated the school into its system, hiring Mr. Aoy and two English-speaking assistants. By 1897, more than 200 students were enrolled. In 1899 the city erected a new building and by 1900 over five hundred children were seeking attendance at the school. In El Paso, as in other southwestern cities, school authorities often operated separate schools or classrooms for children who were non-English-speaking. Children in El Paso, for example, were separated by linguistic ability: "All Spanish speaking pupils in the city who live west of Austin Street will report at the Aoy School, corner of 7th and Campbell. English speaking Mexican children will attend the school of the district in which they live."[39] Because these regulations were often applied broadly, they eventually became the subject of several lawsuits.

From: *Report of the Public Schools of El Paso, Texas, 1905–1906*, pp. 35–36. Library of Congress, Washington, D.C.

I deem it unnecessary to go into details relative to the buildings, equipment, and work of all the ward schools, except to say that each of them is thoroughly equipped and the work is well organized and is progressing with unusual smoothness. But I do wish to call the attention of the Board to the Mexican school. This is a twelve-room building occupied exclusively by Mexican children. It is situated in the heart of the Mexican district of the city. It is well heated with steam heat, well lighted and thoroughly ventilated, furnished with the best new single desks, has the best toilet fixtures, is kept scrupulously clean, and all in all, is one of the most conveniently arranged and best equipped buildings we have. It is impossible to estimate the general good that this school is doing and has done among these benighted Mexican people. Yearly there are over six hundred children who attend regularly this school. They come from the humblest homes, where in years past, knowledge of English and habits of cleanliness and refinement were unknown; from families whose ancestors for ages have been under clouds of ignorance and superstition. To these little fellows, the school building is a veritable palace. It is possibly the most comfortable house, outside the church, that they ever entered. Nice, clean floors, beautiful pictures on the walls, porcelain lavatories, clean towels, brushes, combs, well kept toilet rooms, are luxuries that these little beings never knew. Add to these comforts neatly dressed, cultured, refined, sympathetic teachers, whom these little ones love devotedly, and upon whom they look as beings a little more than human, and you see at once the opportunity for doing good at this building. Among the first lessons instilled into these children when they enter the school room is cleanliness. It is not an uncommon sight here to see a kind hearted school ma'am standing in the lavatory room by one of these home-neglected urchins and supervising the process of bringing about conditions of personal cleanliness as he applies with vigor to rusty hands, dirty ears and neck, unkempt face and head the two powerful agencies of American civilization, soap and water.

These are perhaps secrets intended by the instructors of this building to be kept sacredly within the walls of the school house, but I feel that they are such potent agencies for good that I am justified in revealing them. It has come to me that on several occasions these youngsters have returned home at the close of the day and were not recognized by their parents. I will not vouch for the truthfulness of this statement, but to him who his [*sic*] seen one of these little ones "before and after taking" it is not an incredible story.

Looking back over a period of ten years, since this work began one is astonished at what really has been accomplished . . .

Boys and girls who then were half dressed, half fed waifs of the alleys and streets, now speak English, hold positions as clerk in stores, book-keepers, teachers, interpreters, do all kinds of work where intelligent labor is required, dressmaking, laundrying, cooking, housekeeping, blacksmithing, work in foundries, railroad shops, carpenter shops, factories, etc. and have become so Americanized that the influence they exert for good upon this city in point of sanitation and morals can scarcely be estimated.

DOCUMENT 5.10

"Your American Pupil from Puerto Rico"

Between the years 1949 and 1968 the number of Puerto Rican children in the New York City public schools increased tenfold, from approximately 30,000 to 300,000. State and local school officials conducted numerous surveys and reports in order to determine the best methods to effectively teach English-language learners. In the 1950s and early 1960s the school district published several pamphlets designed to assist teachers and administrators with the influx of new pupils. The following images and captions are from the pamphlet, "Your American Pupil from Puerto Rico," which was tailored specifically for school districts 17 and 18 of the Bronx.

From: Material from publication, *Your American Pupil from Puerto Rico*, NY: Board of Education of the City of New York, 1957 (Vertical file—Education-Elementary). In Archives of the Center for Puerto Rican Studies/Centro de Estudios Puertorriqueños, Hunter College, New York, NY.

In a <u>regular classroom</u>, a bilingual Perto Rican pupil can act as a "buddy" for a non-English-speaking Puerto Rican child.

The <u>Special English Teacher</u> helps small groups of children each day. She plans real experiences to provide practice in purposeful conversation.

The Special English Teacher also helps the classroom teachers meet problems connected with non-English-speaking Puerto Rican children. The Puerto Rican coordinators in the junior high schools and elementary schools perform similar duties.

The <u>Substitute Auxiliary Teacher</u> (S.A.T.), a native of Puerto Rico who speaks Spanish and English fluently, works directly with non-English-speaking Puerto Rican pupils and their parents.

DOCUMENT 5.11

Esmeralda Santiago, from When I Was Puerto Rican

Born in Santurce, Puerto Rico, on May 17, 1948, Esmeralda Santiago came to the United States at the age of 13. In this excerpt from When I Was Puerto Rican, *her first encounters with the U.S. public school system in the early 1960s are illuminated. Ms. Santiago graduated from New York City's High School of Performing Arts and earned her bachelor's degree from Harvard and her M.F.A. from Sarah Lawrence College.*

The first day of school Mami walked me to a stone building that loomed over Graham Avenue, its concrete yard enclosed by an iron fence with spikes at the top. The front steps were wide but shallow and led up to a set of heavy double doors that slammed shut behind us as we walked down the shiny corridor. I clutched my eighth-grade report card filled with A's and B's, and Mami had my birth certificate. At the front office we were met by Mr. Grant, a droopy gentleman with thick glasses and a kind smile who spoke no Spanish. He gave Mami a form to fill out. I knew most of the words in the squares we were to fill in: Name, Address (City, State), and Occupation. We gave it to Mr. Grant, who reviewed it, looked at my birth certificate, studied my report card, then wrote on the top of the form "7–18."

Don Julio had told me that if students didn't speak English, the schools in Brooklyn would keep them back one grade until they learned it.

"Seven gray?" I asked Mr. Grant, pointing at his big numbers, and he nodded.

"I no guan seven gray. I eight gray. I teeneyer."

"You don't speak English," he said. "You have to go to seventh grade while you're learning."

"I have A's in school Puerto Rico. I lern good. I no seven gray girl."

Mami stared at me, not understanding but knowing I was being rude to an adult.

"What's going on?" she asked me in Spanish. I told her they wanted to send me back one grade and I would not have it. This was probably the first rebellious act she had seen from me outside my usual mouthiness within the family.

"Negi, leave it alone. Those are the rules," she said, a warning in her voice.

"I, don't care what their rules say," I answered. "I'm not going back to seventh grade. I can do the work. I'm not stupid."

Mami looked at Mr. Grant, who stared at her as if expecting her to do something about me. She smiled and shrugged her shoulders.

"Meester Grant," I said, seizing the moment, "I go eight gray six mons. Eef I no lern inglish, I go seven gray. Okay?"

"That's not the way we do things here," he said, hesitating.

"I good studen. I lern queek. You see notes." I pointed to the A's in my report card. "I pass seven gray."

So we made a deal.

"You have until Christmas," he said. "I'll be checking on your progress." He scratched out "7–18" and wrote in "8–23." He wrote something on a piece of paper, sealed it inside an envelope, and gave it to me. "Your teacher is Miss Brown. Take this note upstairs to her. Your mother can go," he said and disappeared into his office.

"Wow!" Mami said, "you can speak English!"

I was so proud of myself, I almost burst. In Puerto Rico if I'd been that pushy, I would have been called *mal educada* by the Mr. Grant equivalent and sent home with a note to my mother. But here it was my teacher who was getting the note, I got what I wanted, and my mother was sent home.

"I can find my way after school," I said to Mami. "You don't have to come get me."

"Are you sure?"

"Don't worry," I said. "I'll be all right."

I walked down the black-tiled hallway, past many doors that were half glass, each one labelled [*sic*] with a room number in neat black lettering. Other students stared at me, tried to get my attention, or pointedly ignored me. I kept walking as if I knew where I was going, heading for the sign that said STAIRS with an arrow pointing up. When I reached the end of the hall and looked back, Mami was still standing at the front door watching me, a worried expression on her face. I waved, and she waved back. I started up the stairs, my stomach churning into tight knots. All of a sudden, I was afraid that I was about to make a fool of myself and end up in seventh grade in the middle of the school year. Having to fall back would be worse than just accepting my fate now and hopping forward if I proved to be as good a student as I had convinced Mr. Grant I was. "What have I done?" I kicked myself with the back of my right shoe, much to the surprise of the fellow walking behind me, who laughed uproariously, as if I had meant it as a joke.

Miss Brown's was the learning disabled class, where the administration sent kids with all sorts of problems, none of which, from what I could see, had anything to do with their ability to learn but more with their willingness to do so. They were an unruly group. Those who came to class, anyway. Half of them never showed up, or, when they did, they slept through the lesson or nodded off in the middle of Miss Brown's carefully parsed sentences.

We were outcasts in a school where the smartest eighth graders were in the 8–1 homeroom, each subsequent drop in number indicating one notch less smarts. If your class was in the low double digits, (8–10 for instance), you were smart, but not a pinhead. Once you got into the teens, your intelligence was in question, especially as the numbers rose to the high teens. And then there were the twenties. I was in 8–23, where the dumbest, most undesirable people were placed. My class was, in some ways, the equivalent of seventh grade, perhaps even sixth or fifth.

Miss Brown, the homeroom teacher, who also taught English composition, was a young black woman who wore sweat pads under her arms. The strings holding them in place sometimes slipped outside the short sleeves of her well-pressed white shirts, and she had to turn her back to us in order to adjust them. She was very pretty, with almond eyes and a hairdo that was flat and straight at the top of her head then dipped into tight curls at the ends. Her fingers were well manicured, the nails painted pale pink with white tips. She taught English composition as if everyone cared about it, which I found appealing.

After the first week she moved me from the back of the room to the front seat by her desk, and after that, it felt as if she were teaching me alone. We never spoke, except when I went up to the blackboard.

"Esmeralda," she called in a musical voice, "would you please come up and mark the prepositional phrase?"

In her class, I learned to recognize the structure of the English language, and to draft the parts of a sentence by the position of words relative to pronouns and prepositions without knowing exactly what the whole thing meant.

The school was huge and noisy. There was a social order that, at first, I didn't understand but kept bumping into. Girls and boys who wore matching cardigans walked down the halls hand in hand, sometimes stopping behind lockers to kiss and fondle each other. They were *Americanos* and belonged in the homerooms in the low numbers.

Another group of girls wore heavy makeup, hitched their skirts above their knees, opened one extra button on their blouses, and teased their hair into enormous bouffants held solid with spray. In the morning, they took over the girls' bathroom, where they dragged on cigarettes as they did their hair until the air was unbreathable, thick with smoke and hair spray. The one time I entered the bathroom before classes they chased me out with insults and rough shoves.

Those bold girls with hair and makeup and short skirts, I soon found out, were Italian. The Italians all sat together on one side of the cafeteria, the blacks on another. The two groups hated each other more than they hated Puerto Ricans. At least once a week there was a fight between an Italian and a *moreno*, either in the bathroom, in the school yard, or in an abandoned lot near the school, a no-man's-land that divided their neighborhoods and kept them apart on weekends.

The black girls had their own style. Not for them the big, pouffy hair of the Italians. Their hair was straightened, curled at the tips like Miss Brown's, or pulled up into a twist at the back with wispy curls and straw straight bangs over Cleopatra eyes. Their skirts were also short, except it didn't look like they hitched them up when their mothers weren't looking. They came that way. They had strong, shapely legs and wore knee socks with heavy lace-up shoes that became lethal weapons in fights.

It was rumored that the Italians carried knives, even the girls, and that the *morenos* had brass knuckles in their pockets and steel toes in their heavy shoes. I stayed away from both groups, afraid that if I befriended an Italian, I'd get beat up by a *morena*, or vice versa.

There were two kinds of Puerto Ricans in school: the newly arrived, like myself, and the ones born in Brooklyn of Puerto Rican parents. The two types didn't mix. The Brooklyn Puerto Ricans spoke English, and often no Spanish at all. To them, Puerto Rico was the place where their grandparents lived, a place they visited on school and summer vacations, a place which they had complained was backward and mosquito-ridden. Those of us for whom Puerto Rico was still a recent memory were also split into two groups: the ones who longed for the island and the ones who wanted to forget it as soon as possible.

I felt disloyal for wanting to learn English, for liking pizza, for studying the girls with big hair and trying out their styles at home, locked in the bathroom where no one could watch. I practiced walking with the peculiar little hop of the *morenas*, but felt as if I were limping.

I didn't feel comfortable with the newly arrived Puerto Ricans who stuck together in suspicious little groups, criticizing everyone, afraid of everything. And I was not accepted by the Brooklyn Puerto Ricans, who held the secret of coolness. They walked the halls between the Italians and the *morenos*, neither one nor the other, but looking and acting like a combination of both, depending on the texture of their hair, the shade of their skin, their makeup, and the way they walked down the hall.

One day I came home from school to find all our things packed and Mami waiting.

"Your sisters and brothers are coming," she said. "We're moving to a bigger place."

Tata and I helped her drag the stuff out to the sidewalk. After it was all together, Mami walked to Graham Avenue and found a cab. The driver helped us load the trunk, the front seat, and the floor of the rear seat until we were sitting on our bundles for the short ride to Varet Street, on the other side of the projects.

I'd read about but had never seen the projects. Just that weekend a man had taken a nine-year-old girl to the roof of one of the buildings, raped her, and thrown her over the side, down twenty-one stories. *El Diario*, the Spanish newspaper, had covered the story in detail and featured a picture of the building facing Bushwick Avenue, with a dotted line from where the girl was thrown to where she fell.

But Mami didn't talk about that. She said that the new apartment was much bigger, and that Tata would be living with us so she could take care of us while Mami worked. I wouldn't have to change schools.

The air was getting cooler, and before Delsa, Norma, Héctor, and Alicia came, Mami and I went shopping for coats and sweaters in a secondhand store, so that the kids wouldn't get sick their first week in Brooklyn. We also bought a couch and two matching chairs, two big beds, a *chiforobe* with a mirror, and two folding cots. Mami let me pick out the stuff, and I acted like a rich lady, choosing the most ornate pieces I spotted, with gold curlicues painted on the wood, intricate carving, and fancy pulls on the drawers.

Our new place was a railroad-style apartment on the second floor of a three-story house. There were four rooms from front to back, one leading into the other: the living room facing Varet Street, then our bedroom, then Tata's room, then the kitchen. The tub was in the bathroom this time, and the kitchen was big enough for a table and chairs, two folding racks for drying clothes washed by hand in the sink, and a stack of shelves for groceries. The fireplace in the living room, with its plain marble mantel, was blocked off, and we put Tata's television in front of it. The wood floors were dark and difficult to clean because the mop strings caught in splinters and cracks. The ceilings were high, but no cherubs danced around garlands, and no braided molding curled around the borders.

On October 7, 1961, Don Julio, Mami, and I went to the airport to pick up Delsa, Norma, Héctor, and Alicia. Papi had sent them unescorted, with Delsa in charge. The first thing I noticed was that her face was pinched and tired. At eleven years old Delsa looked like a woman, but her tiny body was still that of a little girl.

In the taxi on the way home, I couldn't stop talking, telling Delsa about the broad streets, the big schools, the subway train. I told her about the Italians, the *morenos*, the Jewish. I described how in Brooklyn we didn't have to wear uniforms to school, but on Fridays there was a class called assembly in a big auditorium, and all the kids had to wear white shirts.

Tata prepared a feast: *asopao*, Drake's cakes, Coke, and potato chips. The kids were wide-eyed and scared. I wondered if that's the way I had looked two months earlier and hoped that if I had, it had worn off by now.

All my brothers and sisters were sent back one grade so they could learn English, so I walked to the junior high school alone, and my sisters and brothers went together to the elementary school on Bushwick Avenue. Mami insisted that I take the long way to school and not cut across the projects, but I did it once, because I wanted to find the spot where the little girl had fallen. I wondered if she had been dead when she fell, or if she had been still alive. Whether she had screamed, or whether, when you fall from such a great height, you lose air and can't make a sound, as sometimes happened to me if I ran too fast. The broad concrete walkways curved in and around the massive yellow buildings that rose taller than anything else in the neighborhood. What would happen to the people who lived there in case of fire? I imagined people jumping out the windows, raining down onto the broad sidewalks and cement basketball courts.

The walls of the projects and the buildings nearby were covered with graffiti. I didn't know what LIKE A MOTHER FUCKER meant after someone's name. Sometimes the phrase would be abbreviated: "SLICK L.A.M.F." or "PAPOTE L.A.M.F." I had heard kids say "shit" when something annoyed them, but when I tried it at home, Mami yelled at me for saying a bad word. I didn't know how she knew what it meant and I didn't, and she wouldn't tell me.

"Mami, can I get a bra?"

"What for, you don't have anything up there." She laughed.

"Yes, I do. Look! All the girls in my school . . ."

"You don't need a bra until you're *señorita*, so don't ask again."

"Mami," I said a couple of weeks later as she changed out of her work clothes. "I'm going to need that bra now."

"What?" she stared at me, ready to argue, and then her face lit up. "Really? When?"

"I noticed it when I came home from school."

"Do you know what to do?"

"*Sí.*"

"Who told you?" Her face was a jumble of disappointment and suspicion.

"We had a class about it in school."

"Ah, okay then. Come with me, and I'll show you where I keep my Kotex." We walked hand in hand to the bathroom. Tata was in the kitchen. "Guess what, Tata," Mami said. "Negi is a *señorita!*"

"Ay, that's wonderful!" She hugged and kissed me. She held me at arms length, her eyes serious. Her voice dropped to a grave tone. "Remember, when you're like that, don't eat pineapples."

"Why not?"

"It curdles the blood."

In the bathroom Mami showed me her Kotex, hidden on a high shelf under towels. "When you change them, wrap the soiled ones in toilet paper, so no one can see. Do you want me to help you put the first one on?"

"No!"

"Just asking." She left me alone, but I could hear her and Tata giggling in the kitchen. The next day Mami brought me a couple of white cotton bras with tiny blue flowers between the cups. "These are from the factory," she said. "I sewed the cups myself."

While Mami worked in Manhattan, Tata watched us. As the days grew shorter and the air cooler, she began drinking wine or beer earlier in the day, so it wasn't unusual for us to come home from school and find her drunk, although she still would make supper and insist that we eat a full helping of whatever she had cooked.

"My bones hurt," she said. "The beer makes the pain less."

Her blood had never thickened, Don Julio explained, and she had developed arthritis. Tata had been in Brooklyn more than fifteen years, and if her blood hadn't thickened by then, I worried about how long it would take.

We complained about being cold all the time, but Mami couldn't do anything about it. She called *"el lanlor"* from work, so that he would turn on the heat in the building, but he never did.

On the coldest days, Tata lit up the oven and the four burners on the stove. She left the oven door open, and we took turns sitting in front of it warming up.

One evening as we all sat grouped around the stove I told the kids a fairy tale I'd just read. Don Julio crouched in the corner listening. Like my sisters and brothers, he frequently interrupted the story to ask for more details, like what color was the Prince's horse, and what did the fairy godmother wear? The more they asked, the more elaborate the story became until, by the end, it was nothing like what I had started with. When it was over, they applauded.

"Tell us another one," Hector demanded. "Tomorrow."

"If you tell it now," Don Julio said, "I'll give you a dime."

"For a dime, *I'll* tell a story," Delsa jumped in.

"I'll do it for a nickel," challenged Norma.

"Everyone quiet! It's my dime. I'll tell it."

Edna and Raymond huddled closer to my feet. Delsa and Norma, who had sprawled on the linoleum floor wrapped in a blanket, argued about who had to move to give the other more room.

"Let me get another beer," Don Julio said, and he lumbered to the refrigerator.

Tata lay on her bed in the next room. "Get me one too, will you Julio?" she called out. "Negi, talk louder so I can hear the story."

"Would anyone like some hot chocolate and bread with butter ? " Mami offered.

There was a chorus of "Me, me, me, me."

"Do you want me to tell the story or not?"

"Yes, of course," Don Julio said. "Let's just get comfortable."

"Go ahead and start, Negi," Mami said. "The milk takes a while to heat up, and I have to melt the chocolate bar first."

"All right. Once upon a time . . ."

"One minute," Alicia interrupted. "I have to go to the bathroom. Don't nobody take my place," she warned.

The fluorescent fixture overhead buzzed and flickered, its blue-gray light giving our faces an ashen color, as if we were dead. Don Julio's face looked menacing in that light, although his small green eyes and childlike smile were reassuring. My sisters and brothers were huddled together as close to the open oven door as they could manage without getting in Mami's way as she melted a bar of Chocolate Cortés and kept adjusting the flame on the pan of milk so that it wouldn't boil over. The room looked larger when we were all together like this, leaning toward the warmth. The walls seemed higher and steeper, the ceilings further away, the sounds of the city, its constant roar, disappeared behind the clink of Mami's spoon stirring chocolate, the soft, even breathing of my sisters and brothers, the light thump each time Don Julio set his beer can on the formica table. Brooklyn became just a memory as I led them to distant lands where palaces shimmered against desert sand and paupers became princes with the whush of a magic wand.

Every night that first winter we gathered in the kitchen around the oven door, and I embellished fairy tales in which the main characters were named after my sisters and brothers, who, no matter how big the odds, always triumphed and always went on to live happily ever after.

"Come kids, come look. It's snowing!" Mami opened the window wide, stuck out her hand, and let the snow collect on her palm. It looked like the coconut flakes she grated for *arroz con dulce*. The moment it fell onto our hands, it melted into shimmering puddles, which we licked in slurpy gulps.

"Can we go down and play in it, Mami?" we begged, but she wouldn't let us because it was dark out, and the streets were never safe after dark. We filled glasses with the snow clumping on the fire escape then poured tamarind syrup on it to make *piraguas* Brooklyn-style. But they tasted nothing like the real thing because the snow melted in the cup, and we missed the crunchy bits of ice we were used to.

The next day schools were closed, and we went out bundled in all the clothes Mami could get on us. The world was clean and crisp. A white blanket spread over the neighborhood, covering garbage cans and the hulks of abandoned cars, so that the street looked fresh and full of promise.

When schools opened again, kids ran in groups and made snowballs, which they then threw at passing buses, or at each other. But as beautiful as it was, and as cheerful as it made everyone for a while, in Brooklyn, even snow was dangerous. One of my classmates had to be rushed to the hospital when another kid hit him in the eye with a rock tightly packed inside a clump of snow.

Every day after school I went to the library and took out as many children 's books as I was allowed. I figured that if American children learned English through books, so could I, even if I was starting later. I studied the bright illustrations and learned the words for the unfamiliar objects of our new life in the United States: A for Apple, B for Bear, C for Cabbage. As my vocabulary grew, I moved to large-print chapter books. Mami bought me an English-English dictionary because that way, when I looked up a word I would be learning others.

By my fourth month in Brooklyn, I could read and write English much better than I could speak it, and at midterms I stunned the teachers by scoring high in English, History, and Social Studies. During the January assembly, Mr. Grant announced the names of the kids who had received high marks in each class. My name was called out three times. I became a different person to the other eighth graders. I was still in 8–23, but they knew, and I knew, that I didn't belong there.

DOCUMENT 5.12

Migrant Children in U.S. Public Schools

Mexican Americans began migrating to the Midwest in the early 1900s when demands for labor in manufacturing and agriculture brought them northward from Central Mexico. The children of migrant workers, particularly those too young to work in the fields, experienced even greater hardships than resident Latino children. The U.S. government recognized the needs of migrant children in the mid-twentieth century and provided special resources as a result of the Elementary and Secondary Education Act (ESEA) of 1965. Prior to that date, state and local agencies such as those illustrated in this 1961 demonstration summer school often took responsibility for the care of migrant children.

From *Education on the Move. Part II. Report of a 1961 Demonstration Summer School for Migrant Children in Manitowoc County, Wisconsin,* Governor's Commission on Human Rights, Madison, WI. Located in the Wisconsin State Historical Society, Madison, WI: Box 13, Folder 11, "Migrant Education," Equal Rights Division, Bureau of Community Services, 1971–1975, Series 2164.

Mrs. Shirley Mecha, regular teacher at Meadow Brook School, helps migrant young-sters to use resources of the "library corner" which can make their travels a more meaningful experience.

Teacher Richard Vaughan answers a question for one of the six resident children who took the opportunity offered by the 1961 Manitowoc Summer School to learn about the larger world from the migrant children.

CHAPTER SIX
Cuban Arrivals, 1959–1980

The Committee's conclusion is that though resident children must share their schools and community services with Cuban Refugee children, it can be done amicably and with deep understanding if the reasons why are handled skillfully. Actually, Florida children have a star role in creating an "Image of America" for the world to see. They have much to gain spiritually and educationally. Difficult as it is, they are privileged to be a part of a historical moment.

—"Report to the Governor of Florida on the Cuban Refugee Problem in Miami," 1961[1]

Cuba's proximity to the United States, only 90 miles to the tip of Florida, has facilitated migration between the two countries since the nineteenth century. Cuban independence movements against Spain from the mid-1800s until the 1898 Spanish-American War sparked social and political turmoil. As a result, several cigar manufacturers established factories in Key West and Tampa, drawing thousands of workers from the island to Florida.[2] By 1900, an estimated 20,000 Cubans and Spaniards had left Cuba for the United States. Among tourists, business men, and workers, travel between Cuba and the United States was also popular. Key West, New Orleans, New York City, and Tampa were the primary destinations of Cuban visitors and settlers.[3] The brief period of educational Americanization in Cuban from 1898–1902 (as discussed in chapter four) and Cuba's status as a protectorate of the United States under the Platt Amendment (1903–1933), further cemented political and economic ties between the two nations. In 1934 President Franklin Delano Roosevelt abrogated the Platt Amendment as part of his Good Neighbor Policy towards Latin America. However, U.S. business interests in Cuba, particularly in sugar plantations, continued from the 1930s to the 1950s, and tourism also flourished between the two countries.

The well-established political and economic relationship between Cuba and the United States ended abruptly with Fidel Castro's overthrow of the Batista regime in 1959. Furthermore, the Cuban Revolution of 1959 was set against the backdrop of cold war political tensions between Communist and non-Communist nations. As Alex and Carol Stepick point out, cold war ideology contributed to the warm reception and generous financial assistance the U.S. government offered to Cuban refugees fleeing communism. No other Latino immigrant group had before—or has since—been allotted similar resources and aid. In short, this mid-twentieth century Cuban exodus is linked to this political context and is both a result of the already high education levels and the reason for subsequent educational gains in the United States.[4]

The history of Cuban migration to the United States in the post–World War II era is typically divided into three waves. The first wave consisted of 250,000 Cubans who arrived between 1959 and 1962. Sometimes called the "Golden Exiles," they disproportionately represented the urban, educated professional and business classes. For instance, although only 4 percent of all Cubans in 1959 had received high school diplomas, 36 percent of exiles during this time had completed high school.[5] The Cuban Missile Crisis in October 1962 suspended flights between the two countries.

A second wave of Cubans, totaling almost 300,000, began during the "freedom flights" of 1965 and lasted until 1973.[6] The social and economic composition of this second wave has been described as more working class and less educated than the first wave Golden Exiles. Furthermore, refugees were restricted in the amount of funds they could bring to the United States.

The third wave began in 1980. Cubans involved in this migration were pejoratively named the "*marielitos*" (Mariel, Cuba, was the port of departure). Castro, responsible for their negative reputation, announced that he had sent by boats the worst elements of Cuba. An additional 125,000 Cubans came to the United States during this third wave.[7]

The Cuban refugees from 1959 to 1962 included children and adults with little knowledge of the English language, and the arrival of so many placed a severe strain on services in the greater Miami area—especially the education system. The enrollment of Cuban children in the public and private schools tested the flexibility and resources of school officials. Superintendent of Dade County Public Schools, Joe Hall, provided detailed information on children arriving in 1960 to the federal government. Superintendent Hall reported that a total of 3,127 "nonimmigrant" children (defined as "those who are admitted into the country as political exiles or as persons on student visas") had been admitted to the public schools between August and November of 1960.[8] By January of 1961 the number of nonimmigrant Cubans in the public schools had increased to 3,500 and the Catholic parochial schools had enrolled 2,650 new children.[9] These numbers, according to Sylvia Carothers, director of Florida Children's Commission, had the parochial schools "stretched to the breaking point"; Superintendent of Public Education Thomas Bailey explained that financial assistance was "urgently needed at the earliest possible moment, not only to reimburse Dade County School System for extraordinary expenses already incurred, but, also, for such expenses as long as the situation continues to be critical."[10]

Funds from the federal government arrived quickly in south Florida. President Eisenhower authorized the immediate transfer of one million dollars to the Department of Health, Education, and Welfare (HEW) for assistance to South Florida.[11] Eisenhower appointed lawyer Tracy Voorhees, who had headed the Hungarian refugee crisis of the 1950s, to spearhead the relief efforts. In December 1960 the federal Cuban Emergency Refugee

Center was created to serve as a reception, processing, and relief agency.[12] (See document 6.1.) From this center the U.S. government eventually created the large-scale Cuban Refugee Program (CRP), which cost almost one billion dollars between 1965 and 1976.[13]

The response of local school officials to the influx of over six thousand children was creative and prompt. Furthermore, assistance from both local groups, such as the P.T.A., and distant philanthropists, such as the Rockefeller Foundation, eased the educational crisis. State Superintendent of Education for Florida Thomas D. Bailey noted some of the conditions facing the schools and the innovations created to meet the needs of English-language learners. Bailey found that at Riverside Elementary School, "69% [of 1,027 pupils] are 'Latins' with varying degrees of competence in speaking English. Sixty-two new children entered within the week prior to our visit and 14 on the day of our visit. Upon entering the school almost all of these spoke very little or no English."[14] Bailey reported that the public schools had developed a guide, "Planning for Non-English Speaking Pupils," that offered suggestions for schools. In this guide, administrators recommended that Cuban refugees be assigned "directly to a grade and class without any prior sessions for learning English or general orientation." Then, an orientation teacher worked 45 minutes each day with groups of ten children. Bilingual visiting teachers were also hired at Riverside Elementary School and the P.T.A. sponsored a Mother's Club for the Spanish-speaking parents.[15] At Citrus Grove Elementary School in 1961, one-third of the 900 students were non-English speaking. At this school, Spanish-speaking children were pulled out for an hour a day to work with orientation teachers whose purpose was to "help children learn to speak and understand English." Furthermore, the P.T.A. at Citrus Grove employed a bilingual teacher to teach a six-week English course to a group of 32 parents. According to Bailey, the English course for parents "serves also to help the Cuban parents understand the elementary school and its program."[16]

At the junior high school level, Dade County schoolteachers utilized different strategies than were used in the elementary schools. At Ada Merritt Junior High School in 1961, one-half of the 800 students was Cuban. New students were placed in a regular homeroom but assigned a "bilingual buddy." Each student attended the four major subjects of English, mathematics, science, and social studies in addition to a two-hour daily orientation class that focused on English skills. At Ada Merritt, federal resources provided funds to hire a bilingual secretary, a bilingual counselor, and an orientation (English) teacher.[17]

At the high school level the strategies varied by school. New students at Miami Senior High were placed in orientation programs (curiously, never called English classes). The curriculum of these one- or two-hour daily classes consisted of studying "English, customs, school regulations, and general citizenship." High schoolers also continued to take the regular academic curriculum and were encouraged to graduate from the orientation program as soon as possible. The strategy at Coral Gables Senior High School was modeled on a more traditional "sink or swim" immersion method of language acquisition that had been used for immigrants in many turn of the century public schools. According to Superintendent Bailey's 1961 report, of the 200 Cuban students in a school of 2,670 students, the nonimmigrants were placed "directly in the regular high school classes." At that date no special English classes were provided even though some teachers "believe[d] that a 'crash program' for teaching English might be valuable before students are placed in regular academic courses." Because of the numbers arriving to Coral Gables Senior High, Bailey pointed out that it would be "difficult to adhere to the county guideline of placing no more than 3 to 5 non-English speaking children in a single classroom."[18]

At the parochial schools in Miami, a different language strategy emerged. At Gesu Elementary School, where nearly half of the 450 students were Spanish-speaking, those who spoke only Spanish were placed in separate "first grade" orientation classes until they gained a "working knowledge of English so that when they are assigned to a regular grade class they have a chance to keep up."[19] The parochial schools in Miami also proposed an experimental bilingual secondary school. Housed at the Centro Hispana Catolico, the curriculum was modeled on a "typical Cuban secondary school with the addition of English and American history." The purpose of the school was transitional: "it will help Cuban students maintain their academic level until they return to Cuban schools. Spanish will be spoken for teaching purposes. The staff is drawn from Catholic sisters who have taught in Cuba and from Cuban refugee teachers."[20]

Eventually the public schools also began experimenting with fully bilingual education. In 1963 Coral Way Elementary School piloted a voluntary bilingual program in grades one through three. Children received instruction in Spanish one-half of the day and English for the other half. Bilingual programs such as the one at Coral Way received substantial federal assistance and were the first bilingual public schools in the post–World War II era. Between 1960 and 1972 Dade County public schools received more than $130 million for bilingual education programs.[21] According to James and Judith Olson, "the field of bilingual education was born in those Dade County schools when it became clear that when Cuban-American children were able to become literate in their mother tongue as well as learn English, their success rates in school were dramatic."[22] Careful observers note that while Puerto Ricans had encountered barriers to receiving bilingual schooling in New York City, such programs at Coral Way Elementary School and at other South Florida sites were opened without controversy and received adequate staffing and funding.[23]

The bilingual programs in Dade County's schools also provided jobs for refugees who had taught in Cuba. Before Cuban teachers (and other professionals seeking certification or U.S. credentials) could be hired, numerous logistical and political obstacles had to be overcome. Some of the first problems school officials in Florida encountered while trying to hire teachers resulted from cold war politics. For example, the Florida legislature had passed a law (Fl. Statutes, Section 231.17) requiring that teaching certificates only be issued to U.S. citizens or "citizens of other nations not antagonistic to democratic forms of government."[24] Because the United States had severed ties with Cuba, after December 1960 Cubans could no longer be certified to teach in the public schools. James L. Burnsted, director of adult and vocational education, Broward County, demanded that the state superintendent of education find an administrative solution to the dilemma. He pointed out that the state was "slamming a door in the faces of the people who could help us during this emergency and who, by their very actions, have indicated that they are not antagonistic to a democratic form of government. These people, at tremendous personal, professional, and financial risk and loss have sought refuge in Florida . . . It seems incredible that these people could be certificated [*sic*] and employed as teachers as long as their application was processed prior to December. Nothing about these people has changed. Their attitudes concerning the Castro government and our democratic government are the same as they were when they fled their homeland."[25] In response, Superintendent Bailey cautioned "it has been made crystal clear that many Cubans now in the United States are not exiles from the Castro government but are supporters of Castro." Thus, according to Bailey, the Department of Education would be charged with trying to "classify an individual Cuban as an exile from or supporter of the Castro regime."[26]

The administrative red tape was eventually lifted, which opened the door for professional work among the exile community's educators. One of the HEW-funded programs included hiring and training Cuban professionals to work as bilingual teacher aides in the Dade County public schools. Another sizeable HEW grant allowed the University of Miami to open the Cuban Teacher Training Program in 1962.[27]

The first years in Miami were difficult and confusing for both new Cuban teachers and children. Dr. Mirta R. Vega, a retired Dade County school administrator, recalled her early experiences: "I started working as a Cuban aide in May 1962 . . . My students were all Cuban refugee children and they were very frightened. Many of the American teachers were also frightened. I could understand that. Each day two or three or four new refugee children arrived at the school . . . every day. New classes had to be created all the time. Some of the teachers just couldn't handle it. One day, two American teachers came up to me and said, 'Don't take it personally, but we've decided to take early retirement. The changes here are just too much for us.'"[28] Some Cuban teacher training programs in South Florida continued to receive federal funding as late as 1972.[29]

The special forms of assistance to Cuban refugees also extended to higher education. In 1961 HEW's Office of Education created a college loan program for Cuban students. By 1966, María Cristina García noted that more than 5,500 Cuban Americans had taken advantage of this special perquisite.[30] A total amount of $34 million was distributed to Cuban American college students between 1962 and 1976 through the Cuban Refugee Program.[31] The combination of direct and indirect assistance to Cuban pupils and teachers during the 1960s and into the 70s translated into positive educational and economic outcomes that created a successful Latino enclave in south Florida.[32]

In addition to the strain placed on local public and private schools, the United States also faced a unique situation in providing for not only the educational but the welfare needs of over 14,000 Cuban children who arrived unaccompanied by family members. The program was called "Operation Pedro Pan," because the first child sponsored was named "Pedro" and the flight to Miami was reminiscent of Peter Pan's flight to Neverland in J. M. Barrie's classic book. Father Bryan O. Walsh, director for the Catholic Welfare Bureau, headed up the cause from Miami and was tireless in securing temporary homes for children.

Cuban families made the decision to send their children ahead amid rumors that Castro was planning to remove children from their families. As one of the Catholic agencies reported in 1960, "Of grave concern to the Church and local welfare authorities is the increasing number of children who are being sent by themselves into South Florida. A rumor, which is persistent in Havana, whispers that parental rights are going to be terminated and that children between the ages of five and 16 will be taken from their parents and sent to government-controlled schools for military training and indoctrination."[33] Indeed, the new government did send youth to the countryside as literacy teachers and to Russia for military training; and it nationalized both public and private schools.[34]

The Pedro Pan children first arrived slowly and, as Miami became overwhelmed, were eventually placed not only in private homes and Catholic shelters, but in orphanages and foster homes in 100 cities in 35 states.[35] As the memoirs of two former Pedro Pan children testify (see documents 6.2 and 6.3), the trauma of leaving Cuba without family members was often painful and long-lasting. Some families believed they would only be separated from their children for a few weeks. However, political situations such as the October 1962 Cuban Missile Crisis suspended flights for almost three years, causing considerable hardship for children and their stranded parents. Although one study of Pedro Pan children suggests that they felt the sacrifice was worthwhile for their long-term futures, others believed it was an unnecessary hardship. Today, Pedro Pan children have formed their own organization for mutual support and to raise money for children's relief work.[36]

The 1980 exodus of Cubans to the United States occurred in a very different social-historic context than the exodus of the cold war 1950s and 1960s. In 1980, Dade County, Florida, which had once provided exceptional hospitality to the first wave of Cubans, was in the midst of a political backlash against immigrants. An English-only amendment was on the ballot and angry Anglos felt resentful over the widespread use of Spanish in public spaces.[37] Politicians demanded that the U.S. government not burden south Florida with a second inundation of Cubans, particularly ones who Castro had called his "refuse." Rumours that the *Marielitos* (from the name Mariel—the port of departure from Cuba) were overwhelmingly composed of criminals and mentally ill patients spread through the media. As María Cristina García points out, less than 4 percent of the total number of entrants were criminals and many "criminals" had been jailed in Cuba for crimes "not recognized in the United States," but the public was convinced otherwise and feared the latest arrivers.[38]

Approximately 13,000 new Cuban children needed adjustment in the Dade County public schools, prompting another crisis among local educators. Fortunately, many federal programs created initially for the first waves of Cuban refugees were already in place. Furthermore, one school official pointed out the abundance of "sufficient qualified bilingual personnel. That instant pool of skilled, qualified teachers, teachers' aides and clerical employees is unique to South Florida."[39] In June 1980 the federal government approved a one-million-dollar grant under the Bilingual Education Act (Title VII of the Elementary and Secondary Education Act of 1965) for a summer program. Designed for 5,000 students, over 9,000 boatlift children were served.[40] The rapid entrance and sheer number of Cuban refugee children in the Mariel boatlift overwhelmed south Florida's schools: "When school opened, the day after Labor Day in September, 1980, there were over 13,000 refugee students enrolled who had not been in the United States five months before."[41] Although the Dade County public schools had traditionally incorporated English-language learners into regular neighborhood schools, the short-term segregation of Mariel children was utilized for the first year of acculturation to the United States. Afterward, refugee children were placed throughout the general student population.[42]

Pressure from Floridian politicians to disperse the burden of the new refugees throughout the country resulted in the creation of three receiving centers outside of south Florida. Military bases in Fort Chaffee, Arkansas; Indiantown Gap, Indiana; and Fort McCoy, Wisconsin, were opened to receive refugees, many of whom did not want to relocate outside of south Florida. In addition, the local communities were not welcoming of their arrival. Fort McCoy, Wisconsin, hosted a large number of unaccompanied Cuban adolescents sent ahead by family members and without family members to receive them. Many of these children were placed in at-risk, poorly supervised situations resulting in additional trauma to the migrant process.[43] As youth from Fort McCoy and other refugee centers were placed in nearby communities, those school districts faced new challenges. In Milwaukee, school officials prepared a proposal for the U.S. Department of Education for special assistance of Cuban refugee adolescents (see document 6.4). As captured in sympathetic portraits of Cuban refugees, the military bases proved to be harsh adjustment centers for many youth.[44] (See documents 6.5a, b, and c.)

The Cuban experience in the United States represents a unique aspect of the immigrant story. Although the socioeconomic status of the third wave of Cubans (who spent their formative years in Communist Cuba) was considerably lower than that of the first two waves, their predecessors had created a strong community with political clout for them to enter. At this date, of all

Latino groups, the Cuban population has the highest educational levels. This is in part a legacy of the substantial resources and governmental commitment to the Cuban refugees of the cold war era.[45]

DOCUMENT 6.1

Excerpts from a Statement Made by President Eisenhower and Tracy S. Voorhees from the White House on January 18, 1961

As a result of the cold war, the United States opened its doors to individuals fleeing communist countries, including Hungarian refugees between 1956 and 1957. Tracy S. Voorhees, chairman for the President's Committee for Hungarian Refugee Relief (1956–1957), also served as the President's Personal Representative for Cuban Refugees from 1960 to 1961. In this report Voorhees illuminates the hardships placed upon Miami, Florida, with the arrival of a quarter of a million Cubans. The schools, both public and private, were particularly impacted by children for whom English was not their native tongue.

> From: Tracy S. Voorhees, *Report to the President of the United States on the Cuban Refugee Problem* (Washington, D.C.: United States Government Printing Office, 1961). In Rockefeller Archives Center, Rockefeller Foundation Archives, Sleepy Hollow, NY, Box 67, Folder 565.

"I am releasing herewith the final report on Cuban refugee problems by Tracy Voorhees who has been acting as my personal representative in this matter.

"In appointing Mr. Voorhees for this task last November, and in giving him funds and added powers on December 2, I sought to express by effective action the interest which, as President of the United States, I felt in these troubled people, as well as my deep sympathy for them and desire to be of help to them.

"This latest exodus of persons fleeing from Communist oppression is the first time in many years in which our Nation has become the country of first asylum for any such number of refugees. To grant such asylum is in accordance with the long-standing traditions of the United States. Our people opened their homes and hearts to the Hungarian refugees 4 years ago. I am sure we will do no less for these distressed Cubans."

REPORT TO THE PRESIDENT ON THE CUBAN REFUGEE PROBLEM

Foreword

This report supplements my Interim Report of December 19th. In the latter I stated that the influx of Cuban refugees to the United States had approached almost 40,000 in number and was continuing at a rate of more than 1,000 a week. The principal port of entry for these refugees was—and is—Miami.

Although there has been a large spillover to the New York Metropolitan area—including Newark—and to a relatively small extent to other cities, the majority remain in the Miami area. There, an ever-mounting Cuban population quite obviously has overrun the community's capacity to cope with it.

The problem is now a national one . . .

Education

Below College Level

Over 6,500 Cuban students are going to the Miami public and parochial schools. About 93 percent of those in the public schools have been exempted by necessity from the $50 fee provided under Florida State law; 18 percent have even been exempted of necessity from school charges for instructional supplies; and 6 percent even from the lunch charge. Cafeteria supervisors report that many students are getting their one hot meal per day at school.

Teachers and administrators describe these young people as "generally a very high type of student from ambitious, education-minded families."

Classes range in size up to 42 in the public schools and 60 in the parochial schools. Sometimes in a class as many as two-thirds are unable to speak English. The public school system is developing special orientation sessions given by Spanish-speaking teachers, but the public schools need 25 more of these teachers plus up to 50 additional regular classroom teachers. The parochial schools also need help.

These students have seen and heard much of fear and violence. They are eager to adjust and to learn. There is a dramatic contrast between the frightened faces of little children at the registration desk each day and the purposeful student in a first-grade reading group or a ninth-grade civics class—which sometimes includes arrivals in Miami of only a few days before.

A detailed report on the school situation has been prepared by a member of my staff, concurred in by the Florida State Superintendent of Education and by the Superintendent of Schools of Dade County—the Miami area. At the request of the State and local offices, the U.S. Department of Health, Education, and Welfare now has a team in Florida to work out a program of possible help for schools in Dade County.

One school principal summed up the challenge: "There are a lot of very fine people coming into our schools. They *can* be fine ambassadors for us, if they return sometime to Cuba; or fine potential citizens of the United States, if we handle things right here in the schools."

University Level

As to the needs of the college student our facts are as yet inadequate. A study has been prepared but further analysis is necessary as to the requirements for the second semester. One report expresses the belief that as many as 900 students are in financial difficulty and may have to leave college.

Giving these boys and girls a chance is a challenge which the people of the United States will not fail to meet.

Temporary provision from a private gift has been made for 20 of the most deserving and needy cases of students now at the University of Miami.

The Crux of the Refugee Problem

This situation has been created by evil events which at least for the present have subjugated Cuba to communism.

Although it has only been recently—and as yet inadequately—realized, the Cuban refugees present us with a national problem following from our nation's traditional humane policy of granting asylum as long as they need it to people fleeing from oppression, however they come in and whatever their status . . .

Respectfully submitted,
Tracy S. Voorhees

Washington, D.C.
January 18, 1961.

DOCUMENT 6.2

Legacy of Operation Pedro Pan

In the following two documents Cuban refugees who were a part of Operation Pedro Pan reflect on their memories of their passage and adjustment to the United States. Researchers estimate that over 14,000 children were quietly removed from Cuba between 1959 and 1965 in order to escape an uncertain fate in the newly communist Cuban society. Many of these unaccompanied children were placed in foster homes, with Catholic relief agencies, and in orphanages until they could be reunited with family members. The trauma of being separated from family members impacted their educational experiences, particularly for those who were sent to live in communities, such as Kansas and Wisconsin, unaccustomed to Latinos.

SELECTIONS FROM ``THIRTY-TWO YEARS LATER FOR MY SISTER, ALICIA"

by Flora Gonzalez Mandri

Flora González Mandri is the mother of Rachel Werner Baldwin. Born in Havana but raised in Camagüey, she left Cuba as an adolescent with her sister, Alicia, as part of Operation Pedro Pan and was relocated to California. She now lives in Cambridge, Massachusetts.

From: "Thirty-Two Years Later: Florence González Mandri" pgs. 107–112 from REMEMBERING CUBA: LEGACY OF A DIASPORA edited by Andrea O'Reilly Herrera, Copyright © 2001.

When my sister, Alicia, and I landed at the Havana airport, thirty-two years after we had left in 1962, she insisted that I take a picture of her right after she stepped on Cuban soil:

"I can't believe we're in Cuba, together," my sister said. "Why don't you take a picture of me in front of the Cubana airplane?"

In 1980 I, alone, had returned to the Island after eighteen years of absence, looking for the little girl I had been back in the fifties and early sixties. I had also had my picture taken then, but I was so nervous that I managed to lose my camera before developing that memorable photograph.

On that first return, I was carrying with me all the carefully forgotten memories of having left with my sister, who had been eleven; I had been thirteen. My parents had said goodbye to us at the airport. "Just for a short while," they said. The separation would be temporary, they assured us, but my unspoken fears told me otherwise. We ended up waiting to leave at the airport for hours, inside the *pecera*, that memorable glass-enclosed waiting room. Our parents were on the other side of the glass, feeling the anguish of sending their daughters away for the first time in their lives, hoping against hope that they had made the right decision. They had to protect us from the threat of communism. We were so young and susceptible . . .

In order to protect myself from the mixed emotions of that goodbye, I had erased my parents' anguish from the map of my soul. So in 1980, when I felt I was an adult—like my parents had been—with a daughter of my own, I had returned to recover my lost past, my Cuban past. Was losing my camera in Havana a weak gesture representing my failed attempt to recover a memory? Because that first return failed to recapture those forgotten hours in the pecera, I persisted and persisted by returning. I kept boarding planes alone from Miami to Havana in 1985, 1987, and 1988. On each trip, I gradually recovered the streets and the houses of my childhood; I forgave my parents for letting us go so gently; and I began to reconstruct my ties to the red soil of my native Camagüey.

In 1994, while returning with my sister, I asked myself, "What memories and emotions would this trip bring?" With my Kodak camera, I did capture the image of my sister smiling. But what of my identity? Would I find it on the way to Camagüey with Alicia?

After several days in Havana, Alicia and I boarded a plane again, this time to accomplish the real purpose of our trip: to visit our cousin, Mario, and his daughter, Gloria, in Camagüey.

. . . Even though this was the fourth trip back to the land where I grew up, the mask of the knowledgeable traveler began to show gargantuan cracks. As a child, I was always guided around the streets of Camagüey by my mother, aunt, or grandmother; I seldom went out alone, other than the short walk to and from school. During previous trips, I assumed the persona of the family member who had returned after having made it big back in the United States. I often relied on the knowledge of the taxi drivers to get me around. This trip was different because I did not feel alone; I did not have to play the role of the provider. When we left Cuba, two sisters alone, I had stood as the parent to my younger sibling. Now, I thought, it was time to reverse the roles. So I began to rely on my sister, who had the advantage of a psychology degree and many years of practice reading my emotional well-being. This was not a casual trip to Cuba for me—if there is such a thing as that for me, or anyone who had left as a child, "never to return." Many of us who had left as children, the so-called Pedro Pan generation, did go back looking for their relatives and their roots. This time, however, I was returning with my sister, just as in '62 I had left with her. The circle now felt complete, even though on this trip she was the one in charge.

My fantasy had been that like that first exit, this reentry would put me, the older of the two, in the driver's seat. After all, Mami had told me, her thirteen-year-old daughter, "You take good care of your little sister now, you hear?" And I had taken that sentence as a command, as my destiny until "that wonderful moment when the family of four would again reunite." And like the "good" girl I had grown up to be, I had taken good care of my sister, always following the examples of my mother, grandmother, and aunts, who had given each of us the clear sense that we were a whole person, our limbs firmly secured to our torsos, our hair nearly combed, and our heads held high. A whole person definitely connected to a very large extended family.

But our goodbye at the Havana airport January 11, 1962, had shaken all that. And the feeling of taking one's own body for granted in all its integrity had disappeared. As if in fear that I might wake up one day with my arms and legs disjointed and dancing away from me to the sounds of some mysterious New World Symphony, I acquired the bad habit of unconsciously tensing my muscles to retain physical integrity. Only when I returned to the neighbor-hoods of my childhood did my body feel that it owned the space that surrounded it. Then the tension in my muscles melted and flowed through my toes into the cobblestones of the colonial streets in Havana, or into the rich, compressed dirt of the Casino Park in Camagüey. For the first time after my many returns, I could share these feelings with Alicia, who knew exactly what I was talking about.

"If we close our eyes and hold hands, we can pretend we never left, can't we?" she said, as if to bridge the chasm between Miami and Casino Park. As soon as she made the statement, we held hands so as to let the power of our shared past invade our present. But our pragmatic American selves knew better. I now know that remaining Cuban as a citizen of the United States means hard work. Hard work to retain my command of the Spanish language. For my sister, it is hard work to understand that being Cuban in Miami is not the same as being Cuban in Camagüey. Hard work to learn about the reality of Cubans who stayed. Hard work to imagine, "What if I had never left?"

My present intellectual work has me obsessed with the artistic production of Cuban women around my age who live and write their poetry, produce films, and paint canvases in a land with severe economic limitations, rather than in the "land of opportunity" where I reside. Their cultural expressions cross over to the mainland, terra firma, and become a lifeline for my sister and me, who left unexpectedly, unaware. During each trip to Havana, I collect more engraved images of the Cuban culture I experience, vicariously, as they stare down at me from the walls of my living room, adding color and warmth to the cold Boston winters.

DOCUMENT 6.3

Reflections on Adolescent Migration from Cuba

"LIFE AL REVÉS"

by Alicia Serrano Machirán Granto

Alida Serrano Machirán Granto was born in 1949 in Santiago de Cuba, Oriente Province. She left the Island in February 1963 and now lives in Buffalo, New York.

From: "Life al revés: Alicia Serrano Machirán Granto" pg. 116–121 from REMEMBERING CUBA: LEGACY OF A DIASPORA edited by Andrea O'Reilly Herrera, Copyright © 2001.

I left my childhood behind in Cuba at the age of fourteen. I still have a terribly vivid memory of my brother and me in the *pecera* [the fishbowl] at the Havana Airport, where they placed children leaving the country to keep them separated from their relatives as they waited to board the plane that would take them away from everything and everyone they had known since birth.

A couple of weeks after the disastrous Bay of Pigs Invasion, our family home was confiscated. My parents were given forty-eight hours to remove their belongings and go *"donde carajo pudieran,"* [wherever the hell they could] which turned out to be my grandmother's house. I, along with my brother and sister, was bewildered and in a daze; I was so full of pain that I could not even talk about it.

Together, my parents, Adela Machirán Ortiz and Diómedes Serrano Cala, came to the painful conclusion to send my brother and me out of Cuba when my father discovered that I was writing anti-Castro graffiti on the walls of a beach clubhouse that we frequented. After having overheard a *militario* say that if he caught the "bastard" who was writing things against *"El Comandante,"* he or she would regret it, my father became frantic with fear when my older cousin, who had already spent three months in jail for distributing pro-freedom-of-speech flyers, told him that I was the culprit. The deciding factor, however, came when my parents saw my name posted on a list at the Conservatory of Music of "promising piano students" who were chosen to be sent to Moscow on an art exchange scholarship.

In order to leave as a family, my father would have had to resign his job, which would have severed the means he needed to get us out of the country.

With the same epidemic desperation caught by hundreds of parents, who were fearful that their children's brains would be "washed by communism," Papi and Mami sent my brother and me out of Cuba in January 1963; they remained behind with my sister, who was mentally retarded. We were sent to my maternal uncle and aunt, who lived in Miami; they had been exiled for a couple of months. I remember being in a daze for days. It was not culture shock yet—being surrounded by the family softened that blow somewhat—it was trauma, which thanks to my immaturity manifested itself in deep denial. In effect, I convinced myself that I was visiting the States as a tourist, and soon the vacation would be over and I'd go home to my parents.

In addition to the emotional stress, I went from lacking nothing in Cuba to living a provisional life in Florida. For example, because I had almost nothing to wear, other than the clothes on my back, each night I would wash the armpits of my dress with Ivory soap (I came to hate that soap) so I would have something clean to wear to school the next day. Despite the material poverty we faced, I still have some very fond memories, such as the occasional evenings spent with our Cuban neighbors playing bingo with pennies.

A few months after our arrival in the States, my uncle was relocated to Wilmington, North Carolina; and my brother and I went with him. (He was my mother's oldest brother, and he decided to take on the responsibility of caring for his sister's children.) This is when culture shock *really* hit me. Not only were there no Cubans around, but in the town where we lived, Blacks walked on one side of the street, and Whites walked on the opposite side. We were regarded with either curiosity or suspicion by both. I remember feeling completely displaced and not belonging; the possibility that I might not see Cuba again suddenly occurred to me, and it triggered a defensive reaction in me. As a result, I submerged myself in my studies so intensely that I became one of the top five students at the Catholic school I attended. Once in a while, I would wake up in the middle of the night with anxiety attacks. "Oh my God," I would ask myself, "am I ever going to see my parents again? Am I ever going back to Cuba?"

I skipped the tenth grade and moved up to the eleventh grade at a nearby public high school; and again, most of the unpleasant experiences that occurred stemmed from ignorance on the part of my classmates, or resentment and/or jealousy on the part of those who could not accept the fact that "a boat refugee" could excel academically. Things began to change, however, in January 1965, with the arrival of Papi and Mami.

After years of trying to gain passage out of Cuba, my parents were finally granted a visa, thanks to my cousin, who lived in Mexico and had Mexican citizenship. When they arrived in Mexico, they applied for entry into the U.S., certain that the worst of their trials were over and that we would soon be reunited as a family. To their great dismay, they were denied passage to the United States—a denial due, in part, to my sister's handicap. In desperation, my parents and my sister, along with several other Cuban refugees and Mexicans, crossed the Rio Grande in a canoe, were arrested upon their arrival in Brownsville, Texas, and were tried as illegal aliens. Although the judge at their trial announced that they were going to be deported—a pronouncement that caused my mother to faint right in the middle of the courtroom—my parents discovered soon afterwards that the trial had been a sham, staged in order to deport the Mexicans who had crossed into the States. As a result, all of the Cubans were granted entry, and my parents were flown to Miami, where my brother and I were waiting at the house of Adela Babún, who had generously offered to share her rented one-bedroom house with all of us until my parents could get back on their feet.

After only a few months in Miami, my parents announced that we would be moving, a decision based on the fact that most of the better jobs available to Cubans in south Florida had already been taken. At the Cuban Refugee Center, which was relocating Cuban professionals all over the United States, they learned that teaching positions were available in New York and New Jersey. (My father refused to allow my mother to work at any job other than teaching, the profession she had held in Cuba.) Although my parents had the option to go to New Orleans, they thought it best to move further away, in order to resist the temptation of returning to Miami. As a result, we moved to Niagara Falls in September of 1965; and my mother, at the age of forty-seven, went back to school to get her certification to teach in the States.

Despite our isolation (we were the only Cuban family in the area at the time), my parents worked very hard to preserve our Cuban culture. When I became engaged to a very old-fashioned Italian American, it was very important to us that the wedding ceremony reflect our heritage. As a result, I wore a *mantilla* as my wedding veil, and we insisted on paying for the reception meal so that we could serve the traditional Cuban *puerco asado* to our guests.

Soon after my wedding, however, my parents decided to move the family to New York City. Due in part to his inability to speak English, my father was unable to find work in Niagara Falls. In addition to feeling culturally isolated, he felt emasculated, for he was unable to fulfill his traditional role as provider and "man of the house." Knowing that there were teaching positions available to my mother in New Jersey, my father accepted a job in Merrill Lynch's Latin American division, a position that did not require him to speak English.

My family's departure, coupled with my father's sudden death in an automobile accident in 1969 and the reality of both marital and maternal responsibilities, brought my *cubanismo* to yet another stage. I remember one particular Saturday morning, while I was getting ready to do housework, I put on records of songs that were hot around the time I left Cuba, such as "*Sabora mi*" and "*Imágenes*" (just writing about it now brings tears to my eyes); as I listened to the music, I succumbed to an enveloping cloud of nostalgia that would physically paralyze me for hours at a time. This would happen so very often that I started playing a little tune over and over in my head at night:

> Espavílate muchacha, que no estás,
> haciendo na', te has quedado sin na' de allá, y
> no estás creando na' por acá.

Wake up, girl; you're not
doing anything; you're left with nothing from there, and
you're not starting anything here.

Of course, the *allá* was Cuba, and the *acá* was my new life (which was supposed to be filled with marital bliss!) in Niagara Falls. Finally, I woke up one morning, gathered all my wonderful Cuban records together, and took them to Goodwill. No more Cuban music to listen to and get immobilized by! I also stopped trying to teach my (then) three babies Spanish, and I joined and became very active in an Italian lodge, becoming the first non-Italian officer to serve in it.

In the meantime, my husband was unintentionally helping me define myself even more as a *cubana*. Soon after our marriage, I realized that we were separated by some very fundamental cultural differences. For instance, whereas he tended to be negative and shortsighted, I tended to be positive and always see a light at the end of the tunnel—whenever he focused on the dark side of things, my optimistic approach would cause him to accuse me of always wearing "rose-colored glasses." I was also misunderstood by others, such as my new extended family and my colleagues at the university, several of whom referred to me in a negative sense with terms such as indomitable, ambitious, nonconformist, cocky (more often than not), and the like. "Aren't you ever satisfied?" I would be asked. "Why don't you stay home where you belong?" Others still would ask, "Is the way you behave typically Cuban, or is it you alone, or, perhaps, the way that you were brought up?" In self-defense, I would always respond by saying, "I don't know . . . I don't know how to be any other way. I don't think I would know how to behave differently." The fact that I didn't stay home raising my family and taking care of the housework, coupled with my professional success, defied their stereotypical notion of how a Latina should behave. In other words, I wasn't subdued and submissive and docile. Both personally and professionally, I have always felt driven to excel; I have tried to do my best and be as productive as possible, so as not to mar that prevalent image of my people as hard workers. In the same vein, I do not share *mis problemas y pesares* with the entire world (a typically Cuban trait!), and I grow weary of the manner in which many of my non-Cuban co-workers whine and moan and groan about their working conditions. I find myself thinking that rather than complaining, most of the Cubans I know would be moved to action, either by trying to change things or by looking for another alternative (perhaps even another job).

In addition to my own career and my work in the community, I wanted to pass on to my children the empowering legacy of their heritage. Oftentimes, I would wake up in the middle of the night and worry that my children were being deprived of the family warmth and caring that I witnessed during my own childhood. So surrounded were they by Italians while growing up, I feared no matter how hard I tried, I would fail. As a result, my most desperate effort throughout the last twenty years has been to instill in my children a positive frame of mind and an optimistic perspective (what they now refer to as "the Cuban mentality we got from our mom"). My interaction with my own children has always been guided by an attempt to replicate a collage of experiences from my early years growing up in Cuba—the caresses and hugging, the tolerance of everyone talking at the same time, the abundance of food (which I love to prepare), the warmth and company of family and friends. To have them feel that way is the greatest reward for my tenacity in adhering to teaching them by action and example and not by simply talking and preaching.

I can honestly say that every aspect of my life is impacted by my heritage, sometimes more than I would like. Externally, I have adjusted very well to American life. Internally, however, I often feel schizophrenic in my dealings outside of my Cuban "circle." In some sense, this other person inside of me sees and reacts to most things differently than my American twin. Whenever I am mistaken for something other than a *cubana*, or even called an American-Born Cuban, I have made it a point to explain that I am Cuban, born and raised. Why is this so? I'm not exactly sure if I can explain, but I suppose it reflects my tremendous pride in my Cuban heritage.

This testimonial, which is based upon several conversations, was co-written by Alicia Serrano Machirán Granto and Andrea O'Reilly Herrera.

DOCUMENT 6.4

Immigrant and Refugee Program, Milwaukee Public Schools

Upset over the continued arrival and financial strain caused by Cuban refugees, the state of Florida requested the federal government's assistance in resettling refugees to other parts of the country. In response, several thousands of Cubans were relocated to Wisconsin, Pennsylvania, and Arkansas in 1980–1981. This document is a copy of a grant proposal submitted to the United States Department of Education by the Milwaukee Public Schools. As the proposal reveals, many Cuban adolescents arriving without family members faced horrific circumstances in their adjustment and adaptation to a new country. Through the proposal of innovative and holistic programs this proposal reveals one school system's attempt to meet the needs of a unique population. Within the grant proposal are included the analyses of psychologists, social workers, and others involved in the assessment and delivery of services to Cuban refugee adolescents.

From: Milwaukee Public Schools, Division of Planning and Long-Range Development, "Immigrant and Refugee Assessment and Academic Support Program Demonstration Project." February 13, 1981. In Hispanics in Wisconsin collection, 1981, M89–116, folder, "Next edition—Hispanic Bibliography," Wisconsin Historical Society, Madison, Wisconsin.

MILWAUKEE PUBLIC SCHOOLS RECENT IMMIGRANT AND REFUGEE ASSESSMENT—BILINGUAL ACADEMIC SUPPORT PROGRAM

The Milwaukee Public Schools is requesting Title VII, E.S.E.A. funds for the establishment and operation of a pilot project designed to demonstrate the feasibility of an assessment and bilingual academic center support for newly arrived immigrant and refugee youths. The program is designed to address the needs of all pupils categorized as newly arrived adolescents, limited in English proficiency and with little or no prior school experience, but most particularly the needs of Cuban refugee teenagers who have been resettled in Milwaukee and neighboring communities. The project proposes to meet the needs of these juveniles by offering three types of services: 1) thoroughly assessing the pupils' social, psychological, linguistic, academic and vocational needs; 2) providing students with access to the best available educational program in the Milwaukee Public Schools; and 3) providing the participating students with access to a bilingual academic support program that includes a bilingual supportive services component.

A. Need

The main client population for this project are the newly resettled Cuban refugee youths placed with sponsors or sponsoring agencies in the city of Milwaukee . . .

The impact of being uprooted from their homeland and the experience of living in a refugee camp for four months have left these youths, most of them young men in their late teens with needs that transcend the current capacity of the Bilingual Education Program of the Milwaukee Public Schools to adequately satisfy. Therefore, in order to understand the needs of these young men, we must review their common backgrounds in terms of their experience in Cuba and in the United States refugee camps in which they were housed while awaiting resettlement.

I. The Refugee Camp Experience

The state of Wisconsin was one of four states chosen by federal officials in April of 1980, as a site for the establishment of a relocation and processing center to house some of the estimated 120,000 Cubans who were arriving daily in Florida as a result of the Freedom Flotilla. The Wisconsin Cuban Refugee Center, located at Fort McCoy, a military reservation in the central part of the state, was open[ed] on May 29, 1980. By the time the center ceased to function four months later, nearly 15,000 Cubans seeking assylum [*sic*] in the United States had been processed.

Most of the refugees were quickly resettled with relatives already in the United States or with American citizens, willing to act as their sponsors. One group of refugees, however, languished in the camp for months unable to leave. They were the 425 Cuban juveniles who came to the United States unaccompanied by their parents. As the deadline for closing the camp neared in late August, only 25 of the youths had been able to leave the camp.

Their difficulty in getting out of Fort McCoy resulted from two main reasons. First, according to federal regulations, the Cuban minors could only be resettled with their natural parents. The regulations would not allow any others, even close relatives from assuming guardianship of the youths. The federal officials were fearful of encouraging a recurrence of a problem encountered in the resettlement of unaccompanied Indochinese children. Some of the children, spirited out of Indochina as alleged orphans, had been placed with American foster parents, who later lost custody of the child when the natural parents arrived in the United States.

Another reason that slowed down the resettlement of the unaccompanied Cuban minors was the bureaucratic confusion caused by different govern-

mental administration of the camp. During the four months that the Relocation and Resettlement Center was in operation at Fort McCoy, responsibility
for administrating the program was shaped by three different government
agencies. At first, the United States Army was in charge of the camp, then
supervision of the center was assumed by the United States State Department,
and finally control of the camp was turned over to the state of Wisconsin. In
the ensuing change of command, confusion was added and delays created.
This long internment period raised the level of frustration, boredom and
despair in the Cuban youths, and these feelings in turn were manifested in
rambunctious and undisciplined behavior on the part of some of the youngsters. Officials at the camp fearing that such negative behavior could lead to
more serious problems, or possibly prompt a recurrence of the disorders that
occurred at the resettlement centers at Fort Chaffe, Arkansas, and Indiantown
Gap, Pennsylvania, decided in mid July to isolate all the teenagers in one area.
The move was also prompted by a sincere desire to protect the youths from
the other more aggressive refugees at the camp. Unfortunately, the area
chosen to house the young Cubans was a section of the compound formerly
used as a detention center to hold wrongdoers at the camp. The teens felt they
were being unfairly punished and treated like prisoners by being placed in a
compound surrounded by several chain link fences that were topped with
barbed wire.

Life for those inside the special compound was full of fear and anxiety, as
many of the youngsters became the victims of older youths and other refugees
who would invade the area at night to rob and sexually abuse those held inside.
Knowledge of these conditions as reported in the media, gave rise to public
concern for the safety and well being of refugees at Camp McCoy and
motivated Governor Dreyfuss to create a Citizen Commission to investigate
the conditions in the camp.

The report issued by the Governor's fact-finding commission, confirmed
the horrors previously reported in the press and voiced by concerned camp
employees. They found substantial evidence of sexual and psychological
abuse of both male and female Cuban juveniles at Fort McCoy. In two cases,
the commission documented, a young woman refugee was raped by 54 men
and another fell victim to nine men when she accidentally went into a barracks
housing single men. Cases of homosexual attacks on younger boys were also
reported. The report also detailed incidents in which some of the youths were
being coerced by others into stealing articles to supply a growing black market
among the refugees. Evidence of sporadic acts of violence were also acknowledged by the commission, including acts of self-inflicted violence caused by
the sense of frustration and hopelessness felt by many of the camp's inhabitants.

The commission's report and recommendations helped to speed up the
resettlement of the unaccompanied Cuban youths. Agreement was reached
with federal authorities for the state of Wisconsin to assume legal responsibility of the youths and to find them sponsors. As a result, 295 of the Cuban
juveniles were placed with foster families or in group homes in Wisconsin,
with a substantial number of youths being resettled in the Milwaukee area . . .

2. The Homeland Experience

To gain additional insight into the educational and psychological needs of these pupils, we must also study the adolescent's experience in modern day Cuba. What about the homeland experiences of these youths would shed some light on the difficulties they are likely to face as students in the Milwaukee Public Schools? Dorita R. Marina, a staff member of the Spanish Family Guidance Clinic in Miami, Florida, has attempted to answer this question in an article published in the fall 1980 issue of *Clinical Psychologist* entitled "Predicting the Cuban Entrant Acculturation Rate." Dr. Marina suggests that a comparison of characteristics of the new arrivals with that of the earlier Cuban exiles is helpful in attempting to forecast the problems they are likely to encounter in adjusting to their new environment. She points out four key factors that differentiate the newer arrivals whom she calls "Marielitos" because most left for the United States from Mariel Bay in Cuba, from other Cubans in the Unites [*sic*] States.

A. Work Values an [*sic*] Experience

"Work values and experience was crucial in the relative easy adjustment of Cuban Americans to America. The results of research on this factor indicated that both male and female Cuban Americans had a high need for vocational (and academic) achievement. The Cuban Americans, many of whom were middle class and professionals, were able to adapt successfully to their new economic situation because they had the requisite job skills and high motivation to succeed."

"In contrast, the Marielitos' achievement motivation has been less developed in Cuba in the last 20 years. Further, even for those Marielitos who are motivated to work and are not mentally or physically disabled, finding jobs in the already cluttered American job market will be difficult."

B. Education

"Cuban Americans place a high value on academic achievement. Many were middle class immigrants and had attained a high level of achievement in Cuba. It appears that many of the Marielitos are not highly educated, further, their chances are certainly less than ideal."

C. Family

"The cohesiveness, the displayed affection, and strong parental control of the Cuban American family have been important factors in maintaining traditional cultural values which simultaneously fostered successful adaptation to life in the United States by providing the Cuban American with a strong sense of personal identity, integrity and self-esteem."

"Many Marielitos' families were disrupted during the exodus. Many Marielitos arrived in the United States unaccompanied by any family members. Life in the refugee camps for the unaccompanied Marielito minor is specially difficult. The unaccompanied minor has only feelings about and memories of his or her family for support. Many of the unaccompanied Marielitos come from families unlike Cuban Americans. In a large number of the unaccompanied minor's families, the mother had abdicated her role and left the children with family relatives. Still other adolescents lived at home and felt very close to family members, especially their mothers. This last group is the most depressed of the youngsters ... The unaccompanied minors, lacking the emotional support of a family are more likely to have problems in the acculturation process."

D. Race

"Race is another factor that may influence the Marielitos' acculturation rate. One third of the Marielitos are black. Historically, the assimilation of blacks into the American mainstream has been an extremely slow process. Thus, the race factor could delay the acculturation process."

Because of all these factors, the author ends by concluding that the newer Cuban arrivals will face a relatively slow and difficult adjustment period in the United States. As a result, she predicts a sustained need for supportive social services will be needed by the teenagers of the Freedom Flotilla.

3. Statement of the Problem

Many of the new Cuban pupils attending the Milwaukee Public Schools have had a difficult time adjusting to their new cultural, social and academic environment. Among the examples of their difficulty to cope with their new school surroundings, are the following:

A. Absenteeism from Certain Classes

Some of the students were not accustomed to attending school all day. In Cuba, schooling after the sixth grade usually involves half day vocational or occupational training coupled with some trade related work. Those pupils now in Milwaukee mostly desire to attend the English as a Second Language classes, and regularly skip the other classes of lesser interest to them. They would thus often be found roaming the school corridors, where they would be apprehended by a school aide or faculty member; when the student is admonished in a language he yet does not understand, a confrontation results.

B. Shoplifting

Unlike their experience in Cuba, where even the most ordinary consumer items are severely rationed, the wealth of goods displayed, and stores nearby some of the schools have proved too tempting for a few students. This behavior also shows in part the influence of the refugee camp atmosphere where the currency of exchange was often stolen, cigarettes, radios, blankets and blue jeans.

C. Fights

The newcomers were not readily accepted by their peers. Distrustful of all others after their bitter experience at Fort McCoy, the Cuban adolescents have sought each other out for mutual support and protection. Unwittingly, they have become a part of the juvenile gang rivalry already existing in some high schools. Even minute arguments like what should a space traveler be called, an astronaut or a cosmonaut, have created unnecessary friction and resulted in some serious conflicts.

D. Loss of Self Esteem

Press reports about disturbances in some of the refugee camps fueled the negative perception of the refugees already held by many Americans who opposed the whole idea of allowing the Cubans into the United States. The incidents of truancy, hall walking, fighting and shoplifting, although only involving a few members of the group, have unfortunately fed these very negative stereotypes. The Cuban youngsters have internalized some of these views which, coupled with their previously held feelings of inadequacy in an English speaking society, have had a damaging effect to their sense of self esteem.

E. Severe Academic Retardation

The academic skills of the majority of the Cuban pupils is much lower than that of their fellow pupils the same age, in part, because most have had limited formal schooling after the sixth grade. Those who were institutionalized in Cuba received little academic instruction. The others only went to school part-time, and had frequent vacation periods that interrupted the schooling process. Vacations were declared by the state when extra labor was needed to help with the sugar harvest or to work in public works projects. The Cuban educational system also appears to be less advanced than the public school system in America. Teachers and others have reported that the youngsters are behind in reading, writing and mathematics skills . . .

THE CUBAN REFUGEE EXPERIENCE IN WISCONSIN

A Chronology of Events From Press Reports

1980

April 20   Fidel Castro announces anyone who wishes is free to leave Cuba.

May 23	Camp McCoy prepares for as many as 12,000 Cuban refugees.
May 25	Tomah, Wisconsin radio station declares 60% of respondents in phone survey opposed to Cubans being resettled in Wisconsin.
May 26	76,000 Cubans reported to have landed thus far in Florida. Committee of previous Cuban refugees formed to assist newcomers. 3,000 estimated already in Wisconsin. Handful of Cuban refugees escape from processing center at Elgin Air Force Base in Florida.
May 27	Officials declare Camp McCoy may not receive any Cuban refugees after all.
May 28	Government official declares less than 1% of refugees have criminal records.
May 29	The first Cuban refugees arrive in Wisconsin; 172 men. Rumors that 25,000 instead of 15,000 refugees are expected at McCoy surface. Gallup poll results reveal 97% of Americans against allowing Cubans into the United States.
May 31	Another 1,200 Cubans arrive at McCoy, including first women and children. McCoy officials told to prepare to receive 25,000 refugees.
June 1	A total of 2,000 refugees have arrived at Fort McCoy. 94,249 refugees have arrived thus far in boats at Key West, Florida. Harris poll claims 75% of those [Americans] questioned opposed Carter decision to let the Cubans in.
June 2	95,000 refugees have reached Florida. Some refugees reportedly set fires and battled with soldiers and police at Fort Chaffe, Arkansas. First Fort McCoy refugee released to sponsors. More than 3,000 refugees now at McCoy. Front page headline in the Chicago tribune "1,000 Cubans Riot—Refugees Seize Camp in Arkansas."
June 6	Refugees at Fort McCoy oppose actions of rioting refugees in Arkansas. 6,000 refugees now at McCoy. New arrivals raise number of refugees landing in Florida to 100,000.
June 10	Thirteen days after the refugees started arriving, there are now 12,818 at Fort McCoy. Number of total refugees from Cuba now stands at 112,500. Secretary Muskie declares fewer than 1,000 of the Cuban refugees are felons.
June 12	Governor Dreyfuss tours Fort McCoy. A total of 13,343 refugees have arrived at McCoy.
June 22	114,000 Cubans have arrived in Florida on the "Freedom Flotilla."
June 25	700 refugees abandoned by their American sponsors found homeless and destitute in the streets of Miami, Florida.
July 4	Cuban refugee mother gives birth to first child born at Fort McCoy.
July 13	A total of 14,159 refugees have arrived at Fort McCoy since the center opened.

July 19	425 unaccompanied Cuban minors are moved to a separate compound.
July 24	More than 14,000 Cubans have up to now been processed at Fort McCoy. Only 7,000 still remain at the camp. Sponsor pool is diminishing slowly.
July 28	Plans to close Fort McCoy by September are announced.
July 30	State of Wisconsin asked by federal officials to assume guardianship of 180 Cuban teenagers without relatives in the United States.
August 6	Disorders erupt at the Fort Indiantown Gap Pennsylvania Refugee Center. 118,000 Cuban refugees have come ashore in Florida to date. 18,000 refugees still remain at the refugee centers.
August 12	250 refugee teenagers moved to a separate compound at the Indiantown Gap Refugee Center.
August 13	Five refugees escape from Fort McCoy, are found and returned.
August 14	Camp psychiatrists are concerned that rising feelings of anxiety and depression are responsible for several suicide attempts at Fort McCoy.
August 15	The fear of being transferred to Fort Chaffe and Indiantown is blamed for causing disturbances among the Cuban refugees at the Resettlement Center. Class action suit filed against federal officials [—the suit is] protesting the treatment of juveniles at Camp McCoy.
August 19	The state of Wisconsin assumes responsibility for the resettlement of 295 unaccompanied Cuban youths at Fort McCoy. An agreement with the federal government will allow placing the youths with foster families.
August 20	The Governor of Wisconsin creates a Citizen Commission to investigate changes of poor treatment of refugees at Fort McCoy.
September 1	The Governor's Cuban Resettlement Fact-Finding Commission issues their report of findings to Governor Dreyfuss.
September 15	The Cuban Refugee Resettlement Center at Fort McCoy is closed.

Submitted by The Milwaukee Public Schools, Division of Planning and Long-Range Development, February 13, 1981, to the U.S. Department of Education for funding under Title VII of the Elementary and Secondary Education Act, Catalog Number 84.003N.

DOCUMENT 6.5

Marielitos at Fort McCoy, Wisconsin, 1980

In 1980 Fidel Castro announced that all residents who wished to leave Cuba would be granted permission. Between April and October 1980, 124, 776 Cubans arrived in the United States from the Port of Mariel, Cuba. The arrival of so many refugees strained local, state and federal resources in South Florida to the breaking point. In the midst of a recession, President Jimmy Carter was criticized for permitting the Cuban government to take advantage of the U.S. largesse. One solution to the strain placed on South Florida was to relocate refugees to three military centers in Pennsylvania, Arkansas, and Wisconsin. On July 4, 1980 photographer Dave Archibald visited the Fort McCoy Refugee Center and captured the various experiences of refugee children and youth in transition.

From: Reproduced by permission of The Wisconsin Historical Society, Dave Archibald Collection, Photographs and Digital Images. Madison, Wisconsin. Image numbers (in order): Whi-10522; Whi-10520; and Whi-10521.

Young Cubans with petitions. In this photograph Archibald captured the smiling faces of Cuban youth who had created their own governance association to establish order in the often unruly camps. One of the refugees is holding a petition signed by members of Fort McCoy protesting poor conditions and long waiting periods for resettlement.

Interpreter Steve Gunderson, center, with four children and a member of the United States military services. The placement of refugees at military camps, including penitentiary sections of camps, often angered newly arrived Cubans who felt they were being punished and treated like criminals.

Children sitting on steps. Life in refugee camps was often long and boring for both adults and children. Military camps rarely possessed recreational facilities suitable for young children.

CHAPTER SEVEN

In Search of Educational Opportunity and Access, the 1960s and 1970s

Whiteness, Blackness, and Brownness are not modes of thinking or feeling, nor principles of human understanding; and accordingly, they cannot provide the proper basis of any academic curriculum or administrative structure.

—*Students of Claremont Men's College to the Board of Trustees, 1969*[1]

The Chicano movement . . . is forcing education to look closer at itself, to make important changes in the social sciences in order to relate more realistically and comprehensively to contemporary society. For too long the archaic, ivory tower social sciences have been turning out "professionals" who in reality are nothing more than bookish robots with little or no understanding of minority community problems. Now, through political and social pressure, often producing either ethnic studies courses or changes in existing curricula, the social sciences are beginning to change for the better.

—*Ysidro Ramon Macias, "The Chicano Movement," 1970*[2]

INTRODUCTION

Frustration on the part of the Latino community at the slow pace of educational improvements for Mexican American and Puerto Rican children catalyzed rebellion and reform in the latter part of the twentieth century. *El movimiento*, as it is often called, did not solely focus on education but was part of overlapping social revolutions occurring in the 1960s and 1970s. Vietnam War protests, the Free Speech Movement, Women's Rights, the Black Power movement, and the call for rights for numerous disenfranchised groups such as gays and lesbians spread throughout the nation, but they concentrated primarily in urban areas and on college campuses. Discriminatory labor practices, persistent oppression and denial of rights guaranteed under the

1848 Treaty of Guadalupe Hidalgo, political disenfranchisement, and other issues sparked the Latino community into forming a Mexican American *Chicano* movement and its counterpart among Puerto Ricans, the *Boricua* movement Both *Chicano* and *Boricua* were terms adopted by the movement that reflected the indigenous origins of Mexican Americans and Puerto Ricans and symbolized a shift in civil rights strategy from assimilation to cultural nationalism.

Several individuals tower over the long list of activists who led the crusade for Puerto Rican and Mexican American justice. Notably: César Chávez and his close associate, Dolores Huerta, leaders of the United Farm Workers Association; Rodolfo "Corky" Gonzalez, founder of the civil rights organization Crusade for Justice in Denver, Colorado, 1965; José Angel Gutiérrez, founder of the Mexican American Youth Organization (MAYO) and La Raza Unida (LRUP) political party; Reies Lopez Tijerina, founder of the Alianza de los Pueblos Libres (Alliance of Free City-States), organized to revive the struggle over Spanish and Mexican land grant titles; and Miguel "Mickey" Melendez, founder of the New York branch of the Young Lords.[3]

Isolated efforts for acquiring civil rights, including state school desegregation cases in the 1930s and 1940s and the formation of mutual aid associations such as the League of United Latin American Citizens (LULAC) and the American GI Forum, seemed too slow and ineffective for the new generation of activists. As Juan Gonzalez explained in *Harvest of Empire*, Latin youth concluded: "their parents' attempt at integration within the political system had failed. Only through massive protests, disruptive boycotts, and strikes or even riots, the new generation decided, could qualitative (some called it revolutionary) change be accomplished."[4] Inspired by—and often in cooperation with—the more militant aspects of the African American civil rights movement, Mexican American and Puerto Rican youth turned to more aggressive means for achieving long-denied civil rights in the areas of employment, housing, political enfranchisement, and education. High school and college youth led many of the educational movements of the 1960s and 1970s. These movements ultimately gained the respect and support of the older generation.

Although Latinos had agitated individually and collectively for their civil and political rights since the 1800s and some improvements were made in the 1950s, the educational status of Latinos vis á vis Anglos and African Americans in the early 1960s was cause for serious concern. Overall, in the Southwestern states Latinos averaged three to four years less schooling than Anglos. Approximately one-half of Mexican Americans over the age of 25 had not completed eighth grade and less than 6 percent had some college education. College participation among Anglos was four times greater than among Latinos.[5]

In addition, equal access to schooling seldom translated to equal opportunity. Mexican American children often used worn-out textbooks and outdated science equipment. Because inequities also persisted within the school system, only one-half of Mexican American students who entered the Los

Angeles public high school system graduated. Role models were few—in early 1960s East Los Angeles, which had the most heavily concentrated population of Mexican Americans in the United States, only 3 percent of the teachers and 1.3 percent of the school administrators had Hispanic surnames.[6] The lack of bilingual education for the waves of new immigrants arriving in the 1950s and 1960s hindered educational advancement. High school student Patricia Delgado testified before the U.S. Civil Rights Hearings in 1967 on the language issue, stating, "if you come up from the Latin countries, they don't have any special classes for you. They will stick you in anywhere. How can you make it through if you don't understand what the teacher is talking about? If you go into a math class and you don't know one word of English and she is trying to explain to you, what can you do about it? You just sit there and then they say the Latin's don't know anything"[7] (see document 7.2). Compounding language difficulties (and often as a result of English language limitations instead of inferior intelligence), Latino students in California's public schools were disproportionately represented in special programs for the "mentally handicapped." Although Latinos comprised 14 percent of the state's K-12 population in 1968, Latinos were 40 percent of all students in the separate "mentally handicapped" programs.[8] Chicano youth of the 1960s spearheaded protests against this intolerable situation.

HIGH SCHOOL WALKOUTS IN THE CIVIL RIGHTS ERA

Crowded conditions, the practice of tracking of Latino students into vocational programs, and insensitivity on the part of Anglo teachers toward Latino culture and the Spanish language contributed to the famous 1968 walkouts (sometimes called "blowouts" by students) in Los Angeles. In the context of Latino educational history, the walkouts represented the first time that students—rather than parents or other stakeholders—directly confronted school administrations. Furthermore, the walkouts united high school students with the newly emerging college-educated Chicano activists to work toward social change.

On March 1, 1968, 300 students at Wilson High School walked out when the principal cancelled the student production of *Barefoot in the Park*. The cancellation of the play symbolized the failure of teachers, administrators, and school board members to address student demands. Among numerous demands at Wilson, Garfield, Roosevelt, and Lincoln (the four East L.A. high schools with majority Chicano populations), students petitioned for smaller classes, more rigorous college preparatory curricula, greater representation of Mexican-origin faculty and administrators, parental input, and the end of the "no Spanish" language rules.[9] Throughout the Southwest, Mexican-origin youth presented lists of demands to school officials (see document 7.3). The demands were often ignored or marginalized, and students staged massive walkouts during the spring of 1968.

The initial walkout at Wilson High on March 1, 1968, was followed by the departure of 2,000 students at Garfield High and 4,500 at Lincoln and Roosevelt High Schools. By mid-March, more than 15,000 students had walked out of Los Angeles–area schools.[10] College students who were members of the newly formed United Mexican American Students (UMAS) chapters in Los Angeles, members of the newly formed youth organization the Brown Berets, and local activists provided the strike committee with support and ideas. One of the few Mexican-origin teachers at Lincoln High School, Sal Castro, actively supported the student strikes. Adult supporters formed a committee called the Educational Issues Coordinating Committee (EICC) and met with school board members to follow up on student demands.

The local power structure responded to the walkouts and demands with repressive measures. The Los Angeles Grand Jury presented indictments against 15 members of EICC for conspiring to incite riots, disrupting the public schools, and disturbing the peace.[11] Thirteen EICC members were arrested (the LA Thirteen), including teacher Sal Castro, who was dismissed from his position at Lincoln High School. The community galvanized to reinstate Mr. Castro and appeal the indictments. Only after two years of legal battles were the LA Thirteen cleared of charges.[12]

The student walkouts in Los Angeles added a new level of militancy to the emerging Chicano movement. Mexicans had typically been described as "docile" and "passive" because of their subordinate social, political, and economic condition and seeming acquiescence to the status quo. The walkouts, which spread throughout the Southwest to Colorado and Texas, permitted students to see their collective power. Despite the arrests of the LA Thirteen, school board members eventually gave in to some demands from students, particularly regarding curricular issues.

Student protests represented one aspect of the multiple strategies to fight for Latino rights. Another tool, legislative measures at the federal level, built potential for legal recourse to violations of civil rights. The 1964 Civil Rights Act required "full and equal enjoyment . . . of any place or public accommodation" for individuals in the United States regardless of "race, color, religion, or national origin." Most importantly, the federal government could suspend funds to any entity (such as a school district or college) receiving federal monies if it violated the spirit or letter of the Civil Rights Act. Created the next year, the 1965 Voting Rights Act declared no person or organization shall "deny or abridge the right of any citizen of the United States to vote on account of race or color." Long denied access to the political or legal system as voters or jurists, Mexican Americans were finally empowered to register to vote, which allowed them to select candidates who would promote their political interests. Although both federal Acts were originally designed to combat discriminatory measures against African Americans, the rights extended to Latinos.

Among the several organizations created during the 1960s and 1970s that were committed to expanding the legal rights of Latinos, the Mexican

American Legal Defense and Education Fund (MALDEF) was the most prominent. Mexican American lawyer Pete Tijerina is credited with founding MALDEF, but it was the influence of the NAACP that led to the financial backing that made the organization viable. Responding to a funding request in 1968, the Ford Foundation awarded the "sum of $2.2 million to be spent over five years on civil rights legal work for Mexican Americans; $250,000 of the grant should go for scholarships to Chicano legal students."[13] With the necessary financial resources, MALDEF began to combat discrimination against Mexicans in the Southwest through litigation.

In the 1970s MALDEF was involved in key education lawsuits. A groundbreaking case in the civil rights history of Mexican Americans was *Cisneros v. Corpus Christi Independent School District* (1970). For years, Mexican American school desegregation advocates had charged that separate schools for Mexican children violated the equal protection clause of the Fourteenth Amendment by arguing that Mexicans were "white," and thus not subject to segregation clauses for blacks or Indians. This tactic was reversed in *Cisneros*. The plaintiffs now desired that Mexicans be classified as a separate minority group that could benefit from rulings such as *Brown v. Board of Education* (1954). State Judge Owen Cox ruled in their favor, stating that "on the basis of their physical characteristics, their Spanish language, their Catholic religion, their distinct culture, and their Spanish surnames, Mexican Americans were an identifiable ethnic minority group for desegregation purposes."[14]

Because *Cisneros* was ruled only at the federal district court level it could not be broadly applied. Thus, as Guadalupe San Miguel, Jr., chronicles in his work, *Brown Not White: School Integration and the Chicano Movement in Houston*, school districts in cities like Houston continued to classify Mexican American students as "white" to achieve desegregation of black schools.[15] MALDEF was instrumental in pushing the federal court system to classify Mexican American as a legally distinct group. In *Keyes v. School District Number One, Denver, Colorado* (1973) the U.S. Supreme court recognized Mexican Americans as having the constitutional right to be recognized as an identifiable minority. After *Cisneros*, MALDEF shifted its resources to other measures, chiefly establishing bilingual education. Since Latino children were no longer officially being sent to separate "Mexican" schools, MALDEF sought to ensure equal educational opportunities within desegregated settings.

Many of the demands of Latinos were met through legislative and judicial measures passed by the U.S. government in the 1960s and 1970s. As part of President Lyndon B. Johnson's War on Poverty, the U.S. government passed the Elementary and Secondary Education Act (ESEA) of 1965. ESEA contained several programs beneficial to Latinos, including migrant education, English for adult learners, and early childhood education.[16] Furthermore, ESEA was amended in late 1967 to create the Bilingual Education Act. Section 702 of the act stated "in recognition of the special educational needs of the large numbers of children of limited English-speaking ability . . . Congress hereby declares it to be the policy of the United States . . . to meet these special educational needs." The government thus authorized millions of dollars to create special programs for language education in local school districts across the nation.[17] (See document 7.4.) The Bilingual Education Act of 1968 was further reinforced through the judicial branch in the U.S. Supreme Court case *Lau v. Nichols* (1974) and legislatively in the amended Bilingual Education Act of 1974. In *Lau* the U.S. Supreme Court ruled that the civil rights (but not constitutional rights) of non-English speaking children were being violated in schools that did not address their linguistic/educational needs.[18] According to legal scholars Herbert Teitelbaum and Richard J. Hiller, the ruling in *Lau* "raised the nation's consciousness of the need for bilingual education, encouraged additional federal legislation, energized federal enforcement efforts . . . aided the passage of state laws mandating bilingual education, and spawned more lawsuits."[19] (See document 7.4.)

Even before the *Lau* ruling, the Office of Civil Rights (OCR) had viewed Title VI of the Civil Rights Act as a measure protecting language minority children. Title VI specified "No person in the United States shall, on the ground of race, color, or national origin be excluded from participation in, be denied the benefits of, or be subjected to discrimination under any program or activity receiving federal financial assistance." Since public school districts received substantial federal funding by 1970, the OCR required school districts to file compliance plans showing that language minority children were not being excluded from any programs nor being tracked into programs that "operate as an educational dead-end or permanent track."[20] *Lau v. Nichols* provided the legal reasoning (and federal enforcement) that the rights of non-English-speaking students under the equal protection clause of the Fourteenth Amendment were indeed violated when public schools failed to provide adequate programs to facilitate bilingual education or English language acquisition. The rulings in *Lau* and the subsequent guidelines—called the "Lau Remedies"—did not definitively end the public controversy over government policies for language instruction in public schools—an issue that has beleaguered the long history of American education. Rather, contentious debates in academia, between politicians, and among the citizenry over the benefits and disadvantages of bilingual education and best practices in English language acquisition characterized the 1970s and continued into the 1980s and 1990s.[21]

K-12 EDUCATION AND THE PUERTO RICAN DIASPORA

The Puerto Rican child and the Negro children share many humiliations, not the least of which is a system of even-handed injustice dispensed by big-city school administrations throughout the North. But the Puerto Rican child carried additional burdens all his own, his status as a stranger in our midst being perhaps the heaviest, and these have been sufficient to keep him at the very bottom of the educational pyramid.

The observer can but dimly discern the everyday frustrations which many Puerto Rican school children have come to take for granted: their imperfect grasp of English, which often seals both their lips and their minds; their confusion about who they are (what race? what culture?), a confusion compounded by the common ravages of white prejudice; their sense of being lost, or traveling through a foreign country with a heedless guide and an undecipherable map.

—*Richard J. Margolis, The Losers: A Report on Puerto Ricans and the Public Schools, 1968*[22]

By the 1960s Puerto Ricans were established in barrios not only in New York City but in other northeastern cities such as Hartford, Connecticut, and Boston. Furthermore, they settled in the midwestern cities of Chicago, Milwaukee, and elsewhere, seeking jobs in the industrial labor market. No longer first-generation by the 1960s, many Puerto Ricans viewed themselves as second-generation mainlanders and were frustrated with the limited opportunities available to them as American citizens, particularly in the areas of housing, employment, and education.

Similar to that of Mexican Americans, the educational status of Puerto Ricans on the eve of the Civil Rights Movement remained low compared to Anglos. In 1960, only 13 percent of adult Puerto Ricans had finished high school and over one-half had less than eight grades of schooling. The pipeline to college was effectively blocked at the high school level. In 1963 a modest 331 (1.6 percent) of the 21,000 academic diplomas granted to New York City teenagers were granted to Puerto Ricans. Of those 331 students, only 28 pursued higher education. Typically, only academic diplomas permitted high school graduates to continue on to higher education. In contrast, the majority of New York Puerto Ricans who attended high school were clustered in the vocational diploma track. Puerto Rican youth represented 23 percent of all youth in vocational high schools and 29 percent of those attending "special" schools.[23]

Inferior educational conditions and the slow response to desegregation mandated in *Brown v. Board of Education* (1954) united African Americans and Puerto Ricans against the New York City public schools. Boycotts and strikes in the late 1950s and early 1960s brought leaders from both communities together.[24] Although African American and Puerto Rican parents came together on issues of mutual concern (crowded schools, poor facilities, underqualified teachers, and lack of parental input), their priorities for equal opportunity diverged. The Puerto Rican community emphasized bilingual education and matters of cultural sensitivity. Puerto Ricans also believed (as Mexicans had in Houston, Texas) that white school officials were wrongfully utilizing Puerto Rican children to satisfy federal racial integration orders.[25]

The post–World War II Puerto Rican leadership shifted strategies in the late 1950s and early 1960s. The immigrant assimilationist frame of thinking was discarded in favor of a nationalist separatist identity as mainland ethnic minorities.[26] For example, the U.S. government's designated agency to help islanders adapt to mainstream American culture and language, the Migrant

Division, decreased in favor among Puerto Ricans. The new generation rejected these impositionist efforts and formed grassroots organizations focused on issues of particular concern to their everyday needs—chiefly, poverty, schools, housing, and civic participation.[27]

The pre–World War II Puerto Rican community had created numerous social clubs and advocacy groups.[28] The make-up and missions of organizations formed in the 1960s reflected concerns of a new generation no longer willing to quietly assimilate to American culture and language or to socialize with hometown clubs from the island.[29] Community groups such as the Puerto Rican–Hispanic Leadership Fund (1957); ASPIRA (from the Spanish word for "to aspire"), founded in 1961 by educator Antonia Pantoja; and the Puerto Rican Community Development Project (1964) provided forums for Puerto Ricans seeking social justice and change.

Organizations founded in the 1960s and 1970s utilized diverse approaches to secure reforms. For example, the United Bronx Parents, Inc. (UBP), founded in 1965 by Dr. Evelina Antonetty, focused on empowering parents to improve the public schools. The grassroots organization viewed its mission as "targeting and correcting the unresponsiveness of New York City public schools" to the needs of Puerto Rican children.[30] As part of UBP's focus on training parents, they received a Department of Health, Education and Welfare (HEW) grant to create a Parent Leadership Training Program. Other groups, such as Casa de Puerto Rico, Inc.—established in Hartford, Connecticut in 1967—and the Puerto Rican Legal Defense and Education Fund (1972), targeted community participation or legal avenues to bring about change.[31] The most militant group was the Young Lords, founded in 1969 in Chicago. With a strong base in both Chicago and New York, the Young Lords tapped into the energy of Puerto Rican youth. After 1972, the Young Lords became the Puerto Rican Revolutionary Workers organization.[32]

ASPIRA and the Puerto Rican Legal Defense and Education Fund were instrumental in expanding the rights of Puerto Rican children.[33] As a result of activism in the community and the creation of federal programs with funding, bilingual education was implemented in New York City's Ocean-Hill Brownsville district and at P.S. 25, the "first bilingual school in the northeast, established in 1968."[34] ASPIRA also sponsored an important conference in 1968 that examined the educational levels of both Puerto Ricans and Mexican Americans. For this conference, ASPIRA commissioned social scientist Richard J. Margolis to investigate the conditions of Puerto Rican children in public schools around the country. Entitled *The Losers: A Report on Puerto Ricans and the Public Schools*, the study provided damning evidence of how public schools in the Northeast and Midwest sabotaged the Puerto Rican children's opportunities for equitable schooling. One particularly disturbing finding of Margolis's report was that as a result of uninformed and unsympathetic teachers and administrators and low expectations, "the longer Puerto Rican children attended public school in the United States, the less they learned."[35] Student attendance rates decreased, they were sidetracked to vocational programs, and discouraged from academic work leading to college.

In response to such reports, ASPIRA and the Puerto Rican Legal Defense Education Fund sued the New York City Board of Education with the hopes of instating bilingual education. The lawsuit was finally resolved in August 1974. A significant victory, the ASPIRA Consent Decree of 1974 mandated bilingual education in the New York City public schools.[36] The efforts of both Puerto Rican and Mexican American community organizations to improve K-12 public education contributed to the development of a new Latino presence and leadership on college campuses.

EL MOVIMIENTO IN HIGHER EDUCATION, 1960–1980

It is a fact that the Chicano has not often enough written his own history, his own anthropology, his own sociology, his own literature. He must do this if he is to survive as a cultural entity in this melting pot society which seeks to dilute varied cultures into a gray upon gray pseudo-culture of technology and materialism. The Chicano student is doing most of the work in the establishment of study programs, centers, curriculum development, and entrance programs to get more Chicanos into college. This is good and must continue, but students must be careful not to be co-opted in their fervor for establishing relevance on the campus. Much of what is being offered by college systems and administrators is too little too late.

—*Manifesto of El Plan de Santa Barbara, 1969*[37]

Reckard [Chaplain, Claremont Colleges] commented on the dilemna [sic] all of us have in dealing with the Mexican-American subculture; for the most part his group is saying that they would prefer their traditional culture, the maintenance of strong family ties, etc. than it would to have their young people become doctors and lawyers. In spite of this dilemma, one knows that the seams of the old culture are cracking or will crack and that in spite of the difficulties it would be worthwhile to move bright young individuals, who hopefully can be leaders in the future, into the main stream of American higher education.

—*Rockefeller Foundation Report on proposed grant to the Claremont Colleges, 1967*[38]

The Latino pioneers of the pre-1960 era demonstrated that Latinos could succeed in college if they were granted access. Furthermore, the first generation of Latino faculty appearing on campuses served as role models. Civil rights activism, Vietnam War protests, and various expressions of anti-authoritarianism converged on college campuses in the 1960s and 1970s. This era brought unprecedented numbers of Latinos into academe and clearly represents a watershed in the history of higher education. No longer isolated individuals and small groups on campus, the new Puerto Rican and Chicano college students across the nation organized to demand their rights.

The parallel Chicano and Puerto Rican activism movements emerged with a concrete agenda: to increase the attendance and success of Latino youth in college. These radicalized youth demanded a broad range of services to aid access and retention for Chicano and Puerto Rican students. Often connected to organizations that pushed for reform in K-12 schools and community improvements, the Latino student youth movement nonetheless had its own leadership and momentum. The histories of the Chicano (Mexican American), and Boricuan (Puerto Rican) movements in the 1960s will be examined separately as regional and sociohistoric contexts varied these experiences.

The early 1960s were a period of identity development for many Mexican Americans as they looked for inspiration from their indigenous roots rather than following pathways of assimilation. As the black civil rights movement for access to white colleges and universities gained visibility in the national media, Latino students examined their own status and numbers in the academy.[39] Available data suggest that on the eve of the turbulent 1960s Latino students were still on the margins of academe. Valdez noted that in 1960 Michigan, 40 percent of all 17–24 year olds in the state were attending some form of postsecondary institution, compared to only 7 percent of its Mexican population.[40] In Houston, only 3 percent of Mexican American adults older than 25 had completed four years of college that same year.[41] Researcher Herschel T. Manuel found in 1958 that only 5.7 percent of freshmen in 146 Southwestern colleges had Hispanic surnames.[42] By the mid-1960s however, numbers were increasing. Finding strength in numbers, Latinos organized el movimiento.

One of the issues that engaged students was the small number of Latino faculty. Muñoz pointed out that at the beginning of the movimiento's heyday (1968–1973), "fewer than one hundred scholars of Mexican descent held doctorates in the US. Of these, most held doctorates in education (Ed.D.s), which located them in a distinctly different research network with a very different emphasis from those scholars holding a Doctor of Philosophy."[43] The low number of faculty who might be interested in teaching or researching Latino issues was a principal issue addressed in the youth movement's conferences and lists of demands.

Several simultaneous external events also sparked the creation of a militant Chicano movimiento that spread through the Southwest. The national attention given to César Chávez and the farmworkers' movement, the politicization of Latinos through Viva Kennedy clubs, and a growing desire to take advantage of federal War on Poverty programs and monies from private foundations contributed to the formation of student groups. The fall of 1967 witnessed the birth of several Mexican American student organizations. The Mexican American Youth Organization (MAYO) began at St. Mary's College in San Antonio, Texas and then at the University of Texas at Austin. In Los Angeles the United Mexican American Students (UMAS) formed several chapters in area institutions including UCLA and Loyola. At the large East Los Angeles Community College, the Mexican American Student Association (MASA) was formed.[44] Frustrated with the lack of attention to Latino demands for additional faculty and more relevant courses, the students began a series of conferences and protests to change the academy. José Angel Gutiérrez's (1944-) work for this cause through MAYO and PASO (Political Association of Spanish-Speaking Organizations) illuminates the risks such activism posed to individuals wishing to change the status quo. (See document 7.1.)

The Chicano movement received inspiration and training from African American organizations, particularly the Black Panther Party and the Student Non-Violent Coordinating Committee (SNCC).[45] However, campus dynamics between Chicano and black protest organizations were often tenuous. Both Muñoz and Olivas note that Chicanos often felt slighted by colleges' emphasis on the recruitment of black students. As a result "bitter and intense conflicts between Mexican Americans and Blacks on several campuses, [made] viable coalition politics difficult, if not altogether impossible."[46]

The first Latino protest activity on a college campus occurred at San José State College (CA) in 1968. Approximately two hundred graduating seniors and members of the audience walked out of commencement exercises to protest the under-representation of Chicano students and lack of bilingual and cultural training for professionals (teachers, social workers, and policemen) who worked in Latino communities. The San José State walkout was the beginning of a series of events that changed the image of Mexican Americans from a "passive" group to that of a visible, militant group. Dolores Delgado Bernal's study of the East Los Angeles blowouts also illustrates the significant leadership roles of Chicanas in the movement.[47]

Among the many conferences and strikes that students led, the University of California–Santa Barbara conference of 1969 holds the most significance for Latino higher educational history. Emerging from the conference was El Plan de Santa Barbara; the plan presented the clearest and most detailed articulation of the demands of Latino college youth. Although El Plan Spiritual de Aztlan, created at the National Chicano Youth Liberation Conference in Denver, March 1969, was a pivotal "magna carta" of the general Chicano movement, El Plan de Santa Barbara focused specifically on higher education.

The Chicano Coordinating Council on Higher Education, a coalition of students, faculty and staff from California institutions of higher education, organized the Santa Barbara conference to develop a unified platform for higher education reform in the areas of Chicano studies programs, access, and retention. The conference resulted in the call for individual student groups to forego their original names and unite to become the Movimiento Estudiantil Chicano de Aztlan (MEChA). Many groups joined MEChA, but others retained their individuality. The manifesto "El Plan de Santa Barbara" focused on three key areas. First, it emphasized the obligation of college and university Chicanos to maintain ties with the barrio community. Second, it stressed the importance of changes to institutions of higher education that could open their accessibility to Chicanos. The hiring of Chicano faculty, administrators, and staff was viewed as a key step in achieving this objective.

Lastly, the Santa Barbara plan called for the alteration of traditional European white interpretations of history, literature, and culture to incorporate Third World viewpoints and particularly Chicano perspectives.[48] El Plan de Santa Barbara served as a model that was circulated throughout the nation as Chicanos and Puerto Rican students not only entered the academy in the 1960s and 1970s in significant numbers but also worked to change the academy itself. (See document 7.6.)

A significant aspect of higher educational history during the 1960s and 1970s concerns the impact of both federally and privately funded programs to increase access and retention of minority students at colleges and universities nationwide. Although little has been written on these programs, the Rockefeller Foundation, the Carnegie Corporation, and the Danforth Foundation, to name a few, contributed millions of dollars to support Latino educational success. Given optimistic names such as "Project Open Future," "Upward Bound," and "Horizons Unlimited," these programs also received support from the U.S. Office of Economic Opportunity as part of the War on Poverty. For instance, in 1968, after a year of discussion and debate, the Rockefeller Foundation granted the Claremont Colleges in California $650,000 to create a "Program of Special Directed Studies for Mexican-American and other economically and socially disadvantaged students."[49] Although some administrators were initially skeptical of the abilities of Mexican American, African American, and other socially disadvantaged youth, the five-year program significantly increased the numbers of such students at a cluster of colleges that had been elite enclaves for decades. (See documents 7.7. and 7.8.)

While Chicano students on the West Coast and in the Southwest forged a largely Chicano identity on campuses, Puerto Rican students in the urban Northeast and Midwest also joined with community-based groups or created their own organizations. The population of Puertorriqueños swelled in the mid-twentieth century. In New York, the Puerto Rican population rose tenfold from 70,000 in 1940 to 720,000 in 1961. The work of ASPIRA to establish bilingual and bicultural educational programs in New York City set the groundwork for future militancy. Founded by educator and leader Antonia Pantoja in 1961, ASPIRA focused on raising educational levels and preparing Puerto Rican youth for leadership. At the time of its founding, few Puerto Ricans were graduating from high school and among those who did graduate few received academic diplomas; instead, most received vocational high school diplomas that limited options for higher education.[50] The numbers of Puerto Rican students in the city colleges were correspondingly low. Traub found that in 1964, City College, although located in Harlem, did not enroll more than 2 percent of its students from the black community and enrolled "a much smaller number of Puerto Ricans."[51]

The Puerto Rican youth movement did not emerge from a vacuum; it was built upon decades of Puerto Rican American associations that had raised awareness of deficiencies in the community and local governments' response to Puerto Rican poverty, discrimination, and lack of bilingual assistance.[52] Similar to Mexican American youth who shifted their identities to embrace both the spiritual homeland and signify a new activist stance, Puerto Ricans used "Boricua" as a term symbolic of their ideologies and goals. The name "Boricua" appears in the titles of numerous 1960s student organizations and Boricua College was founded in New York City in 1973, along with the Universidad de Boricua (a university without walls founded by Antonia Pantoja). "Boricua," explained Santiago, stands for the word "Borinquen" or "Land of the Brave Lord," which the Arawak Indians called the island prior to Columbus's arrival.[53] In contemporary times, "Boricua" is a term of endearment used among the Puerto Rican community. In the 1960s and 1970s, "Boricua" served as a rallying cry for activist youth.

The radicalization of Puerto Rican youth is attributed to the small but significant numbers of Puerto Rican college students who benefited from early 1960s programs such as City University of New York's (CUNY's) Search for Education, Elevation, and Knowledge (SEEK), College Discovery, and SUNY Educational Opportunity Programs (EOP).[54] These pioneer students created several organizations, including the Puerto Rican Student Movement (PRSM) for ASPIRA members in the CUNY system. In addition, Puerto Ricans for Education Progress (PREP) was a network of private college students who worked to get more students into prestigious private schools such as Princeton and Yale.[55] Although these early organizations were not long lasting, they provided a fundamental leadership for the more radical movement that followed.

At Yale College for example, Puerto Rican students organized into a small group called Boricuas Unidos which pushed for more Latino enrollment. Utilizing data gathered from college records, sending open letters to the president of the college, and through other means of activism, Puerto Rican students were determined to increase their numbers and clout.[56] (See document 7.9.)

One strategy Boricuas Unidos pursued was encouraging Yale recruiters to target New York City area Catholic high schools. As a result of these initiatives, ASPIRA compiled information on Catholic high schools in order to determine the level of cooperation with college recruiters. For example, at the girls' school Mother Cabrini High School it was recorded: "Sister Patricia is very interested in her students. She is aware of ASPIRA, and, if the high school does not have an ASPIRA club, many of the Puerto Rican students participate in [the New York area] ASPIRA. The school is top-notch academically speaking and draws to it many Latinas. Sister Patricia is ready to meet anyone whom she feels will help her students."[57] The pivotal turning point in the Puerto Rican student movement was the successful fight for open admissions at City College, generally considered the flagship campus of the CUNY system. In 1969 conditions were ripe for a showdown between minority students and administrators. Black and Puerto Rican students organized a strike and campus shutdown that is credited for pressuring administrators to enact the Open Admissions policy. The controversial Open Admissions policy at CUNY resulted in a flood of Puerto Rican and black students. By 1975 Puerto Rican undergraduates had increased to 18,570 or

8.3 percent of the CUNY system student population. In 1969 Puerto Rican students had only numbered 5,425 or 4 percent.[58] Similarly, at Brooklyn College students from the Black League of Afro-American Collegians (BLAC) and the Puerto Rican Alliance (PRA) occupied the president's office and issued a series of demands. Throughout the greater New York area during the tumultuous years of 1969 and 1970, public and private colleges experienced demonstrations and strikes as underrepresented groups fought their way into college.[59]

In addition to increasing college enrollment for Puerto Rican, then Dominican, and other Latino groups in the greater New York area during the 1970s and beyond, the movement created Puerto Rican research centers critical to the rewriting of Puerto Rican history and a renewed interest in Puerto Rican culture. In 1971, after two years of battles, CCNY finally approved the creation of a Department of Puerto Rican Studies. Dr. Frank Bonilla of Stanford was hired to assist in the design of the program. Furthermore, CCNY agreed to open an office of Puerto Rican development to increase Latino admissions and bolster retention rates. At Hunter College, the Puerto Rican Student Union (PRSU) facilitated the creation of the Centro for Estudios Puertorriqueños (Center for Puerto Rican Studies) in 1972, an archival and scholarly research center with a continuing presence in the research community.

Students at the Chicago Circle campus of the University of Illinois (now UIC) successfully fought for the inclusion of a Latino curriculum and assisted in the creation of the Rafael Cintron-Ortiz Latino Cultural Center. Similarly, the Center for Chicano-Boricua Studies at Wayne University in Detroit was founded in 1971 and continues its presence as a midwestern research center.[60] The peak of the Puerto Rican student movement occurred between 1969 and 1973. After 1973 many student organizations declined in numbers and militancy. The end of the open admissions experiment at CUNY in 1975 affected Puerto Rican youth. The combination of new tuition charges—the first ever in the institution's history—and the general economic recession of the 1970s led to a decline in the steady Puerto Rican enrollment growth of the early 1970s.[61]

Throughout the country, the student youth movement began to wane as government programs, private foundations, and institutions implemented many of the students' demands. Affirmative action initiatives among many institutions of higher education in the 1970s brought Latinos into colleges, many who are now faculty members.[62] In the long continuum of Latino higher educational history, the 1960s and 1970s was a permanent watershed, although Latinos remained clustered overwhelmingly in public two-year institutions. Although Mexican Americans and Puerto Ricans had not been barred from white colleges and universities on account of racial policies, their participation prior to the mid 1960s was negligible. The Latino student movement introduced the possibility of not only access to white institutions but the creation of uniquely Hispanic colleges. The U.S. government's growing acknowledgement in the 1970s that African Americans were not the only "minority" significantly impacted subsequent state and federal court rulings and policies. President Richard Nixon's O.M.B. Statistical Directive 15 of 1973 finally acknowledged Hispanics as a separate federally identified group. As the controversial Richard Rodriguez wrote in *Brown, The Last Discovery of America* (2002), the result of this federal directive was that "several million Americans were baptized Hispanic."[63] The intervention of the federal government in shaping policies toward Hispanic higher education and political dynamics accompanying the ever-increasing Latino immigration are explored in the following chapter.

DOCUMENT 7.1

A Mexican American Youth Coming of Age during Civil Rights

Born in 1944 in Crystal City, Zavala County, Texas, José Angel Gutiérrez and his crusade for justice epitomized the work of the new Chicano generation of the 1960s and 1970s. Similar to many first-generation Latino college students, Gutiérrez found the pathway to higher education through a community college, Southwest State in Uvalde, Texas, and then received his B.A. at Texas A and I University in Kingsville. He continued his graduate studies at St. Mary's University in San Antonio. While in college he founded the Mexican American Youth Organization (MAYO) and later created the political party La Raza Unida Party (LRUP).

In this excerpt Gutiérrez describes the community censure he faced as a result of his political activities, even as a young man. Within his high school and community college we learn about the unwritten codes governing social relations between Mexican boys and Anglo girls.

From: José Angel Gutiérrez, "The Education of a Chicano," chapter 3 in his *The Making of a Chicano Militant: Lessons from Cristal* (Madison: University of Wisconsin Press, 1998). Reprinted by permission of The University of Wisconsin Press.

I graduated from high school in May, 1962, just as PASO was beginning to organize in Cristal, and there was talk of Los Cinco Candidatos for the 1963 election. I welcomed the possibility of social change that PASO seemed to be trying to bring about because I was frustrated with my inability to understand why there were two worlds during the school day, one of privilege for Anglos and another of exclusion and discrimination for Chicanos. My feelings were jumbled with regard to who *could* be my friends, who *should* be my friends, and who *were* my friends after school while in the barrio. As a Chicano I felt proud that we as a community would rise and take from the Anglos what had been rightfully ours to begin with, our self-respect and equality. My thoughts about these concepts had evolved during my high school years. I became the declamation champion for medium size schools in the state of Texas in 1961. No prior champion had been a Mexican American. In track and field events, another classmate, Richard Gallegos, also won the mile run state championship in 1961, another first for Mexican Americans in our school and in the entire state of Texas. At a school assembly where all athletes and interscholastic league forensic winners for the year who had "lettered" were given their varsity jackets, Richard and I, among others, were honored by the administration and student body. His trophy and my plaque for those honors were not displayed in the awards case of our high school in any prominent fashion; however, our awards were placed away in the lower rear shelf of the case, nearly hidden. There have been no other state champions for Cristal in either of these two categories since.

My winning state championship speech, ironically, was on the meaning of democracy. For demonstration I used a copper penny. With the coin I would elaborate on the contributions made by President Lincoln and on the meaning of the engraved words, *E Pluribus Unum*. The speech was very moving and emotional, and I bowled the audience over with both the message and my delivery. Here was this young Chicano kid from some dinky little school in South Texas who spoke English without an accent, was very poised and natural, and declaimed better than Anglos. The use of the penny was also most effective: I had studied the coin and found a unique feature not found in other coins. The profile of the personality on the nickel, dime, and quarter all face left, while Lincoln's profile on the penny looks toward the right. I found symbolism in this pattern: only Lincoln looked forward into the future of the nation as one people while the others turned away from that reality. Everyone could relate to the coin and speech. While today, you cannot buy anything for a penny, it was the most frequently used coin by the poor during my youth.

WHITE GIRLS AND SPEECH

I actually got interested in public speaking because of my budding interest in girls. As I passed the auditorium in the high school during my freshman year one afternoon, practice for interscholastic league activities was underway. On stage reciting were some of the school's prettiest white girls and only a white boy or two. I gawked at the white girls through an open window until the speech teacher, Aston Pegues, yelled at me. "Keep moving or come in and recite. You're a distraction to the girls." I am not one to play the coward before authority, then or now, so I walked into the auditorium and up to him. Mr. Pegues was a slight, small man with a big greasy nose and thinning hair who was effeminate in his gestures. He gave me the once over and pushed a mimeographed sheet into my hand, saying, "Read it. Be ready to read aloud in a few minutes. We're practicing for interscholastic league forensic competition." I didn't comprehend the big words he uttered but did begin reading the words on the sheet. And then I did exactly what he told me to do, I read it aloud. He loved my voice projection, poise, and delivery, and said so. (Little did Aston Pegues know that I had learned my oratory in Spanish in kindergarten with Suse Salazar, the barrio teacher my family had sent me to when I was five years old, and who had trained me very well.) I knew I was going to be in "speech" from then on. It was easy, and I was right there next to the prettiest white girls in the school—all of them.

This was a turning point for me because Chicanos did not try out for speech as an extracurricular activity. Chicanos took the speech class in high school because it was an easy elective course, but we didn't take this stuff seriously, much less try out for competition. For the Chicano boys, however, the speech class gave us access to the Anglo girls, some of whom liked to be with Chicano boys after dark.

In my time Chicanos didn't read poems, give speeches, partner in a debate team or act in plays. Chicanos also didn't hang around with Anglos, and Anglos avoided Chicanos as study partners, social friends, and dates. One's own Chicano group of friends would also apply severe peer pressure not to become white-acting, or agavachado. The only one who I knew of who did become agavachado by always hanging out with Anglos and entering the speech competitions was a boy named Hector, who was a year or two ahead of me. Chicanos called him *joto* (queer) because of his effeminate ways, and because he hung around during school hours with two other Anglo classmates of his who were considered "queer" looking. He also became our first Chicano drum major, which was a big no-no for any male Chicano. In our teenage view of the world, girls, not boys, became twirlers and drum majors!

I had tried out for the football team my freshman year but the coach used the big Chicano kids like me as practice dummies for the scrawny, puny Anglo kids. The coach would place us to be tackled by them over and over. We never got to practice tackling them. Chicanos were placed mostly on the B team, not the varsity A team. All the Anglo punies, however, made the varsity squad, along with a few Chicanos. The average Chicano player like me never seemed good enough to make the varsity or traveling team, though we were always good enough to take hits from the Anglos every afternoon. Given the alternative of the Anglo girls in the auditorium for speech practice, I lost interest in football very quickly.

Risking all of the accusations and ridicule that was sure to come from my Chicano peers, I gave up football and tried out for the forensics team. After brief practice sessions of poetry reading and extemporaneous declamation, I was good enough to travel with the group to out-of-town competitions.

Most of the time, I would bring home the first- or second-place ribbons; I was getting really good at public speaking. I was often given a motel room to myself on these trips. I thought the special treatment was because I was a winner, a "star"; little did I know that my speech teacher was only following the local rules against Anglos mixing with Mexicans, let alone sleeping in the same room together. The good side of this segregation eventually was that I could have female company—from either our group or from a rival team— during the night. No one knew of these escapades except me and my female companion.

SPEECHES ABOUT THE CONDITIONS OF LA RAZA

The next year at the first Chicano political rallies at La Placita for Los Cinco Candidatos, I honed my speaking ability not in English but in Spanish.

I would rail against the discrimination in Cristal that was obvious to me. Chicano barrios had unpaved and unlit streets, no sidewalks, no sewers, and no parks. My block on West Edwards between West Zavala and North Avenue A was the only one in my barrio with pavement, sidewalks, and a street light, and only because my father and the Galvan family across the street had paid for these improvements. The De Zavala elementary school, like the county, was named after the first Vice President of the Republic of Texas, Lorenzo de Zavala. The school was across the street from my house and occupied an entire block of its own. That block had street lights on two sides only, but no sidewalks. The playground, where the Martinez boys and I played baseball, had no grass. What it *did* have was broken glass, plain dirt, rocks, and a rickety backstop made of chicken wire mesh nailed to a wood (two by four frame). This was our "athletic field." The classroom buildings were old Army barracks, except for one that had been removed from a Japanese internment camp that had been located in Cristal from 1941 to 1947. It had held mostly Japanese Americans forcibly brought from California and the West Coast, and a few German and Italian prisoners of war, and was located over in another barrio, *el Avispero en el Campo*, the "hornet's nest by the camp." I don't know why Chicanos called the neighborhood that. Sometimes the camp area was also called "*el* airport" because the municipal airport was adjacent to it. The Army barracks near the airport location had also been made into a junior high school for "migrants," meaning Mexicans.

The Anglo elementary school, on the other hand, had lights on each corner, sidewalks, green grass, real backstops and a baseball diamond, and extensive playground equipment such as merry-go-rounds, seesaws, monkey bars, and many sets of swings. I don't think it had a name other than "the grammar school"—at least that is the name that Chicanos gave it—and it was nothing like the Mexican schools. This school was a two-story monument made of red brick with two gleaming metal slides for fire escapes attached from the top of the second floor. My neighborhood buddies and I would sneak over without permission from our parents to that school at night because we loved to climb up the slides and come down. The school authorities tried to keep us out with their "KEEP OFF" signs, chains across the top and bottom of the slides, and even with lights shining on the tops of the slide platforms. But to no avail: this was our Chicano amusement park at night.

The dilapidated conditions of the schools, streets, and La Placita; the social distance between whites and browns in Crystal City, specifically, and South Texas generally; the presence of the Texas Rangers and their intimidation tactics; and the increased wages and price per piece work offered to Chicano labor during election times were the themes in my early political speeches. The democracy I spoke of in my state championship speech with the penny as the symbol of all that was good and right with my country was long forgotten because of its irrelevance to Chicano reality. My Chicano friends and I would always mock the ending of the Pledge of Allegiance, reciting " . . . with liberty and justice for some." Democracy didn't exist for Mexicans in Cristal, only for gringos. This is so because the *gringos* thought democracy was reserved only for white people, that they "fought" for it at the Alamo, and they would tell you so in school. I said this repeatedly in public at La Placita. I would also admonish la raza who would rather earn money than pay money for the poll tax in order to go vote. Then, I did not understand why Chicanos turned away from the poll tax. Later on I learned that Chicanos had very limited choices on the ballot: with the poll tax and voter registration certificate, the only choice was still to vote for or against the only white Democrat on each ballot for all local, and most statewide, contests.

. . . A small group of us enrolled at the local two-year college, Southwest Texas Junior College, some forty miles north in Uvalde, Texas. Juan Patlan from Carrizo Springs, and Hector Alva, Francisco Rodriguez, Mike Delgado, Rudy Palomo, Esequiel Romero, and Eddie Avila from Cristal were among those who went to junior college with me . . .

JUNIOR COLLEGE

Doing political work at the community college in Uvalde was much harder than in Cristal. The story of my kidnapping [by local officials because of his outspoken political views] was spread by my friends and I was looked at with awe by other Chicano students. The area Chicano students attending this commuter school were mostly reluctant to put up a challenge, let alone a fight. The odds seemed impossible: the whites outnumbered us by about four to one. I soon learned that the Chicano students on the campus were easily divided and controlled by contacts with their former, hometown Anglo classmates. The Anglos knew each other, even across the region, because their families socialized together. We, as Chicanos, did not know each other from one town to the next. The Anglos have always been in solidarity with each other because they have lived with a siege mentality among us forever, since their ancestors took our land. Anglos were a minority in most of the South Texas towns, so they had to stick together to keep us under control. And to maximize their strength they had to keep us divided. I learned early that when divided, the many become few.

Other Chicanos viewed those of us from Cristal as being too radical. They believed all of the negative press coverage of PASO, the Teamsters, Los Cinco Candidatos, and the general Chicano struggle in Crystal City. Because of this persistent and extremely negative press, most Chicanos avoided affiliation with Cristal. An Anglo had only to ask of a Chicano, "Do you want another Crystal City here?" to get a resounding "No!" Chicanos actually learned to become ashamed of us having taken power in Cristal, and they felt compelled to apologize for that exercise in democracy. Their colonial mentality was more pervasive than I wanted to admit at that time.

The battle of the Alamo of Mexican-era San Antonio is negatively treated in the press and in both high school and college courses in much the same way. So much so that to this very day, most Chicanos will shun identification with General Antonio Lopez de Santa Ana and the Mexican troops in that battle.

The extremely sad irony is that Chicanos *did* win in Cristal, as their ancestors had won at the Alamo. And we were *entitled* to win.

At the junior college the relations between Chicanos and Anglos were not much better than they had been in my earlier years. I had decided to seek the office of student body president, and met with a lot of resistance, to say the least. There were very few Mexicans attending the junior college—I dare say a little under two hundred—and most of them were terribly intimidated by the all-Anglo administration and teaching staff (except for the Spanish teacher, Dr. Roberto Galvan) and the overwhelmingly Anglo student body. Needless to say, I lost the election. Additionally, I was rebuked by the dean of students, Jerald Underwood for having run a very "anti-Anglo" campaign. This was another extremely sad irony that I encountered. Here I was championing a Chicano cause: open admissions, more Chicano faculty, more Chicano scholarship money, and the right to run for office. I knew I couldn't win—that was not the point! The educational campaign, the organizational effort, and the challenge to Anglo dominance were what drove me. Rather than interpret these actions as pro-Chicano, I was accused of being anti-Anglo. I refused then, as I do now, to accept that twisted, illogical, self-serving gringo trick, that our only option is to be pro-Anglo. To be pro-Chicano is *not* axiomatically or automatically to be anti-Anglo.

During the campaign I was called a communist, atheist, radical, and militant Chicano. These words were scrawled on my posters. Dean Underwood dismissed the name calling and defacing of my campaign posters. He said they were my own doing, a reaction to my negative campaign. I reminded him of how I was kept from campaigning in the dormitories by racist gringo students. He somehow did not recall my complaint that these *busca burros* (burro seekers), as we called the storefront cowboys, had chased me on one occasion in an attempt to beat me with their wide-strap belts and big buckles. The students who wore the cowboy garb back then were only the white kids; we Chicanos hated the cowboy look because it reminded us of the dress code of the *pinches rinches* and of most of the police. Today the *Tejano* look is to dress in that garb of an earlier time; we have a right to reclaim all of our cultural trappings, including this type of dress and country style.

  . . .

SOCIAL CONTACTS AND DATING

There were a few Anglo girls who favored Mexicans, but this had to be on the sly, just as it had in my high school. I saw that this is the way it was between Anglo girls and Chicano boys: you met after sundown and kept the relations top secret. I had learned by then that relations with white girls were taboo. The only difference at the junior college in Uvalde was that it was easier to conceal the relationships because Anglos only knew each other, not the Mexicans from other cities. And, unlike high school in Cristal, where people were known by the car they drove, a person's car at the commuter college was not recognizable, many cars were alike. The nearby hill country offered a private retreat; cabins could be rented in many spots within thirty miles of the campus. Even San Antonio, a very large city, was but an hour's drive away from Uvalde. During this era the legal speed limit was seventy miles per hour, and a person could zip over the miles quickly.

George Ozuna, the Cristal city manager, on occasions lent me his car to enjoy female companionship. Aside from a couple of Anglo girls, I had my eye on Ernestina Rodriguez, a beauty from Uvalde. She was the most beautiful Mexican American girl I had ever seen and she liked me. The problem was that she did have a local boyfriend and I had competition from another local boy, Rafael Tovar Jr. Hence it was not easy to drive off with Ernestina to Garner State Park, twenty miles away. It was easier to hit on the Anglo girls and trip out to San Antonio, or to any number of camping areas with cabins near Uvalde, any morning or afternoon in a borrowed car.

It was still, however, very dangerous for a Chicano to be seen alone with an Anglo girl in a car. Schoolmates of mine in high school found this out the hard way. They were almost killed by irate Anglo fathers and their friends because they were dating Anglo girls. The usual technique, then, was to get some of the gay Anglo guys to supposedly call on the Anglo girl for a date, take her out away from the home, and deliver her to the waiting Chicanos. The Chicanos reciprocated by getting someone to do favors for the gay Anglos. Some of these girls got pregnant by Chicanos and had abortions, which was unheard of at that time.

DOCUMENT 7.2

Hearing Before the United States Commission on Civil Rights

In response to the Los Angeles Watts Riots of 1965 and other civil disorders, the U.S. Commission on Civil Rights began a series of hearings around the country to investigate racial and ethnic tensions, particularly in urban areas. In this testimony, the concerns of Latino high school students are articulated: crowded schools, limited knowledge about college access, and a lack of bilingual education are emphasized as barriers to educational equity.

From: Hearing before the United States Commission on Civil Rights, hearings held in San Francisco, California May 1–3, 1967 and Oakland, California, May 4–6, 1967 (Washington: U.S. Government Printing Office, 1967), pp. 406–421.

TESTIMONY OF MISS PATRICIA DELGADO, SAN FRANCISCO, CALIFORNIA, AND MR. GABRIEL VICARIO, SAN FRANCISCO, CALIFORNIA

CHAIRMAN HANNAH. Now, Mr. Sandoval, will you introduce the new people? You can share the microphones. They [the microphones] are both alive so that each of you can take two. We regret, Mr. Sandoval, the necessity of interrupting your testimony. So we would first like to have you introduce the young people who have not been introduced.

MISS DELGADO. Gentlemen of the Commission, ladies and gentlemen of the Commission. I am here to tell you about some of the problems—

MR. GLICKSTEIN. Would you just identify yourself now? Give us your name and address and we will address questions to you a little later.

MISS DELGADO. I am Pat Delgado from 1723 Palou. I am a Horizons Unlimited graduate.

MR. GLICKSTEIN. Would the young man next to you please give us his name and address?

MR. VICARIO. Gabriel Vicario, 3140 Army Street, graduate of Horizons Unlimited for Mission District.

MR. GLICKSTEIN. Mr. Sandoval, would you tell us—

CHAIRMAN HANNAH. I think the young lady has a statement that she wants to give. Is that correct? The understanding was that we would let them do this the way they started to and then let's have the questioning. They have a program they have arranged.

MISS DELGADO. I am here to tell you about some of the problems, but first I would like all the students, all the fellow students from Horizons Unlimited that came with us, to stand. Will you please stand?

CHAIRMAN HANNAH. Members of the project? Very good. Welcome.

MISS DELGADO. They are all kind of mad today because we all spent all afternoon making signs and everything, and the police took them away. So they are kind of upset about it.

And, well, you called us here. You subpenaed [*sic*] us here to hear our problems, because I came here yesterday and I heard that you haven't heard our problems, you haven't seen our leaders and you don't know what our organizations are. Well, you were talking to some of the leaders last night and you are going to be talking to us today. We are leaders. Maybe we are young leaders, but we are leaders and we are going to be the future leaders, and we are tired of all of you older people trying to tell us, you know, what is good for us when who should know better but us? And I would like to say that we had a Negro boy speaking with us. He was supposed to be here on the committee, but you wanted all Latin people. But it is not all Latin people's problem. It is all minorities' problem. It is a lot of the Spanish-speaking's problem, but it has the same to do with the Negroes and the same problem. And if you want to see leaders, you see us and I could take you down to the ghettos and I could show you all the leaders of the gangs if you want to see leaders.

And if you want to hear riots and if you want to hear us making noise, you say you can't see our groups and everything, you want to see us making noise, we will have some riots for you and you can see the noise that we will make, because there are problems here.

Maybe you are looking on the outside and you are not on the inside where we are and you can't just tell by letting us talk for five or 10 minutes what our problems are. You want to hear facts, you want to hear our problems, and you want to hear all the proof of it. You didn't give us enough time to show you all the proof of it.

You come here for three days and you let us, the kids, who know the problems more because we have to live with them for the rest of our lives, you give us a half hour to talk with you, and we split it between us, and we got a few minutes. So I am going to try to tell you what some of the problems are.

O.K. You say our schools—we go to school and we all want to go to college. We want a good education and we set out for it. O.K. We try to go to school and we go to our homerooms and the first thing we do is stand and we say the pledge. The pledge says that this is a free country and everything. O.K. We go in our history class and we learn what a free country it is. Then we walk out of the school doors and it is the end of it. You go back to your crummy little house and try to get a crummy little job and all you know is that your accent is different from everybody else's and so you just can't make it.

Then the schools, if you come up from the Latin countries, they don't have any special classes for you. They will stick you in anywhere. How can you make it through if you don't understand what the teacher is talking about?

If you go into a math class and you don't know one word of English and she is trying to explain to you, what can you do about it? You just sit there and then they say the Latins don't know anything. Sure, we don't know anything in that language and you see the kids coming out of school and you will pull one out of them and you say, "What did you learn today in school?" And they will look at you like you're crazy. "What do you mean, what did I learn in school?"

Most of the Latins that I know that just came recently from their country that came here and been to school, all the stuff they know they have learned in their own country. They haven't learned here. And then the jobs you give us. The little jobs with the $1.40 an hour that you take all the taxes out of, you know. We are paying our taxes for the jobs that you give us. Then you take all the taxes back out of it.

CHAIRMAN HANNAH. May I ask the members of the class—We appreciate your enthusiasm. You have been invited. You are guests of the Commission, and if we have any more applause we are going to ask you to leave the room, because we are here for serious business.

MISS DELGADO. Well, when you're 18, if you are not a citizen or you are a citizen, you go and fight in the war. You won't let them vote. You're under 18, you can't vote. Well, if you're 18 you got to wait until you're 21 to vote. So you have no say in your government and you don't even know what you are doing in Vietnam. All you know is that when you get drafted you just go, and then you won't even give them a measly job when they're 16 or 17, but you will send them to war when they're 18 if they have an education or not, and you won't even let them finish their education, push them out of school from their education.

Then—or sometimes you give us a little program, you know, like, you know, like we are sick, you know; we have a disease and we are going to you. You're the doctor because we are telling you our problems and we are going to you with our disease, and we have a bad disease like cancer and you're our doctor, and we finally get an appointment with you like right now.

So we got this appointment with you and we tell you what is wrong with us. So you give us this little pill. We don't know what it is, you know. We go home and we take it. It is a little aspirin, you know. So we come back, you know. We have this big problem. You gave us this little pill, you know, to solve our problems, and we come back and, you know, we go out and there is a receptionist, you know. Well, it didn't help any. Well, we will be back in another seven years, because that it is how long it took you.

You came here seven years ago. Our problems were here seven years ago. Our leaders were here seven years ago. And you went back and it took you seven years to come back. What made you come back now after seven years? The same problems are here. You didn't do anything. I don't know if you are going to do anything now, but I am telling you what the problems are. O.K.

And another things, we have—There is an E.O.P. program [Educational Opportunities Program] at the University of California. It is supposed to be for Spanish-speaking peoples for opportunity to go to college. It is a program for opportunity to go to college, and yet Mission High School has the most

Spanish-speaking people and they won't even send one lousy person to come to Mission High School and tell us about the program. Not one lousy person, and so nobody knows about the program at the university. So what are we supposed to do? I want to go to college and I don't hear about any of these programs because they won't bother to tell us.

And so that's all I have to say and—Oh, another thing. I would say thank you, but I don't think it is a privilege to be here. I think it is a right and I think you should have came a long time ago.

MR. GLICKSTEIN. Do either of you fellows want to make a statement?

MR. VICARIO. Well, we have lots to say.

MR. GOLDBERG. [Citizen and high school student in San Francisco, CA] Absolutely.

MR. GLICKSTEIN. I wonder if you could each tell us, tell the Commissioners, where you were born. Were each of you born in the United States?

MR. VICARIO. Well, I was born in the South.

MR. GLICKSTEIN. In what part of the United States?

MR. VICARIO. Louisiana.

MR. GLICKSTEIN. And where were your parents from?

MR. VICARIO. From the Philippines and some part of Mexico. They were born out there.

MR. GLICKSTEIN. Miss Delgado, where were you born?

MISS DELGADO. I was born in San Francisco.

MR. GLICKSTEIN. And your parents?

MISS DELGADO. Spain.

MR. GLICKSTEIN. And Mr. Goldberg, where were you born?

MR. GOLDBERG. I was born in San Francisco. My mother is from Panama and my father is from back East.

MR. GLICKSTEIN. Your father is from where?

MR. GOLDBERG. Back East.

MR. GLICKSTEIN. Did you want to make a statement, Mr. Goldberg?

MR. GOLDBERG. Yes, I will. I am going to bring up the topic today of the high school dropout and why we are having this big problem today.

In the past you people have been looked upon as the big problem, but today it is—I feel it is one of the biggest problems in this community—

And when a student enrolls in the high schools in the Mission District, you know, they tell them a bunch of stuff, you know, about the school. It is a good school, all this and that, but then, you know, it all just changes towards the student. It makes him feel that he has come in there and they are not teaching him nothing. He is getting bored and either he drops out, if he feels this way, or either he tries to stick with it. If he does, he has many problems. And another thing about—another thing about this dropout thing is that not all students are dropping out. Most of the students are being kicked out. When you turn 18 years old at my school, they kick you out. You cannot go to any other school in San Francisco except night school or a day adult school or something like this. And I feel that we are being deprived of our education. We are being—actually being cheated out of it. We are not getting the proper books and the material and things like this, and I feel that even the slop they feed us at lunchtime, that is all a part of this, too.

MR. GLICKSTEIN. Are there Spanish-speaking teachers at Mission High School?

MR. GOLDBERG. Very few. They are very rare.

MR. GLICKSTEIN. How about Spanish-speaking counselors?

MR. GOLDBERG. None.

MR. GLICKSTEIN. None at all?

MR. GOLDBERG. No.

MR. GLICKSTEIN. What percentage of the students in the school, would you say, are Spanish-speaking?

MR. GOLDBERG. About 30.6.

MR. GLICKSTEIN. 30.6. What year are you in high school?

MR. GOLDBERG. I'm a high 11.

MR. GLICKSTEIN. Can you give any impression about what percentage of students at Mission High School drop out before they finish?

MR. GOLDBERG. Well, I will tell you like this. There is more that drop out than graduate.

MR. GLICKSTEIN. And is it largely the language problem that leads to the dropout?

MR. GOLDBERG. Yes, I feel it is the language problem and most of the students are slow in reading, in math, in other subjects, and because they are, you know, are dropping behind in subjects like this, they lose their interest in school and they lose their interest in their education, and they do not come to school.

MR. GLICKSTEIN. Are there efforts being made in the school to interest—

MR. GOLDBERG. No, there isn't.

MR. GLICKSTEIN. —interest students in going to college?

MR. GOLDBERG. No, there isn't, no.

MR. GLICKSTEIN. I have no further questions.

. . .

MR. GOLDBERG. I would like each and everyone of you to come out to our school one day. I mean just to see how our school is being run. Just to see how we are getting our education and how we are being cheated out of our education.

We walk down the halls like packs of cattle. This is just how small our school is and overcrowded. We walk down the halls like packs of cattle, and then you people wonder why we have riots and fights and things like this, and you people sit up here and call us down and tell us that the younger generation is crazy and stuff like this.

I would like for you people to come out there and see what is happening. I feel that you would be amazed how we are getting our education. The people, the staff at our school are encouraging us to drop out. They tell us, you know, "If you come late to school, why not bother to come to school? Why don't you stay home and go to sleep or something like this?" And one thing I want to ask everyone of you right now, I want to ask you what is going to happen to the people that drop out? What is going to happen to the people that drop out? They go to look for a job and the people tell them, "They don't want you. They don't want you because you don't have an education. You don't have no high schol [*sic*] diploma." Then you look on the other side for some of them are going to feel, well, "Man, we can't make it the honest way."

So I feel that because of this and the way the schools are run in the Mission District and other districts in San Francisco, I feel that the school districts are not doing nothing but filling up prisons and gas chambers and things like this. And this is the way I feel about it.

Why should we go to school? Why should we go to school? I want one of you to tell me one reason why we should go to school.

CHAIRMAN HANNAH. I don't think this is the occasion for speeches. We are interested in your presentation. It is a pathetic presentation.

I think the audience needs to recognize as well as you young people that in this country the responsibility for education is a State and local responsibility. If what you are telling us today is an accurate picture of what is happening in the Mexican American schools in the city of San Francisco or elsewhere, it should be a matter of real concern to all of the citizens of this community. The city of San Francisco and the communities around it are responsible for the operation of the local boards of education. The kind of a school that San Francisco has at the kindergarten level through the high school is determined by the people of San Francisco, by the taxpayers, by the Board of Education. The State of California has had a lot of publicity on—

MR. GOLDBERG. Damn all that! Wait a minute. I want to hear what you are going to do about it.

CHAIRMAN HANNAH. Young man, you will give me three minutes, then we are going to adjourn this session. We are here to listen. Let me answer your question. You asked why are you looking for an education.

There is only one purpose of public education or education of any kind and that is to make it possible for young people like you to take whatever God gave them in the way of potential and encourage them to make of it something that will make it possible for you as an individual to make the best possible citizen out of yourself that you can, fit you with some skill so that you can earn a living, give you some awareness of what society is so that you may be able to carve out a life for yourself that will be interesting and satisfying to you, but this notion that an education is something that people can assure you, this is not true. An education is something that every person has to achieve for himself and once achieved it is his. Nobody can take it away from him. You can't buy it, you can't steal it, or anything of that sort.

Now, there is nothing more important in San Francisco or anywhere else in this country than that we have the best possible educational opportunity for every youngster, every color, race, religion, whatever society he may have been born into. An opportunity to make of himself everything that he can, because this is what makes America America . . .

All we know is what you are telling us. If what you tell us is true, and if it isn't being provided, then we have problems ahead of us.

MR. GOLDBERG. Yes, but why don't you come on out there?

MISS DELGADO. I would like to ask you something.

CHAIRMAN HANNAH. Now, we are going to have no more presentations.

MISS DELGADO. You came to listen to us and now you are leaving. See, you won't even finish listening to us.

CHAIRMAN HANNAH. We will give each of you—

MISS DELGADO. You know, we want to do something before you leave and I think Gabriel better do it now.

DEAN GRISWOLD. Mr. Chairman, it has been said that we came to listen. I think we came to learn.

If you have facts that you can tell us, I would be very much interested in having them. If all you want to tell us is that you are unhappy and you are distressed and you are uncertain and you think there are all kinds of problems and somebody ought to do something about it, I don't think it is helpful to—

MISS DELGADO. All right.

MR. GOLDBERG. What do you mean—

COMMISSIONER GRISWOLD. Just a moment, please.

MISS DELGADO. Wait a minute.

COMMISSIONER GRISWOLD. We were told a little while ago that we came out here and denounced you because you were improper, and so forth. We haven't denounced anybody. We haven't said a thing. We are just trying to learn and if you could present us with facts which will help us to learn, it could he helpful to us and to you and I think that despite the fact that the educational system is primarily a responsibility of the city of San Francisco and the State of California, this Commission does have a responsibility with respect to any aspect of it which might involve a denial of equal protection of the laws on the ground of race or color or religion or other grounds, but I repeat, mere expressions of emotion are not helpful, are not relevant, but facts would be, and if you have facts I would be glad to hear them.

MISS DELGADO. O.K. I would like to tell you a few facts. O.K.

You are a Civil Rights Commission, O.K. So you want to know about civil rights. OK. You come to school in the morning and you are late and the principal will come out and he will look up and down the line and he will say, "Here you are, all you Latin boys again." Emphasize on the "Latin," "All you Latin boys, always late. You don't go to church 10 minutes late. So why should you come to school late? Why don't you just go home and sleep?" That's what they say. "If you don't want to come to school, drop out and then you can sleep all you want." And if you call that encouragement, and that is a fact, O.K., you go down, somebody is—O.K.—Like Gabe here. He was getting kicked out of school but he had a legitimate reason. He was looking for a job. He has a kid to support. He is looking for a job. He goes to school, and they tell him—they tell the members of the Horizons that came with him, "The best thing you can do for this boy is get him a full-time job. That will be the best thing you could do."

A full-time job might help him now, but what happens as standards in the United States are going up and up? A full-time job, and he is a dropout? What is he going to do all his life? What's his kid going to go through?

O.K. We want better schools now, but think of the future What about his little boy—I mean his little girl, his little girl. His parents didn't have an education. My parents didn't have an education. My mother went to third grade. So she can't help me in school. I have to help myself. I don't have any outside help until I got with Horizons.

O.K. So they gave me some outside help. A little tutoring, a little project, you know, that you guys sent down to—that somebody gave us. I don't really know who. O.K. So do you think that he wants to drop out of school now and then have a bad job and then have his daughter have to grow up and she get a bad education, too? We want an education. We really do. We are trying to get to college. We want to learn about these things. I want to talk to my counselor and I want to tell my counselor, well, I want to be a social worker. What classes am I supposed to take? He has got a long line of kids. You get five minutes each turn to talk to your counselor. And right outside you have got a long line of kids. They all want to talk to him too. So he says, "Well, I will put you down for anything and later on when you decide what you want you come in and tell me." And so I told him I wanted to be a social worker. "What courses do I take?" Nothing; no help or anything. So we just want to, you know, there is facts. Maybe they are brought out in emotions, but the facts are behind the emotions. You show your emotions because some facts is behind it pushing the emotions out because all these things that are against you, you just get them all inside of you and, sure, they come out as emotions, but if you were here longer maybe you would listen to them and we could show you the facts.

If you would go down—you are an outsider. You can't see what is going on. You are not here living it. You are not learning one thing and going out in the world and it is a whole total different thing. It is a different thing. You learn democracy in school. You learn everything. You learn U.S. history and you go in the South and they have a whole different U.S. history. What is the truth? Who is going to tell us the truth?

. . .

MR. SANDOVAL. Well, Mr. Chairman, I feel this way, to be truthful, the students could be here all evening telling you about their problems, but if you will give me a few minutes I can briefly tell you exactly what they are facing today and what they are trying to communicate to you and we won't have to sit down there again and wait one more time. Will you allow me the privilege?

CHAIRMAN HANNAH. When you say a few minutes, if your few minutes is not more than four, I would say go ahead, but we still have what? Three more witnesses this afternoon?

MR. GLICKSTEIN. Yes, Sir.

CHAIRMAN HANNAH. And a meeting of the Commission after that. So you tell us what you can in four minutes and then we are going to call the next witness.

MR. SANDOVAL. O.K. First of all, members of the Commission, what the students are trying to tell you is the fact that when the Father mentioned about the various experiences in Latin America and the other places and the philosophy he was speaking about, this is exactly what the problem is in San

Francisco. Who cares about Latin America? Who cares about Africa? Our problem is right here right now.

You, yourself, Mr. Chairman, have stated earlier that we are just a Commission that is going around the country listening to problems. Now, I am not directing my attack toward you. I respect your position. I am not insulting you, but this is the problem of being a Mexican American, Spanish-speaking person.

I negotiate with the school board administrators on one level. I fight for these people in high school. I fight for them at the Board of Education. Believe me, I am a Mexican American from the day I was born. There is continuous ugly discrimination and segregation. There always has been. Nobody ever comes up to my face and tells me, "You're a dirty Mexican," but I can hear them when I walk off. I know exactly what they are thinking, but you know what the problem is? It is not our problem. It has been over 200 years. It is a problem of the Anglo society, the monolingual, monocultural society cannot accept, does not realize what the bicultural problem is.

Sure there is a lot of people like Robert B. Hayden and Margaret Clark and Georgia Christiansen writing books about us. To be truthful, what do they know about us? What do they know about us? You know we have a few people who write a few books, but everybody all over the United States is on this big kick about anti-poverty and about helping the poor Mexican American. But all they do is chastise him and they literally crucify him.

Robert Hayden, for example—I don't know the man, but in his stories of Spanish-speaking people in the Southwest, Dr. Manuel Guerra from USC, if you had kept up with his correspondence that David North has been mailing back and forth between these people and interested people, you can see how even a person such as Robert Hayden, who supposedly has a professional knowledge of Spanish-speaking people, is completely out of line. He is right in some respects, but monolingual, monocultural people do not understand the Spanish-speaking bicultural problem.

I will tell you this: I grew up with Anglo people in the little towns in New Mexico. Now, that Anglo person knows and understands the problems of Spanish-speaking people better than anybody that comes from Harvard or Notre Dame or Michigan State or Duke, or what have you. This is absolutely the truth.

I negotiate for the students over there at the Board of Education. "I have been to Latin America. I can speak in Spanish. I spoke Spanish the day I was born." This is fine, but they do not understand. This is all on a superficial level.

What we are concerned about is from the very beginning of the grass roots stages. I can't—you will never understand what I mean by our type of discrimination. It is hell to be a Spanish-speaking person, believe me, and I think it is time that everybody woke up. Even if the Commission can't do nothing about it, they can write to somebody, appeal to the President or do something, because I am one person that is up to my neck, believe me, and those people are out there, too, and the younger people today every day are calling me and coming to me, "Jack, it doesn't matter about my life or them anymore. Let's walk down that street and if anybody gets in the way we will knock them out."

CHAIRMAN HANNAH. Thank you, Mr. Sandoval. You are excused. Mr. Glickstein, will you call the next witness?

DOCUMENT 7.3

Crystal City School Walkout Demands

Starting in East Los Angeles in the spring of 1968, Chicano high school students throughout the Southwest organized walkouts or "blowouts" from their schools to protest inequitable conditions. In contrast to the stereotype of Latino students as "lazy" and not interested in education, a careful read of these petitions demonstrates their desire for a better quality education and opportunity to access college. Teachers who called Mexican Americans derogatory terms were particularly targeted, as were practices such as having students assist in cleaning the school.

From: Armando Navarro, *Mexican American Youth Organization* (Austin: University of Texas Press, 1995), pp. 251–252. Original owned by José Angel Gutiérrez, University of Texas, Arlington, TX 76019.

1. That all elections concerning the school be conducted by the student body. Concerning class representatives, the petition asked that the qualifications such as personality, leadership, grades be abolished. These factors do not determine whether the student is capable of representing the student body. The students are capable of voting for their own representatives. The representatives are representing the students, not the faculty. All nominating must be done by the student body, and the election should be decided by a majority vote.
2. The present method of electing most handsome, beautiful, most popular, and most representative is elected [*sic*] by the faculty. The method of cumulative voting is unfair.
3. National Honor Society—the grades of the students eligible must be posted on the bulletin board well in advance of selection. The teachers should not have anything to do with electing the students.
4. An advisory board of Mexican American citizens should be part of the school administration in order to advise on the needs and problems of the Mexican American.
5. No other favorites should be authorized by school administrators or board members unless submitted to the student body in a referendum.
6. Teachers, administrators and staff should be educated; they should know our language—Spanish—and understand the history, traditions and contributions of Mexican culture. How can they expect to teach us if they do not know us? We want more Mexican American teachers for the above reason.

7. We want immediate steps taken to implement bilingual and bicultural education for Mexican Americans. We also want the school books revised to reflect the contributions of Mexicans and Mexican Americans to the U.S. society, and to make us aware of the injustices that we, Mexican Americans, as a people have suffered in an "Anglo" dominant society. We want a Mexican American course with the value of one credit.

8. We want any member of the school system who displays prejudice or fails to recognize, understand and appreciate us, Mexican Americans, our culture, or our heritage removed from Crystal City's schools. Teachers shall not call students any names.

9. Our classes should be smaller in size, say about 20 students to one teacher to insure more effectiveness. We want parents from the community to be trained as teacher's aides. We want assurances that a teacher who may disagree politically or philosophically with administrators will not be dismissed or transferred because of it. Teachers should encourage students to study and should make class more interesting, so that students will look forward to going to class.

10. There should be a manager in charge of janitorial work and maintenance details and the performance of such duties should be restricted to employees hired for that purpose. In other words, no more students doing janitorial work.

11. We want a free speech area plus the right to have speakers of our own.

12. We would like September 16 as a holiday, but if it is not possible we would like an assembly, with speakers of our own. We feel it is a great day in the history of the world because it is when Mexico had been under the Spanish rule for about 300 years. The Mexicans were liberated from the harsh rule of Spain. Our ancestors fought in this war and we owe them tribute because we are Mexicans, too.

13. Being civic minded citizens, we want to know what the happenings are in our community. So, we request the right to have access to all types of literature and to be able to bring it on campus. The newspaper in our school does not carry sufficient information. It carries things like the gossip column, which is unnecessary.

14. The dress code should be abolished. We are entitled to wear what we want.

15. We request the buildings open to students at all times.

16. We want Mr. Harbin to resign as Principal of Fly Jr. High.

17. We want a Mexican American counselor fully qualified in college opportunities.

18. We need more showers in the boy's and girl's dressing rooms.

DOCUMENT 7.4

Bilingual Education Act of 1968

As part of President Lyndon B. Johnson's War on Poverty, the U.S. Congress passed the Elementary and Secondary Education Act (ESEA) in 1965. Designed to provide federal funds for compensatory and remedial programs to underprivileged children, bilingual education was amended as part of ESEA in 1967.

From: United States Congress, *Elementary and Secondary Education Amendments of 1967 Public Law 90–247; 81 Stat. 783 [H.R. 7819], Laws of the 90th Congress, 1st Session, January 2, 1968* (Washington, DC: United States Government Printing Office, 1968), pp. 877–923.

An Act to strengthen, improve, and extend programs of assistance for elementary and secondary education, and for other purposes.

Be it enacted by the Senate and House of Representatives of the United States of America in Congress assembled, that:

This Act may be cited as the "Elementary and Secondary Education Amendments of 1967."

TITLE VII—BILINGUAL EDUCATION PROGRAMS

Findings of Congress

Sec. 701. The Congress hereby finds that one of the most acute educational problems in the United States is that which involves millions of children of limited English-speaking ability because they come from environments where the dominant language is other than English; that additional efforts should be made to supplement present attempts to find adequate and constructive solutions to this unique and perplexing educational situation; and that the urgent need is for comprehensive and cooperative action now on the local, State and Federal levels to develop forward-looking approaches to meet the serious learning difficulties faced by this substantial segment of the Nation's school-age population.

Amendment to Elementary and Secondary Education Act of 1965

Sec. 702. The Elementary and Secondary Education Act of 1965 is amended by redesignating title VII as title VIII, by redesignating sections 701 through 707 and references thereto as sections 801 through 807, respectively, and by inserting after title VI the following new title:
"Title VII—Bilingual Education Programs
"Short Title
"Sec. 701. This title may be cited as the 'Bilingual Education Act'.

"Declaration of Policy

"Sec. 702. In recognition of the special educational needs of the large numbers of children of limited English-speaking ability in the United States, Congress hereby declares it to be the policy of the United States to provide financial assistance to local educational agencies to develop and carry out new and imaginative elementary and secondary school programs designed to meet these special educational needs. For the purposes of this title, 'children of limited English-speaking ability' means children who come from environments where the dominant language is other than English.

"Authorization and Distribution of Funds

"Sec. 703. (a) For the purposes of making grants under this title, there is authorized to be appropriated the sum of $15,000,000 for the fiscal year ending June 30, 1968, $30,000,000 for the fiscal year ending June 30, 1969, and $40,000,000 for the fiscal year ending June 30, 1970.

"(b) In determining distribution of funds under this title, the Commissioner shall give highest priority to States and areas within States having, the greatest need for programs pursuant to this title. Such priorities shall take into consideration the number of children of limited English-speaking ability between the ages of three and eighteen in each State.

"Uses of Federal Funds

"Sec. 704. Grants under this title may be used, in accordance with applications approved under section 705, for—
"(a) planning for and taking other steps leading to the development of programs designed to meet the special educational needs of children of limited English-speaking ability in school having a high concentration of such children from families (A) with incomes below $3,000 per year, or (B) receiving payments under a program of aid to families with dependent children under a State plan approved under title IV of the Social Security Act, including research projects, pilot projects designed to test the effectiveness of plans so developed, and the development and dissemination of special instructional materials for use in bilingual education programs; and

"(b) providing preservice training designed to prepare persons to participate in bilingual education programs as teachers, teacher-aides, or other ancillary education personnel such as counselors, and inservice training and development programs designed to enable such programs to continue to improve their qualifications while participating in such programs; and

"(c) the establishment, maintenance, and operation of programs, including acquisition of necessary teaching materials and equipment, designed to meet the special educational needs of children of limited English-speaking ability in schools having a high concentration of such children from families (A) with incomes below $3,000 per year, or (B) receiving payments under a program of aid to families with dependent children under a State plan approved under title IV of the Social Security Act, through activities such as—

"(1) bilingual education programs;

"(2) programs designed to impart to students a knowledge of the history and culture associated with their languages;

"(3) efforts to establish closer cooperation between the school and the home;

"(4) early childhood educational programs related to the purposes of this title and designed to improve the potential for profitable learning activities by children;

"(5) adult education programs related to the purposes of this title, particularly for parents of children participating in bilingual programs;

"(6) programs designed for dropouts or potential dropouts having need of bilingual programs;

"(7) programs conducted by accredited trade, vocational, or technical school; and

"(8) other activities which meet the purposes of this title.

"*Applications for Grants and Conditions for Approval*

"Sec. 705. (a) A grant under this title may be made to a local educational agency or agencies, or to an institution of higher education applying jointly with a local educational agency, upon application to the Commissioner at such time or times, in such manner and containing or accompanied by such information as the Commissioner deems necessary.

. . .

"*Advisory Committee*

"Sec. 707. (a) The Commissioner shall establish in the Office of Education an Advisory Committee on the Education of Bilingual Children, consisting of nine members appointed, without regard to the civil service laws by the Commissioner with the approval of the Secretary. The Commissioner shall appoint one such member as Chairman. At least four of the members of the Advisory Committee shall be educators experienced in dealing with the educational problems of children whose native tongue is a language other than English.

"(b) The Advisory Committee shall advise the Commissioner in the preparation of general regulations and with respect to policy matters arising in the administration of this title, including the development of criteria for approval of applications thereunder. The Commissioner may appoint such special advisory and technical experts and consultants as may be useful and necessary in carrying out the functions of the Advisory Committee,

"(c) Members or the Advisory Committee shall, while serving on the business of the Advisory Committee, be entitled to receive compensation at rates fixed by the Secretary, but not exceeding $100 per day, including traveltime; and while so serving away from their homes or regular places of business, they may be allowed travel expenses, including per diem in lieu of subsistence, as authorized by section 5703 of title 5 of the United States Code for persons in the Government service employed intermittently. . .

Approved January 2, 1968.

DOCUMENT 7.5

Lau v. Nichols, 1974

*Despite the passage of the Bilingual Education Act, English language learners (ELL)
were being underserved in their public schools. In this hallmark Supreme Court case,
the federal government affirmed the rights of children for whom English was not
their first language to receive special services. Although the case was based specifically
upon the needs of Asian students in San Francisco, the ruling has since applied to all
English language learners. The U.S. Office of Civil Rights subsequently issued
guidelines known as the "Lau Remedies" to aid school districts in interpreting and
applying the law.*

414 U.S. 563

Lau et al. v. Nichols et al.

Certiorari to the United States Court of Appeals for the Ninth Circuit

No. 72–6520.

Argued December 10, 1973

Decided January 21, 1974

The failure of the San Francisco school system to provide English language
instruction to approximately 1,800 students of Chinese ancestry who do not
speak English, or to provide them with other adequate instructional procedures, denies them a meaningful opportunity to participate in the public
educational program and thus violates 601 of the Civil Rights Act of 1964,
which bans discrimination based "on the ground of race, color, or national
origin," in "any program or activity receiving Federal financial assistance,"
and the implementing regulations of the Department of Health, Education,
and Welfare. Pp. 565–569.

483 F.2d 791, reversed and remanded.

Douglas, J., delivered the opinion of the Court, in which Brennan, Marshall, Powell, and Rehnquist, JJ., joined. Stewart, J., filed an opinion concurring in the result, in which Burger, C. J., and Blackmun, J., joined, post, p. 569. White, J., concurred in the result. Blackmun, J., filed an opinion concurring in the result, in which Burger, C. J., joined, post, p. 571.

Edward H. Steinman argued the cause for petitioners. With him on the briefs were Kenneth Hecht and David C. Moon.

Thomas M. O'Connor argued the cause for respondents. With him on the brief were George E. Krueger and Burk E. Delventhal.

Assistant Attorney General Pottinger argued the case for the United States as amicus curiae urging reversal. With him on the brief were Solicitor General Bork, Deputy Solicitor General Wallace, Mark L. Evans, and Brian K. Landsberg.*

[Footnote*] Briefs of amici curiae urging reversal were filed by Stephen J. Pollak, Ralph J. Moore, Jr., David Rubin, and Peter T. Galiano for [414 U.S. 563, 564] the National Education Assn. et al.; by W. Reece Bader and James R. Madison for the San Francisco Lawyers' Committee for Urban Affairs; by J. Harold Flannery for the Center for Law and Education, Harvard University; by Herbert Teitelbaum for the Puerto Rican Legal Defense and Education Fund, Inc.; by Mario G. Obledo, Sanford J. Rosen, Michael Mendelson, and Alan Exelrod for the Mexican American Legal Defense and Educational Fund et al.; by Samuel Rabinove, Joseph B. Robison, Arnold Forster, and Elliot C. Rothenberg for the American Jewish Committee et al.; by F. Raymond Marks for the Childhood and Government Project; by Martin Glick for Efrain Tostado et al.; and by the Chinese Consolidated Benevolent Assn. et al. [414 U.S. 563, 564]

Mr. Justice Douglas delivered the opinion of the Court.

The San Francisco, California, school system was integrated in 1971 as a result of a federal court decree, 339 F. Supp. 1315. See Lee v. Johnson, 404 U.S. 1215 . The District Court found that there are 2,856 students of Chinese ancestry in the school system who do not speak English. Of those who have that language deficiency, about 1,000 are given supplemental courses in the English language.[1] About 1,800, however, do not receive that instruction.

This class suit brought by non-English-speaking Chinese students against officials responsible for the operation of the San Francisco Unified School District seeks relief against the unequal educational opportunities, which are alleged to violate, inter alia, the Fourteenth Amendment. No specific remedy is urged upon us. [414 U.S. 563, 565] Teaching English to the students of Chinese ancestry who do not speak the language is one choice. Giving instructions to this group in Chinese is another. There may be others. Petitioners ask only that the Board of Education be directed to apply its expertise to the problem and rectify the situation.

The District Court denied relief. The Court of Appeals affirmed, holding that there was no violation of the Equal Protection Clause of the Fourteenth Amendment or of 601 of the Civil Rights Act of 1964, 78 Stat. 252, 42 U.S.C. 2000d, which excludes from participation in federal financial assistance, recipients of aid which discriminate against racial groups, 483 F.2d 791. One judge dissented. A hearing en banc was denied, two judges dissenting. Id., at 805.

We granted the petition for certiorari because of the public importance of the question presented, 412 U.S. 938.

The Court of Appeals reasoned that "[e]very student brings to the starting line of his educational career different advantages and disadvantages caused in part by social, economic and cultural background, created and continued completely apart from any contribution by the school system," 483 F.2d, at 797. Yet in our view the case may not be so easily decided. This is a public school system of California and 71 of the California Education Code states that "English shall be the basic language of instruction in all schools." That section permits a school district to determine "when and under what circumstances instruction may be given bilingually." That section also states as "the policy of the state" to insure "the mastery of English by all pupils in the schools." And bilingual instruction is authorized "to the extent that it does not interfere with the systematic, sequential, and regular instruction of all pupils in the English language." [414 U.S. 563, 566]

Moreover, 8573 of the Education Code provides that no pupil shall receive a diploma of graduation from grade 12 who has not met the standards of proficiency in "English," as well as other prescribed subjects. Moreover, by 12101 of the Education Code (Supp. 1973) children between the ages of six and 16 years are (with exceptions not material here) "subject to compulsory full-time education."

Under these state-imposed standards there is no equality of treatment merely by providing students with the same facilities, textbooks, teachers, and curriculum; for students who do not understand English are effectively foreclosed from any meaningful education.

Basic English skills are at the very core of what these public schools teach. Imposition of a requirement that, before a child can effectively participate in the educational program, he must already have acquired those basic skills is to make a mockery of public education. We know that those who do not understand English are certain to find their classroom experiences wholly incomprehensible and in no way meaningful.

We do not reach the Equal Protection Clause argument which has been advanced but rely solely on 601 of the Civil Rights Act of 1964, 42 U.S.C. 2000d, to reverse the Court of Appeals.

That section bans discrimination based "on the ground of race, color, or national origin," in "any program or activity receiving Federal financial assistance." The school district involved in this litigation receives large amounts of federal financial assistance. The Department of Health, Education, and Welfare (HEW), which has authority to promulgate regulations prohibiting discrimination in federally assisted school systems, 42 U.S.C. 2000d-1, in 1968 issued one guideline that "[s]chool systems are responsible for assuring that students of a particular race, color, or national origin are not denied the [414 U.S. 563, 567] opportunity to obtain the education generally obtained by other students in the system." 33 Fed. Reg. 4956. In 1970 HEW made the guidelines more specific, requiring school districts that were federally funded "to rectify the language deficiency in order to open" the instruction to students who had "linguistic deficiencies," 35 Fed. Reg. 11595.

. . .

It seems obvious that the Chinese-speaking minority receive fewer benefits than the English-speaking majority from respondents' school system which denies them a meaningful opportunity to participate in the educational program—all earmarks of the discrimination banned by the regulations.[3] In 1970 HEW issued clarifying guidelines, 35 Fed. Reg. 11595, which include the following:

> "Where inability to speak and understand the English language excludes national origin-minority group children from effective participation in the educational program offered by a school district, the district must take affirmative steps to rectify the language deficiency in order to open its instructional program to these students."

"Any ability grouping or tracking system employed by the school system to deal with the special language skill needs of national origin-minority group children must be designed to meet such language skill needs as soon as possible and must not operate as an educational deadend or permanent track."

Respondent school district contractually agreed to "comply with title VI of the Civil Rights Act of 1964 . . . and all requirements imposed by or pursuant to the [414 U.S. 563, 569] Regulation" of HEW (45 CFR pt. 80) which are "issued pursuant to that title . . ." and also immediately to "take any measures necessary to effectuate this agreement." The Federal Government has power to fix the terms on which its money allotments to the States shall be disbursed.

Oklahoma v. CSC, 330 U.S. 127, 142 -143. Whatever may be the limits of that power, Steward Machine Co. v. Davis, 301 U.S. 548, 590 et seq., they have not been reached here. Senator Humphrey, during the floor debates on the Civil Rights Act of 1964, said:[4]

> "Simple justice requires that public funds, to which all taxpayers of all races contribute, not be spent in any fashion which encourages, entrenches, subsidizes, or results in racial discrimination."

We accordingly reverse the judgment of the Court of Appeals and remand the case for the fashioning of appropriate relief.

Reversed and remanded.

Mr. Justice White concurs in the result.

FOOTNOTES

[Footnote 1] A report adopted by the Human Rights Commission of San Francisco and submitted to the Court by respondents after oral argument shows that, as of April 1973, there were 3,457 Chinese students in the school system who spoke little or no English. The document further showed 2,136 students enrolled in Chinese special instruction classes, but at least 429 of the enrollees were not Chinese but were included for ethnic balance. Thus, as of April 1973, no more than 1,707 of the 3,457 Chinese students needing special English instruction were receiving it.

[Footnote 2] Section 602 provides: "Each Federal department and agency which is empowered to extend Federal financial assistance to any program or activity, by way of grant, loan, or contract other than a contract of insurance or guaranty, is authorized and directed to effectuate the provisions of section 2000d of this title with respect to such program or activity by issuing rules, regulations, or orders of general applicability which shall be consistent with achievement of the objectives of the statute authorizing the financial assistance in connection with which the action is taken . . ." 42 U.S.C. 2000d-1.

[Footnote 3] And see Report of the Human Rights Commission of San Francisco, Bilingual Education in the San Francisco Public Schools, Aug. 9, 1973.

[Footnote 4] 110 Cong. Rec. 6543 (Sen. Humphrey, quoting from President Kennedy's message to Congress, June 19, 1963).

DOCUMENT 7.6

El Plan de Santa Barbara, 1969

In 1969 Latino students from several California universities came together at the University of California at Santa Barbara to organize their goals for higher education reform. Emerging from this conference was "El Plan de Santa Barbara"—a clear and detailed articulation of the needs and demands of Latino college students. In addition, under the leadership of the Chicano Coordinating Council on Higher Education (CCHE), student groups at the Santa Barbara conference agreed to forego their individual names and become the Movimiento Estudiantil Chicano de Aztlán (MEChA).

From: Appendix, Carlos Muñoz Jr., *Youth, Identity, Power: The Chicano Movement* (New York: Verso, 1989), pp. 191–202. Original owned by Armando Valdez.

MANIFESTO

For all peoples, as with individuals, the time comes when they must reckon with their history. For the Chicano the present is a time of renaissance, of *renacimiento*. Our people and our community, *el barrio* and *la colonia*, are expressing a new consciousness and a new resolve. Recognizing the historical tasks confronting our people and fully aware of the cost of human progress, we pledge our will to move. We will move forward toward our destiny as a people. We will move against those forces which have denied us freedom of expression and human dignity. Throughout history the quest for cultural expression and freedom has taken the form of a struggle. Our struggle, tempered by the lessons of the American past, is an historical reality.

For decades Mexican people in the United States struggled to realize the "American Dream." And some—a few—have. But the cost, the ultimate cost of assimilation, required turning away from *el barrio* and *la colonia*. In the meantime, due to the racist structure of this society, to our essentially different life style, and to the socioeconomic functions assigned to our community by Anglo-American society—as suppliers of cheap labor and a dumping ground for the small-time capitalist entrepreneur—the *barrio* and *colonia* remained exploited, impoverished, and marginal.

As a result, the self-determination of our community is now the only acceptable mandate for social and political action; it is the essence of Chicano commitment. Culturally, the word *Chicano*, in the past a pejorative and class-bound adjective, has now become the root idea of a new cultural identity for our people. It also reveals a growing solidarity and the development of a common social praxis. The widespread use of the term *Chicano* today signals a rebirth of pride and confidence. *Chicanismo* simply embodies an ancient truth: that man is never closer to his true self as when he is close to his community.

Chicanismo draws its faith and strength from two main sources: from the just struggle of our people and from an objective analysis of our community's strategic needs. We recognize that without a strategic use of education, an education that places value on what we value, we will not realize our destiny. Chicanos recognize the central importance of institutions of higher learning to modern progress, in this case, to the development of our community. But we go further: we believe that higher education must contribute to the information of a complete man who truly values life and freedom.

The destiny of our people will be fulfilled. To that end, we pledge our efforts and take as our credo what José Vasconcelos once said at a time of crisis and hope: "At this moment we do not come to work for the university, but to demand that the university work for our people."

POLITICAL ACTION

Introduction

For the Movement, political action essentially means influencing the decision-making process of those institutions which affect Chicanos, the university, community organizations, and non-community institutions. Political action encompasses three elements which function in a progression: political consciousness, political mobilization, and tactics. Each part breaks down into further subdivisions. Before continuing with specific discussions of these three categories, a brief historical analysis must be formulated.

Historical Perspective

The political activity of the Chicano Movement at colleges and universities to date has been specifically directed toward establishing Chicano student organizations (UMAS, MAYA, MASC, MEChA, etc.) and institutionalizing Chicano Studies programs. A variety of organizational forms and tactics have characterized these student organizations.

One of the major factors which led to political awareness in the 60s was the clash between Anglo-American educational institutions and Chicanos who maintained their cultural identity. Another factor was the increasing number of Chicano students who became aware of the extent to which colonial conditions characterized their communities. The result of this domestic colonialism is that the *barrios* and *colonias* are dependent communities with no institutional power base of their own. Historically, Chicanos have been prevented from establishing a power base and significantly influencing decision-making. Within the last decade, a limited degree of progress has taken place in securing a base of power within educational institutions.

Other factors which affected the political awareness of the Chicano youth were: the heritage of Chicano youth movements of the 30s and 40s; the failures of Chicano political efforts of the 40s and 50s; the bankruptcy of Mexican-American pseudo-political associations, and the disillusionment of Chicano participants in the Kennedy campaigns. Among the strongest influences on Chicano youth today have been the National Farm Workers Association, the Crusade for Justice, and the Alianza Federal de Pueblos Libres. The Civil Rights, the Black Power, and the Anti-war movements were other influences.

As political consciousness increased, there occurred simultaneously a renewed cultural awareness which, along with social and economic factors, led to the proliferation of Chicano youth organizations. By the mid 1960s, MASC, MAYA, UMAS, La Vida Nueva, and MEChA appeared on campus, while the Brown Berets, Black Berets, ALMA, and La Junta organized in the *barrios* and *colonias*. These groups differed from one another depending on local conditions, and their varying state of political development. Despite differences in name and organizational experience, a basic unity evolved.

These groups have had a significant impact on the awareness of large numbers of people, both Chicano and non-Chicano. Within the communities, some public agencies have been sensitized, and others have been exposed. On campuses, articulation of demands and related political efforts have dramatized NUESTRA CAUSA. Concrete results are visible in both the increased number of Chicano students on campuses and the establishment of corresponding supportive services. The institutionalization of Chicano Studies marks the present stage of activity; the next stage will involve the strategic application of university and college resources to the community. One immediate result will be the elimination of artificial distinctions which exist between the students and the community. Rather than being its victims, the community will benefit from the resources of the institutions of higher learning.

Political Consciousness

Commitment to the struggle for Chicano liberation is the operative definition of the ideology used here. *Chicanismo* involves a crucial distinction in political consciousness between a Mexican American and a Chicano mentality. The Mexican American is a person who lacks respect for his cultural and ethnic heritage. Unsure of himself, he seeks assimilation as a way out of his "degraded" social status. Consequently, he remains politically ineffective. In contrast, *Chicanismo* reflects self-respect and pride in one's ethnic and cultural background. Thus, the Chicano acts with confidence and with a range or alternatives in the political world. He is capable of developing an effective ideology through action.

Mexican Americans must be viewed as potential Chicanos. *Chicanismo* is flexible enough to relate to the varying levels of consciousness within La Raza. Regional variations must always be kept in mind as well as the different levels of development, composition, maturity, achievement, and experience in political action. Cultural nationalism is a means of total Chicano liberation.

There are definite advantages to cultural nationalism, but no inherent limitations. A Chicano ideology, especially as it involves cultural nationalism, should be positively phrased in the form of propositions to the Movement. *Chicanismo* is a concept that integrates self-awareness with cultural identity, a necessary step in developing political consciousness. As such, it serves as a basis for political action, flexible enough to include the possibility of coalitions. The related concept of La Raza provides an internationalist scope of *Chicanismo*, and La Raza Cósmica furnishes a philosophical precedent. Within this framework, the Third World Concept merits consideration.

Political Mobilization

Political mobilization is directly dependent on political consciousness. As political consciousness develops, the potential for political action increases.

The Chicano student organization in institutions of higher learning is central to all effective political mobilization. Effective mobilization presupposes precise definition of political goals and of the tactical interrelationships of roles. Political goals in any given situation must encompass the totality of Chicano interests in higher education. The differentiation of roles required by a given situation must be defined on the basis of mutual accountability and equal sharing of responsibility. Furthermore, the mobilization of community support not only legitimizes the activities of Chicano student organizations but also maximizes political power. The principle of solidarity is axiomatic in all aspects of political action.

Since the movement is definitely of national significance and scope, all student organizations should adopt one identical name throughout the state and eventually the nation to characterize the common struggle of La Raza de Aztlán. The net gain is a step toward greater national unity which enhances the power in mobilizing local campus organizations.

When advantageous, political coalitions and alliances with non-Chicano groups may be considered. A careful analysis must precede the decision to enter into a coalition. One significant factor is the community's attitude towards coalitions. Another factor is the formulation of a mechanism for the distribution of power that ensures maximum participation in decision making: i.e., formulation of demands and planning of tactics. When no longer politically advantageous, Chicano participation in the coalition ends.

CAMPUS ORGANIZING: NOTES ON MECHA

Introduction

MEChA is a first step to tying the student groups throughout the Southwest into a vibrant and responsive network of activists who will respond as a unit to oppression and racism and will work in harmony when initiating and carrying out campaigns of liberation for our people.

As of present, wherever one travels throughout the Southwest, one finds that there are different levels of awareness on different campuses. The student movement is to a large degree a political movement and as such must not elicit from our people the negative responses that we have experienced so often in the past in relation to politics, and often with good reason. To this end, then, we must re-define [sic] politics for our people to be a means of liberation. The political sophistication of our Raza must be raised so that they do not fall prey to apologists and *vendidos* whose whole interest is their personal career or fortune. In addition, the student movement is more than a political movement, it is cultural and social as well. The spirit of MEChA must be one of "hermandad" and cultural awareness. The ethic of profit and competition, of greed and intolerance, which the Anglo society offers must be replaced by our ancestral communalism and love for beauty and justice. MEChA must bring to the mind of every young Chicano that the liberation of his people from prejudice and oppression is in his hands and this responsibility is greater than personal achievement and more meaningful than degrees, especially if they are earned at the expense of his identity and cultural integrity.

MEChA, then, is more than a name; it is a spirit of unity, of brotherhood, and a resolve to undertake a struggle for liberation in a society where justice is but a word. MEChA is a means to an end.

The Function of MEChA—To the Campus Community

Other students can be important to MEChA in supportive roles; hence, the question of coalitions. Although it is understood and quite obvious that the viability and amenability of coalition varies from campus to campus, some guidelines might be kept in mind. These questions should be asked before entering into any binding agreement. Is it beneficial to tie oneself to another group in coalition which will carry one into conflicts for which one is ill-prepared or involve one with issues on which one is ill-advised? Can one safely go into a coalition where one group is markedly stronger than another? Does MEChA have an equal voice in leadership and planning in the coalition group? Is it perhaps better to enter into a loose alliance for a given issue? How does the leadership of each group view coalitions? How does the membership? Can MEChA hold up its end of the bargain? Will MEChA carry dead weight in a coalition? All of these and many more questions must be asked and answered before one can safely say that he will benefit from and contribute to a strong coalition effort.

Supportive groups. When moving on campus it is often well-advised to have groups who are willing to act in supportive roles. For example, there are usually any number of faculty members who are sympathetic, but limited as to the number of activities they will engage in. These faculty members often serve on academic councils and senates and can be instrumental in academic policy. They also provide another channel to the academic power structure and can be used as leverage in negotiation. However, these groups are only as responsive as the ties with them are nurtured. This does not mean, compromise MEChA's integrity; it merely means laying good groundwork before an issue is brought up, touching bases with your allies before hand.

Sympathetic administrators. This is a delicate area since administrators are most interested in not jeopardizing their positions and often will try to act as buffers or liaison between the administration and the student group. In the case of Chicano administrators, it should not a priori be assumed that because he is Raza he is to be blindly trusted. If he is not known to the membership, he must be given a chance to prove his allegiance to La Causa. As such, he should be the Chicano's man in the power structure instead of the administration's Mexican-American. It is from the administrator that information can be obtained as to the actual feasibility of demands or programs to go beyond the platitudes and pleas of unreasonableness with which the administration usually answers proposals and demands. The words of the administrator should never be the deciding factor in students' actions. The students must at all times make their own decisions. It is very human for people to establish self-interest. Therefore, students must constantly remind the Chicano administrators and faculty where their loyalty and allegiance lie. It is very easy for administrators to begin looking for promotions just as it is very natural for faculty members to seek positions of academic prominence.

In short, it is the students who must keep after Chicano and non-Chicano administrators and faculty to see that they do not compromise the position of the student and the community. By the same token, it is the student who must come to the support of these individuals if they are threatened for their support of the students. Students must be careful not to become a political lever for others.

Function of MEChA—Education

It is a fact that the Chicano has not often enough written his own history, his own anthropology, his own sociology, his own literature. He must do this if he is to survive as a cultural entity in this melting pot society which seeks to dilute varied cultures into a grey upon grey pseudo-culture of technology and materialism. The Chicano student is doing most of the work in the establishment of study programs, centers, curriculum development, entrance programs to get more Chicanos into college. This is good and must continue, but students must be careful not to be co-opted in their fervor for establishing relevance on the campus. Much of what is being offered by college systems and administrators is too little too late. MEChA must not compromise programs and curriculum which are essential for the total education of the Chicano for the sake of expediency. The students must not become so engrossed in programs and centers created along established academic guidelines that they forget the needs of the people which these institutions are meant to serve. To this end, *Barrio* input must always be given full and open hearing when designing these programs, when creating them and in running them. The jobs created by these projects must be filled by competent Chicanos, not only the Chicano who has the traditional credentials required for the position, but one who has the credentials of the Raza. Too often in the past the dedicated pushed for a program only to have a *vendido* sharp-talker come in and take over and start working for his Anglo administrator. Therefore, students must demand a say in the recruitment and selections of all directors and assistant directors of student-initiated programs. To further insure strong if not complete control of the direction and running of programs, all advisory and steering committees should have both student and community components as well as sympathetic Chicano faculty as members.

Tying the campus to the *Barrio*. The colleges and universities in the past have existed in an aura of omnipotence and infallibility. It is time that they be made responsible and responsive to the communities in which they are located or whose members they serve. As has already been mentioned, community members should serve on all programs related to Chicano interests. In addition to this, all attempts must be made to take the college and university to the *Barrio*, whether it be in form of classes giving college credit or community centers financed by the school for the use of community organizations and groups. Also, the *Barrio* must be brought to the campus, whether it be for special programs or ongoing services which the school provides for the people of the *Barrio*. The idea must be made clear to the people of the *Barrio* that they own the schools and the schools and all their resources are at their disposal. The student group must utilize the resources open to the school for the benefit of the *Barrio* at every opportunity. This can be done by hiring more Chicanos to work as academic and non-academic personnel on the campus; this often requires exposure of racist hiring practices now in operation in many college [*sic*] and universities. When functions, social or otherwise, are held in the *Barrio* under the sponsorship of the college and university, monies should be spent in the *Barrio*. This applies to hiring Chicano contractors to build on campus, etc. Many colleges and universities have publishing operations which could be forced to accept *Barrio* works for publication. Many other things could be considered in using the resources of the school to the *Barrio*. There are possibilities for using the physical plant and facilities not mentioned here, but this is an area which has great potential.

MEChA in the Barrio

Most colleges in the Southwest are located near or in the same town as a *Barrio*. Therefore, it is the responsibility of MEChA members to establish close working relationships with organizations in that *Barrio*. The MEChA people must be able to take the pulse of the *Barrio* and be able to respond to it. However, MEChA must be careful not to overstep its authority or duplicate the efforts of another organization already in the *Barrio*. MEChA must be able to relate to all segments of the *Barrio*, from the middle-class assimilationists to the *batos locos*.

Obviously, every *Barrio* has its particular needs, and MEChA people must determine with the help of those in the *Barrio* where they can be most effective. There are, however, some general areas which MEChA can involve itself. Some of these are: 1) policing social and governmental agencies to make them more responsive in a humane and dignified way to the people of the *Barrio*; 2) carrying out research on the economic and credit policies of merchants in the *Barrio* and exposing fraudulent and exorbitant establishments; 3) speaking and communicating with junior high and other high school students, helping with their projects, teaching them organizational techniques, supporting their actions; 4) spreading the message of the movement by any media available — this means speaking, radio, television, local newspaper, underground papers, posters, art, theatres; in short, spreading propaganda of the Movement; 5) exposing discrimination in hiring and renting practices and many other areas which the student because of his mobility, his articulation, and his vigor should take as his responsibility. It may mean at times having to work in conjunction with other organizations. If this is the case and the project is one begun by the other organization, realize that MEChA is there as a supporter and should accept the direction of the group involved. Do not let loyalty to an organization cloud responsibility to a greater force—*la Causa*.

Working in the *Barrio* is an honor, but is also a right because we come from these people, and, as such, mutual respect between the *Barrio* and the college group should be the rule. Understand at the same time, however, that there will initially be mistrust and often envy on the part of some in the *Barrio* for the college student. This mistrust must be broken down by a demonstration of affection for the *Barrio* and La Raza through hard work and dedication. If the approach is one of a dilettante or of a Peace Corps volunteer, the people will know it and react accordingly. If it is merely a cathartic experience to work among the unfortunate in the *Barrio*—stay out.

Of the community, for the community. *Por la Raza habla el espiritú.*

DOCUMENT 7.7

Interview Summation, President of Claremont Colleges, California, 1967

In this document and document 7.8, the administrative point of view toward bringing "different" students to traditionally exclusive and Anglo colleges is revealed. Officers from the Rockefeller Foundation routinely met with officials from potential grant recipients and reported their interviews to the Foundation. In this summary of one officer's interview, we are made privy to the temper of the times. Initially skeptical of the potential of diverse students, President Benezet of Claremont College eventually agrees to allow the creation of a Latino center to attract and retain Mexican American students to the campus.

> From: Rockefeller Foundation Archives, Record Group 1.2, series 200, box 19, folder 158, Interview with President of Claremont College, 1967, Rockefeller Archive Center, Sleepy Hollow, New York.

Interviews: JEB, LCD
Visit to Claremont Colleges
Louis Benizette BENEZET [*sic*]
President, Claremont Graduate School and University Center

May 25, 1967

Limits on Possibilities of Educating Lower-Class Students at Upper-Class Colleges

When Benizette BENEZET joined the conversation he issued some cautions about expecting too much in the way of getting disadvantaged lower-class students into highly selective colleges such as those in the Claremont cluster. He says that economic and social class, not race, is the relevant variable here. Colleges in this country have always been, and are, highly selective by economic class. Only rarely do lower-class students go to rich colleges.

Benizette [*sic*] does not believe that the Upward Bound type program that Reckard has developed in cooperation with the Independent Schools will contribute many successful applications to the Claremont Colleges. Lower-class students simply cannot adjust to, and be happy in, upper-class colleges.

Moreover, most of them are not capable of doing the work in highly selective colleges.

To be sure, there will always be a few "mutations" from families with poor backgrounds. These students usually do so well in high school they have no trouble getting into good colleges. Claremont Colleges are glad to take all such students that they can find, if they are provided with money for the necessary support. It is very easy, however, to exaggerate how many such students can be found and it is a mistake to admit students who cannot do the work.

The bulk of lower-class students who go to college will continue to go to junior colleges or community colleges, and many of them will drop out. Some will continue in these colleges and get into good vocations. Which is clearly the way Benizette [*sic*] feels it should be, heredity being what it is.

DOCUMENT 7.8

Interview Summation, President of Claremont Colleges, California, 1969

From: Rockefeller Foundation Archives, Record Group 1.2, series 200, box 19, folder 160, Interview with President of Claremont College, 1969, Rockefeller Archive Center, Sleepy Hollow, New York.

Interviews: KWT
Louis Benezet
President
Claremont University Center
Claremont, California

February 12, 1969
New York City

Benezet wanted to discuss the Program for Specially Directed Study (PSDS) program being supported by Rockefeller Foundation (RF) at Claremont. He feels the program has been most worthwhile. 39 of the 40 young people admitted are doing satisfactory work. While their motivations differ and their I.Q.'s range from 110 to 160 they appear to be meeting the requirements of the program. Benezet regrets that Claremont cannot put more time into the evaluation of what is going on. The gap for some of the young people is greatest in the mechanical field: reading, paper writing, etc.

Benezet believes that American colleges and universities are in a social revolution. The blacks are in the vanguard even though at Claremont two-thirds of the [minority] students are Mexican-Americans and one-third are blacks. The philosophy of the revolution is quasi-Marxist. Heredity counts for nothing and environment everything. Therefore, any student can make it if the environment is right. Furthermore, such students view the Claremont project as a mere drop in the bucket. In other words, the climate is not good for a fair educational experiment. Outsiders are either involved in or close to open guerrilla warfare. There is traffic in guns and outside money is rumored to be coming on to campuses from Chinese interventionists. The groups who spark the reaction are the Black Panthers and the United Mexican-American Students (UMAS).

Benezet said frankly that the climate in educational institutions is so bad for a fair experiment that he would rather see the RF help in the development of community leaders rather than formal education. There is so much clamor on campuses that you cannot hear yourself think, let alone plan.

The real need in America is to get more mature and responsible black and brown leaders.

K.W.T.

DOCUMENT 7.9

Report to Yale College Administration from Boricuas Unidos

Puerto Rican students at Yale formed the organization Boricuas Unidos in the early 1970s to address Puerto Rican concerns regarding admission and retention. As this letter articulates, Puerto Ricans were severely underrepresented at Yale, despite its proximity to some of the largest concentrations of Puerto Ricans in the United States.

From El Centro de Estudios Puertorriqueños/Center for Puerto Rican Studies Archives, vertical clipping files—higher education—Yale College, "Yale—Puerto Rican Students," Hunter College, New York, NY.

September 27, 1971

To: Mr. Henry Chauncey, Jr.
University Secretary,
Director of Admissions and Financial Aid Policy

ADMISSION OF PUERTO RICAN STUDENTS TO YALE COLLEGE

I. Introduction

The following statement appears in the Yale Introductory Information Bulletin for incoming freshmen:

Students, of course, learn at least as much from each other as they do from their teachers or their books. In Yale College, the opportunity is great. Yale's constituency as a university and as a college is broadly national and international. Undergraduates [are] typically from all 50 states and from more than 50 foreign countries. Each undergraduate class is carefully selected not only for a high level of general intellectual and academic acumen, but also for a diversity of background, interests, and talents . . . Interchange among members of the College sharpens awareness of and tolerance for a wide range of points of view.

It is to review the nature of the Puerto Rican experience at Yale and in light of the above statement, that this memorandum has been written and endorsed by Boricuas Unidos, that is, the Puerto Rican students at Yale.

Before we discuss the admission process as it relates to Puerto Ricans at Yale, we find it imperative to present a clearer picture of the Puerto Rican situation in the United States today. Along the Northeast corridor, the number of Puerto Ricans divides into 1,000,000 in New York City, 250,000 in New Jersey, 150,000 in Connecticut, with the remainder residing in Massachusetts, Pennsylvania, Illinois and Upper New York State. We can estimate the current mainland population at 1.8 million, 56 per cent of whom live in New York City. The breakdown between New York City and elsewhere would be as follows . . .

Thus, the number of Puerto Rican students enrolled at Yale presently is twelve in total. Five of these are from the mainland and seven are from the island. This is the extent to which Yale has been able to commit itself to the Puerto Rican student.

It is a wonder and puzzlement to us that Yale could find only eight students qualified from the mainland worthy of admission. It is a further mystery to note that in the sophomore class not one Puerto Rican neither from the island nor the mainland, is enrolled. When compared to Princeton which has some twenty-odd Puerto Rican students, nineteen of which are from the mainland; or to Columbia University which has some 46 Puerto Rican students, thirty-nine of which are from the mainland; or to Harvard which has some twelve Puerto Ricans, almost all of which come from the mainland; it is a question of controversy to find the Puerto Rican in general and the mainland Puerto Rican in particular so underrepresented—so undistinguished—at Yale . . . It is for these reasons, that we recommend Yale College commit itself to the increased admissions of Puerto Ricans, with the aim of representing the

Table 2
Yale University Admissions
Puerto Rican Students

Year	# of Students Matriculated		Year	# of Students Graduated	
	Mainland	Island		Mainland	Island
Fall, 1966	1	2	Spring, 1970	1	2
Fall, 1967	0	4	Spring, 1971	0	4
Fall, 1968	0 + 1*	1	Spring, 1972	1	1
Fall, 1969	1	2	Spring, 1973	-	-
Fall, 1970	0	0	Spring, 1974	0	0
Fall, 1971	3	4	Spring, 1975	-	-

*transferred into the class of 1972 from Sarah Lawrence

Puerto Rican in the student body, according to their representation on a national scale, as well as the Northeast region ethnic distribution. Thus, we look for and work toward the day when Yale College will have a 2.2 per cent Puerto Rican student body. Yale College has much to gain and benefit from the increased presence of Puerto Ricans on campus. We must remember, however, that the essential point is not one of numbers and percentages, but one of the quality and depth of the education at Yale College.

In line with what has been presented in this memorandum and with faith in Yale's commitment to a truly representative student body, we make the following recommendations for the recruitment, admission, and matriculation of the mainland Puerto Rican student:

1. That Yale College secure and make available lunch slips and free lunches to Puerto Rican and minority students visiting the campus from the College Bound High Schools of New York City. Such a practice has proved quite successful in the recruitment not only of Puerto Ricans but of Blacks and Asians to Princeton University. This would further involve contacting the guidance counselors of the respective College Bound High Schools in New York, an effort in which Boricuas Unidos is quite willing to participate. It is our understanding that the College Bound Programs of New York City have budgeted moneys for the transportation of their students to colleges in the Northeast.

2. In an effort to persuade the Puerto Rican high school senior to become interested in Yale and to consider the possibility of applying to and being accepted by Yale College, we recommend that each area man visiting major urban cities in the Northeast with large Puerto Rican populations be accompanied by a Puerto Rican student from the College.

3. That Yale provide funds so that Puerto Rican students from the College will be able to recruit in New York City's Catholic High Schools heretofore neglected in Yale's recruitment of Puerto Rican students. Recruitment will also be carried on in the College Bound programs of the New York City High Schools as well as the major urban centers of New Jersey, Connecticut, Pennsylvania and Illinois.

4. That Yale College further provide funds for a Puerto Rican brochure depicting the Puerto Rican educational and social experience at Yale, such that the College may be made more meaningful and relevant to the urban orientated high school senior.

5. That Yale provide money for Puerto Rican representatives of Boricuas Unidos to meet with the Latin American Student Organization at Columbia University, Union Latinoamericana at Princeton, AJUA at Wesleyan and other such student organizations involved in the recruitment of latino [*sic*] students.

6. That Yale College work towards and hire a Puerto Rican Assistant Dean of Admissions who will be able to work on the recruitment of Puerto Rican students in the Northeast while also voting on the admission of the same. Boricuas Unidos is prepared to provide assistance and resumes towards this end.

7. That Yale College provide a $500 summer research grant to a member of Boricuas Unidos for the study of Puerto Rican recruitment in New York via community groups, anti-poverty agencies and model cities programs. This is to facilitate for Yale admissions the long-term recruitment of Puerto Ricans in New York City.

8. That the Yale Admissions Office hire two Puerto Rican students on Work-study for a period of ten weeks at salary of $125 a week plus traveling expenses for the continued recruitment and contact of Puerto Rican students in Northeastern High Schools.

9. That Yale College provide money for a gathering of the recently admitted Puerto Rican students to the Class of 1976 in order that they be persuaded to select Yale over competitive colleges since this will be a demonstration of Yale's concern for their matriculation in the College.

10. That the Yale Admissions Office provide stationery, applications, financial aid information, catalogues and stamps to Boricuas Unidos for use while Puerto Rican students are recruiting in the field as well as for follow-up.

11. That sufficiently clear credentials of introduction be provided to Puerto Ricans doing recruitment for the office of admissions.

12. That Yale exercise flexible admissions criteria in the selection of Puerto Rican students especially with regards to recommendations written by community people and organizations knowledgeable of the community relevant activities of these students.

13. That Yale work with local community organizations and Puerto Rican Talent Search Agencies in order to recruit Puerto Rican students such as Puente, Inc. in Boston, Massachusetts and La Junta in New Haven.

14. That a meeting be scheduled with the Black Ivy League Administrators to be held at Yale in order that we may find and compare what other colleges are doing with regard to latino [*sic*] recruitment.

15. That the admissions office direct any interested Puerto Rican student to meet with Boricuas Unidos.

CHAPTER EIGHT

Un paso hacia adelante, y otro hacia atrás (One step forward, one step back)

Latinos and Schooling in the 1980s and 1990s

My mother and I sat down for lunch at a leading department store in Miami— an advertisement that Colombian cuisine would be served in observation of Colombia's Independence Day brought us to the restaurant. In the packed dining room, we ordered from the special menu, but my mother's Spanish accent infuriated the Anglo waitress. She grabbed the menus from our hands, threw them to the floor, and screamed to a stunned and silenced room, "This is America, speak English!"

—*Victoria-María MacDonald, Miami, Florida, 1981*[1]

Overall, several witnesses agreed that this country is in a state of denial about the existence and causes of, and consequently, the needed solutions to racial and ethnic tensions. Gross institutionalized racial injustice is an issue that the country has never faced fully, or committed itself to resolve. Indeed, at crucial points, American society has retreated from addressing the critical subject of race, and learned to tolerate, rather than to eradicate racial inequality.

—*Racial and Ethnic Tensions in American Communities: Poverty, Inequality, and Discrimination, hearing before the U.S. Commission on Civil Rights, 1992*[2]

INTRODUCTION

The tumultuous decades of the 1960s and 1970s resulted in important legislative, judicial, and institutional changes impacting the Latino educational experience. The 1968 Bilingual Education Act and subsequent U.S. Supreme Court rulings like *Lau v. Nichols* (1974) provided federal recognition and reinforcement of the rights of English-language learners. Latino college students entered universities that had established centers for the study of Mexican American and Puerto Rican history and culture. A new generation of Latino scholars broadened the discourse on race and education to include a previously overlooked Latino population. At the close of the 1970s, however, the United States was still recovering from the economic and political consequences of the prolonged and costly Vietnam War. Economic stagnation and further political disillusionment accompanying the Watergate scandal appeared to stall the momentum of the Civil Rights era. At this tenuous juncture in American history, waves of Latino immigrants from both familiar points of origin (Cuba and Mexico) and newer sources (Central America, Latin America, and the Caribbean) transformed the demographic composition of our nation.[3]

Following 1965, when U.S. immigration laws were liberalized, Latinos and Asians comprised the fastest growing immigrant groups to the United States. Barely noticeable at first, thousands of Dominicans, Guatemalans, El Salvadorans, and Nicaraguans began arriving in the 1960s and 1970s. These often undocumented immigrants (referred to otherwise as "illegal aliens") were fleeing political oppression, violence, and poverty. Scholars have concluded that the exodus from these countries was largely the "direct result of [U.S.] military and economic intervention," which had assisted in the overthrow of existing governments and the installation of oppressive new regimes.[4]

The recent influx of Latino immigrants has, at times, ignited a backlash. During the 1980s and 1990s voters in states such as California and Arizona passed ballot measures targeted at immigrants. For example, English-only movements; Proposition 187 in California, denying illegal aliens public welfare benefits; and the rollback of affirmative action in the California and Texas university systems all accompanied the accelerated pace of Latino immigration. Complicating the situation, Latinos have been immigrating to parts of the United States, such as the South, unaccustomed to non-English speakers. Working in poultry processing plants or picking Vidalia onions in Georgia and North Carolina, the new Latino immigrants have disrupted the traditional black-white southern dichotomies.[5]

Earlier generations of Latinos had limited power and resources to combat institutional inequities. However, by the 1980s strong and well-established national organizations such as the Mexican American Legal Defense and Education Fund (MALDEF) and the National Council of La Raza possessed the political power and resources to monitor and preserve some of the gains of the Civil Rights era. The political disenfranchisement and marginalization of earlier generations, particularly prior to the Civil Rights Act of 1964, was largely avoided through a sometimes awkward but necessary union of diverse Latino subgroups—Mexican Americans, Puerto Ricans, Cubans, Central and South Americans—to create a single movement. This allowed for effective action against measures infringing on citizenship, in support of the rights of undocumented immigrants, and in resistance to a cultural assimilation model in the American public school system.

"NEW IMMIGRANTS" AND THE SCHOOLS AT THE END OF THE TWENTIETH CENTURY

The negative backlash toward the recent influx of immigrants is better understood with a historic understanding of reactions to immigrants one hundred years ago. Between 1890 and 1920 the United States experienced an extraordinary surge of immigration, unparalleled in its history. In 1900, 14 percent of the American population was foreign born. Now called the "old immigrants," they came largely from eastern and southern European nations. Italians, Russian Jews, Slavs, Poles, Greeks, and others came to work in the increasingly industrialized and urban United States. The arrival of so many new people of varying ethnicity, religions, and languages sparked an eventual anti-immigrant backlash. Organizations such as the Immigration Restriction League were founded in the 1890s. Xenophobia increased during World War I, particularly toward Germans. The backlash led to a restrictive immigration law passed in 1921 and also the 1924 National Origins Act. The National Origins Act established specific immigration quotas for each country, largely favoring "Nordic" and "Alpine" nations over Asian nations and other countries considered inferior to the Anglo-Saxon race.[6] This pattern of rapid immigration followed by protest and political backlash reoccurred in the United States at the end of the twentieth century with the arrival of the "new immigrants" in the post-1965 era.

After Congress amended immigration law in 1965, liberalizing quotas, the complexion of American immigration changed significantly. Over 80 percent of newcomers since 1965 have come from Asia and Latin America. In some cases fleeing political terror and oppression and in other cases seeking economic improvement, the Latino population swelled drastically. Latino presence in the United States is not a new phenomenon, but the sudden and dramatic influx has brought unprecedented media and political attention. The media proclaimed the 1980s as the "Decade of the Hispanics," reflecting the nomenclature of the times and the growth of a Latino middle-class moving into politically powerful positions, particularly in Miami and areas of New Mexico and Texas.

However, the sheer numerical increase of Latinos has overshadowed smaller political and economic gains. Between 1990 and 2000 alone the Latino population in the United States increased 58 percent—from 22.4 to 35.3 million. During that same decade the total U.S. population experienced 13.2 percent growth.[7] Furthermore, the 2000 census confirmed that Latinos outnumbered African Americans as the largest minority population in the United States—a demographic shift that has provoked tension between the two groups.[8]

The impact of new Latino immigration upon the nation's schools has been enormous and widespread, extending from the furthest northwestern portions of the country to small Southern towns. Prior to 1968, state and federal governments rarely included an "Hispanic" category in their collection of demographic data and were thus unaware of the rapidly increasing numbers.[9] Between 1968 (the first year national data on Latino students became available) and 1998, the number of Latinos in the public schools increased by 219 percent—from 2 to 4.4 million. In comparison, Anglo enrollment fell by 5.6 million during those decades and African Americans experienced a modest increase of 22 percent (6.3 million to 7.9 million).[10] A youthful population,

Latino immigrants and their children are predicted to continue to expand school enrollments, particularly in the urban centers, and reshape the educational landscape.[11]

Even in the post-1965 era Mexico provided the largest number of immigrants to the United States. However, Latinos from Central America and the Caribbean introduced a new population of children with backgrounds particular to their countries of origin. Beginning in the mid-1960s families from the Dominican Republic, Nicaragua, Guatemala, and El Salvador started an exodus from political terror, violence, and economic hardship. The U.S. government has had a long and sustained interventionist role in these countries, partly designed to prevent the creation of another communist nation like Cuba close to our borders.[12]

In 1965 President Lyndon B. Johnson sent troops into the Dominican Republic to protect American interests during an uprising in which citizens attempted to replace the ruthless dictatorship of General Rafael Leónidas Trujillo with a democratically elected president. As a U.S.-backed regime was put in place, dissidents were tortured, jailed, and killed. Between 1966 and 1974, more than three thousand Dominicans lost their lives.[13] Among the thousands of families who fled from the Dominican Republic was that of the author Julia Alvarez. In one of her memoirs she recounts the fear of leaving home—and the adjustments to the urban United States.[14] Dominicans initially benefited from a tacit agreement in the 1960s between the Dominican Republic and the United States that political dissidents and surplus laborers would be allowed to legally enter the country. In 1962 only slightly over 4,000 Dominicans legally migrated to the United States. By the mid-sixties that number had increased to an average of over 11,000 per year. In the 1970s that number increased to 16,000. By the early 1990s over 40,000 Dominicans were entering the U.S. annually, gravitating to the Dominican centers of New York and Washington, D.C.[15] Although many Dominicans in the first wave belonged to the middle and professional classes, later groups were largely of the working class and poor. Furthermore, Dominicans were generally darker than Cubans and Puerto Ricans and encountered American-style racism as they clustered in the Washington Heights area of New York City.[16] In 1992 the shooting of a young Dominican by a policeman triggered a weekend of rioting in Washington Heights, drawing national attention to the Dominican population.[17]

Political interventions in the small Central American countries of El Salvador, Guatemala, and Nicaragua also triggered widespread immigration to the United States in the 1980s and 1990s. In Nicaragua the U.S.-backed Somoza dictatorship came under fire from the leftist Sandinistas. The Reagan administration assisted in the attempt to remove the Sandinistas from power. During the upheaval of the late 1970s and early 1980s, an estimated 40,000 Nicaraguans were killed and more than 200,000 left their country in terror.[18] Similar situations of political terror and violence arose in El Salvador and Guatemala. In 1980 only 94,000 Salvadoran-born people were recorded in the U.S. census. By 2000 that number had increased to over 1.2 million. The number of legally documented Guatemalans also increased from approximately 70 thousand to one-quarter of a million between 1980 and 2000.[19]

Central American and Caribbean immigrants brought unique situations to the United States, impacting their educational opportunities. Unlike the protective status provided to Cubans, many of the Nicaraguans, Salvadorans, and Guatemalans entered the country illegally and with severe trauma backgrounds. For example, researcher Marcelo Suarez-Orozco studied the lives of new Central American immigrants in two high schools. He learned from these children that "escaping a world of political terror and limited economic opportunities has profound patterned effects upon subsequent functioning in the U.S. The psychological energy invested in dealing with this legacy is not conducive to optimal school functioning, yet I observed no efforts on the part of schools to ameliorate the pain of the victims."[20] Suarez-Orozco nonetheless found a deep commitment to education and resilience among many Central American youth whose parents held strong beliefs in the opportunities of their new country

Another feature of Central American immigration, the sending of one or two family members at a time, also contributed to children's insecurity in the U.S. and its schools. A child psychiatrist for the Los Angeles Unified School District explained a typical pattern: "The father [of a Salvadoran family] came alone first. Two years later, the mother followed with her son. The two oldest daughters, aged 17 and 18, arrived several months later. Finally, the two youngest children, girls aged 9 and 13 who had been living with an aunt in El Salvador, joined the rest of the family."[21] Anxiety over the welfare of those left behind impacted the new immigrants' psychosocial adaptation to a new culture.[22]

Changing attitudes toward immigrants, particularly undocumented immigrants, during the 1980s and 1990s contributed to their educational, social, and economic marginalization.[23] As one school official expressed it, the fear of deportation impeded healthy parent-school relationships: "This fear is expressed in a reluctance to interact with schools officials, to receive recognition, or to 'get into trouble' and call unwarranted attention to [themselves] which may lead to discovery of [their] illegal residency status. Such a fear can poison teacher student relationships. Teachers may feel the immigrant child is not 'trying,' or comes from a family that is not interested."[24]

Limited English skills also impaired the ability of parents to be advocates for their children. In testimony gathered before the U.S. Commission on Civil Rights, witness Rosie Escobar testified that her daughter and son had been hit by their teacher in the Washington, D.C. public schools but "At that time, I could not speak to the teacher, because I could not speak English. There was no one there who would speak Spanish. Some time later, I went to the central office and presented the case. [They told me that] I should not let [2 months] go by before complaining."[25] The same witness eventually pulled her children out of the Washington, D.C. schools, "[Later], I realized that the teacher had a place in the back of the classroom where when she felt like it, all the Hispanic children were placed if they make a mistake. If they didn't bring the homework, she would hit them with a ruler on their hands . . . [M]ore than five times I picked [my daughter up at school], and she was crying. The teacher never wanted to talk to me about the child."[26] These linguistic and immigration status issues significantly shaped the educational opportunities of new immigrants, particularly in areas unaccustomed to linguistic minorities.[27]

LATINO SCHOOLING DURING AN ERA OF IMMIGRANT BACKLASH

In response to the rapid increase of Latinos into the country, state legislatures passed restrictive measures targeted against immigrants, legal or illegal. For example, in 1975 a Texas school district passed a law denying state school funds to schools that permitted undocumented children into their classrooms. This law was taken to court and eventually reached the U.S. Supreme Court. In the landmark case *Plyler v. Doe* (1982) the court struck down the law, 5–4, for violation of the equal protection clause of the Fourteenth Amendment (see document 8.1). During the decade following *Plyler*, the case was used to defend the rights of undocumented children to receive public services.[28] For instance, in states such as Oregon, which began receiving large numbers of Latino children in the late 1990s, *Plyler* was cited to remind public school officials of their legal obligations to educate all children, regardless of immigrant status. (See document 8.2.)

California, home to the largest number of Latinos in the nation, experienced a wave of anti-immigration sentiment in the early 1990s. An economic downturn in the previously booming high technology state helped fuel the passage of Proposition 187 in 1994, Proposition 209 in 1996, and Proposition 227 in 1998. Proposition 187, the "Illegal Aliens, Ineligibility for Public Services Verification, and Reporting Initiatives Statute," would have rendered public services such as schooling at all levels (K-20), health care, and other social services unavailable to undocumented immigrants. Furthermore, it required local and state officials to report any suspected illegal aliens to the Immigration and Naturalization Service. Proposition 187 was immediately tied up in courts after civil rights organizations such as MALDEF and the League of United Latin American Citizens (LULAC) sued the state. Proposition 187 was particularly vulnerable because of its conflict with *Plyler v. Doe*, which forbids such discrimination. Eventually Proposition 187 was declared unconstitutional in September of 1999 without ever having been implemented.[29]

Bilingual education, a historically controversial aspect of Latino schooling, continued to draw controversy from the public and between researchers. As part of the late twentieth century immigrant backlash, an English-only movement began in 1980 and spread throughout the country, particularly in areas of rapid immigrant or refugee arrival. For example, shortly after the arrival of another 125,000 Cuban refugees as part of the 1980 Mariel Boatlift in Miami, Florida (refer to chapter six), the Citizens of Dade (County) United placed a controversial ordinance on the ballot. The organization proposed an ordinance that would prohibit "the expenditure of any county funds for the purpose of utilizing any language other than English or any culture other than that of the United States." The passage of the ordinance inflamed the Cuban community but also received widespread national media attention.[30] Building upon the success of the Miami ordinance, U.S. Senator S. I. Hayakawa of California urged passage of an amendment to the U.S. constitution declaring English the official language of the nation. Legislation for this amendment was introduced in both congress and the senate in 1981, 1983, and 1985, but eventually failed.[31] Although the constitutional amendment was not successful Senator Hayakawa created a nonprofit group in 1983 called "U.S. English" that promoted the establishment of a common language, i.e., English, throughout the country. At the state level, English-only measures were more effective. Between 1981 and 1990, 15 states voted to make English the official language. By 1995 that number had increased to 23. A lawsuit in Arizona challenging the law requiring public business only be conducted in English led to its State Supreme Court striking down the measure in 1998.[32]

By far the most divisive and potentially harmful measure influencing Latino educational equity in the 1980s and 1990s era was the passage in 1998 of Proposition 227, the "Unz Initiative" in California. Proposition 227 (see document 8.3) essentially eliminated most forms of bilingual education in the state's public schools, despite the fact that one out of every eight children in California's schools in 1990 were considered of limited English proficiency (LEP). According to Angela Valenzuela, author of *Subtractive Schooling*, the passage of Proposition 227 was one more negative message for Latino children: "unassessed in current scholarship are the academic consequences to many Mexican youth who 'learn' perhaps no stronger lesson in school than to devalue the Spanish language, Mexico, Mexican culture, and things Mexican."[33] Proposition 227 remains in place in California in 2004 and it has received attention in other states with large numbers of Latino children, including Arizona.

FEDERAL RESPONSES TO LATINO EDUCATION

In 1980 the federal government issued its first major report on Latinos and schools, *The Condition of Education of Hispanic Americans.* In this report, factors that have persisted as obstacles to Latino educational achievement were brought to light.[34] High drop-out rates, low college participation, underrepresentation of Latinos in mathematics and science, and disparately lower scores on standardized tests than Anglo peers were among the items examined in the report. As the U.S. Department of Education was called upon for its resources and funds to assist states with a heavy influx of refugees and immigrants, increased federal attention was placed upon Latinos and education.

In 1994 President William Jefferson Clinton signed Executive Order 12900, "Educational Excellence for Hispanic Americans." Among other things, the order created a special taskforce "to advance the development of human potential, to strengthen the nation's capacity to provide high-quality education, and to increase opportunities for Hispanic Americans to participate in and benefit from federal education programs." (See document 8.4a.) One product from the commission was the research and publication of the extensive 1996 report, *Our Nation on the Fault Line: Hispanic American Education.* (See document 8.4b.) The report's key findings were alarming and critical of the nation's attention to Latinos and schooling. Overall, the report concluded, "educational attainment for most Hispanic Americans is in a state of crisis." Furthermore, "although the gap in some measures of educational attainment is narrowing, the disparity in overall achievement between Hispanic Americans and other Americans is intolerable."

A Nation on the Fault Line also reported the findings of researchers Gary Orfield and his colleagues at the Harvard Project on School Desegregation: that Latinos and African Americans were disproportionately clustered in large urban schools with few white pupils. In essence, black and Latino students were de facto residentially "resegregated" in the 1990s.[35] The U.S. Department of Education produced several reports during the 1980s and 1990s on aspects of bilingual education, high schooling of Latinos, and higher education. A new generation of scholars of Latino heritage and of scholars interested in Latino issues also produced an explosion of research on the challenges facing Latinos and education, a trend that will continue to engage the academic and educational practitioner communities.[36]

The release of Census 2000 data on the numbers of Latinos in the United States reveals a 58 percent increase in the number of Latinos in the United States since 1990; the numbers have also impacted research communities and practitioners seeking to improve the educational achievement of Latinos. (See document 8.5.) In 2000, Latinos (of any race) represented from 11.7 to 12.5 percent of the U.S. population, approaching the number of African Americans, at 12.3 to 12.9 percent.[37] Moreover, in an increasing number of locales and states, Latinos comprise the largest minority group.

Latino immigration is also changing the face of Southern racial and ethnic dynamics. The Latino population has tripled and even quadrupled in states such as Georgia, North Carolina, and Arkansas. (Refer to document 8.5.) Long accustomed to the presence of both blacks and whites, but only recently admitting the necessity to educate them in the same classroom, the public school systems of the South are now contending with new issues such as citizenship education for immigrants and the provision of English for speakers of other languages.[38]

At the beginning of the twenty-first century Latinos are also grappling with both the definitions imposed upon them by the federal government and their own quest for self-definition. The term *Latino* is a relatively new invention, perceived as a broad umbrella term that does not overly emphasize Iberian roots nor dismiss indigenous origins of modern day peoples of Hispanic descent. The simmering tensions between Latino groups over definition and identity will most likely continue as their numbers and political presence grow. (See documents 8.6a and 8.6b.)

DEVELOPMENTS IN HIGHER EDUCATION FOR
LATINOS, 1980–2000

The 1980 census revealed that while 20 percent of Californians were Hispanic, only 2.7 percent possessed college degrees. Nationally Latinos received only 2.3 percent of all baccalaureate degrees in 1980–81 yet earned 3 percent of all doctorates.[39] Detailed examination of these statistics revealed that the majority of undergraduate students were clustered in community colleges. With the exception of Native Americans, Latino college students since 1980 have not fared well compared to African Americans and Asian Americans. Furthermore, Puerto Rican students enrolled in universities on the island of Puerto Rico were counted along with Latinos in the United States, creating a misleadingly high figure of collegiate participation.[40] Most disturbing in the early 1980s was the decline in gains made in the 1970s. The percentage of Hispanic high school graduates attending college for example, decreased from 35.4 percent in 1975 to 29.9 percent in 1980.[41] As a result of the severe disconnect between Latino educational achievement, the size of its population, and its relative underachievement compared to other ethnic groups, a series of steps were taken in the 1980s to regain advances begun in the 1970s.

Michael Olivas, through his supervision of the influential 1982 National Center for Education Statistics report, *Condition of Education for Hispanic Americans*, spurred reform in Hispanic education. Chairman of the Subcommittee on Postsecondary Education Paul Simon initiated the Hispanic Access to Higher Education Project after reviewing Olivas's work.[42] A series of hearings on the topic of Hispanics and higher education subsequently took place on several campuses during 1982 and 1983. Simon introduced H.R. 5240, the Higher Education Act Amendments of 1984, which recommended several reforms to aid Hispanic access and retention. These included the modification of Title III to provide direct aid to institutions with high concentrations of Hispanic students; specific monies for Hispanic students in the TRIO Programs[43] (these first three federal outreach programs for promising students, preparing them to do college-level work, were the creation of the 1965 Higher Education Act); federal early outreach and student services programs; Upward Bound, Talent Search, and student support services; a special emphasis on teacher preparation (Title V) programs to train teachers for Hispanic populations; and increased monies for the Graduate and Professional Opportunities Program (G*POP) to channel more Latinos toward graduate and professional schools. Although H.R. 5240 was not approved, subsequent legislation adopted the bill's key points. Furthermore, publication of the Staff Report on Hispanics' Access to Higher Education (1985) provided data to support future reforms.

The creation of the Hispanic Association of Colleges and Universities (HACU) in 1986 brought together Hispanic leaders in business and two- and four-year colleges and universities with large numbers of Latinos into one powerful advocacy organization. The mission of HACU, to improve the access and quality of college education for Hispanics, has been carried out through offices in San Antonio, Texas, and Washington, D.C.[44] HACU's most successful victory was the establishment of Hispanic Serving Institutions (HSI's) as a federally recognized category. Hispanic Serving Institutions, unlike HBCU's or Tribal Colleges, do not necessarily have a specific historic mission toward Latino education. Instead, the Department of Education defines HSIs as postsecondary institutions with at least 25 percent Hispanic full-time equivalent enrollment and also 50 percent or more low-income students. Laden credits HACU with successfully having HSIs recognized in the reauthorization of Title III of the Higher Education Act of 1992, thus securing eligibility for federal funds.[45] Furthermore, in the reauthorization of the Higher Education Act of 1998 HSIs were included with Tribal Colleges and HBCU's under Title V, allowing them a larger slice of the federal pie.

Higher education for Latinos has also been influenced by the anti-immigrant backlash and anti-affirmative action sentiments of the late 1980s and 1990s. In 1996 California's Proposition 209 eliminated preferential treatment on the basis of "race, sex, color, ethnicity, or national origin" in public sectors, including education, K-20. In Florida and Texas "10 percent" plans were implemented in place of affirmative action to recruit minorities into the states' university systems.

From a judicial point of view, conflicting decisions in cases such as *Hopwood v. State of Texas* (Fifth Circuit, 1994) and *Smith v. University of Washington* (Ninth Circuit, 2000), respectively repudiating and permitting the use of race in admissions, called into question the future of affirmative action programs as well as the future of *Bakke v. Regents of University of California (1978)*, the U.S. Supreme Court decision permitting the use of race in college admissions. On June 23, 2003, however, the U.S. Supreme Court affirmed the use of race as a legitimate tool in law school admissions in *Grutter v. Bollinger et al.* The case held that "the law schools narrowly tailored use of race in admissions decision to further a compelling interest in obtaining the educational benefits that flow from a diverse student body is not prohibited by the Equal Protection Clause." The findings in the law school case also affirmed that "diversity is essential to its [University of Michigan Law School] educational mission." *Grutter v. Bollinger*, and its accompanying case concerning undergraduate admissions at the University of Michigan, provided in 2003 a pause in the general trend against affirmative action measures.

Researchers also point to future agendas impacting Latino higher education. One development concerns lawsuits filed to stop the use of Advanced Placement (AP) courses for admissions to college. The American Civil Liberties Union contends that many students in California and Texas simply do not have these offerings at their schools and should not be punished for unequal school resources.[46]

A further issue concerns undocumented Latino college students. Many students were raised in the United States but their parents did not have them naturalized; as a result they are being rejected from college admissions and financial aid packages. The complicated legal status of these students was brought to the attention of legislators who introduced the Development, Relief, and Education for Alien Minors Act (DREAM Act) in the summer of 2002. The bill would grant legal residency to undocumented students with no criminal records who have been U.S. residents for at least five years and graduated from an American high school or received a GED.[47]

Lastly, although promising gains have been made in higher education enrollment among Latinos, recent studies note that students continue to drop out at high rates for financial reasons. Furthermore, Latinos remain clustered in two-year colleges (40 percent of all Latinos at the baccalaureate level are at community colleges) compared to 25 percent of white college students.[48] Overall, researchers predict that Latinos in the eighteen- to twenty-four-year-old range will be underrepresented by 500,000 students by the middle of the twenty-first century, a devastating loss of human potential.

The perilous political climate of the late 1990s for affirmative action and other programs such as bilingual education suggests Latino higher education may again cycle backward instead of forward. As this narrative has highlighted, numerical dominance of Latinos as the largest minority group in the United States does not readily translate into equity. History teaches us that an equitable higher education for Latinos in the future requires vigilance and activism today.

DOCUMENT 8.1

Plyler v. Doe, 1982

Legal and illegal immigration escalated significantly in the decades following 1965, straining public services and spurring nativist sentiments and legislation. In 1975, the Texas legislature passed a law that would withhold state funds from any school district enrolling children not "legally admitted" to the United States. A series of challenges to this law eventually led to an appeal in the U.S. Supreme Court. In Plyler v. Doe, *the Court ruled that there were neither legal nor economic grounds to deny public schooling to undocumented immigrant children. In fact, denial of access to public education was ruled as a violation of the equal protection clause of the Fourteenth Amendment. The majority opinion argued that these children fell within the jurisdiction of the state, that the children could not be "punished" for their parents' decision to enter the country illegally, and that failing to provide education to this large and already impoverished population would exacerbate their economic oppression and virtually ensure the continuation of poverty for a large portion of the population. The dissenting opinion agreed in principle with the decision but disagreed that it is the Court's role to decide policy. In these excerpts of the lengthy case, the justices grapple with the status and rights of illegal immigrants, and the role of public education in the late twentieth century. The full court case, including detailed footnotes, may be accessed through http://web.lexis-nexis.com/universe* (accessed October 27, 2003).

Plyler, Superintendent, Tyler Independent School District, et al. v. Doe, Guardian, et al.

No. 80–1538

Supreme Court of the United States

457 U.S. 202; 102 S. Ct. 2382; 72 L. Ed. 2d 786; 1982 U.S. LEXIS 124; 50 U.S.L.W. 4650

December 1, 1981, Argued
June 15, 1982, Decided

* Together with No. 80–1934, Texas et al. v. Certain Named and Unnamed Undocumented Alien Children et al., also on appeal from the same court.

Subsequent History:

Petition for Rehearing Denied September 9, 1982.

Prior History: Appeal from the United States Court of Appeals for the Fifth Circuit.

DECISION: Texas statute withholding funds from local school districts of education of children not legally admitted into United States and authorizing districts to deny enrollment to such children, held to violate equal protection clause.

SUMMARY: The Texas legislature enacted a statute which withholds from local school districts any state funds for the education of children who were not "legally admitted" into the United States and which authorizes local school districts to deny enrollment in their public schools to such children. After making extensive findings of fact, . . . the District Court held that the Texas statute violated the equal protection clause of the Fourteenth Amendment, concluding that the statute was not carefully tailored to advance the asserted state interest in an acceptable manner (501 F Supp 544), and, apparently on the strength of its earlier decision, the Court of Appeals summarily affirmed the decision of the District Court . . .

Marshall, J., concurring, emphasized his belief that an individual's interest in education is fundamental and that a class-based denial of public education is utterly incompatible with the equal protection clause of the Fourteenth Amendment.

Blackmun, J., concurring, expressed his view that the nature of the interest at stake in the case was crucial to its proper resolution, that when a state provides an education to some and denies it to others it immediately and inevitably creates class distinctions of a type fundamentally inconsistent with the purposes of the equal protection clause, and that whatever the state's power to classify deportable aliens, the statute at issue swept within it a substantial number of children who would in fact, and who may well be entitled to, remain in the United States . . .

Burger, Ch. J., joined by White, Rehnquist, and O'Connor, JJ., dissented, expressing the view that the United States Supreme Court trespasses on the assigned function of the political branches under the stricture of limited and separate powers when it assumes a policymaking role as the court did in the case before it, and that the distinction that Texas had drawn was based not only upon its own legitimate interest but on classifications established by the federal government in its immigration laws and policies and was not unconstitutional . . .

Opinion by: Justice Brennan

The question presented by these cases is whether, consistent with the Equal Protection Clause of the Fourteenth Amendment, Texas may deny to undocumented school-age children the free public education that it provides to children who are citizens of the United States or legally admitted aliens.

I

Since the late 19th century, the United States has restricted immigration into this country. Unsanctioned entry into the United States is a crime, 8 U. S. C. §1325, and those who have entered unlawfully are subject to deportation, 8 U. S. C. § 1251, 1252 (1976 ed. and Supp. IV). But despite the existence of these legal restrictions, a substantial number of persons have succeeded in unlawfully entering the United States, and now live within various States, including the State of Texas.

In May 1975, the Texas Legislature revised its education laws to withhold from local school districts any state funds for the education of children who were not "legally admitted" into the United States. The 1975 revision also authorized local school districts to deny enrollment in their public schools to children not "legally admitted" to the country. Tex. Educ. Code Ann. § 21.031 (Vernon Supp. 1981). These cases involve constitutional challenges to those provisions . . .

In considering this motion, the District Court made extensive findings of fact. The court found that neither § 21.031 nor the School District policy implementing it had "either the purpose or effect of keeping illegal aliens out of the State of Texas." 458 F.Supp. 569, 575 (1978). Respecting defendants' further claim that § 21.031 was simply a financial measure designed to avoid a drain on the State's fisc, the court recognized that the increase in population resulting from the immigration of Mexican nationals into the United States had created problems for the public schools of the State, and that these problems were exacerbated by the special educational needs of immigrant Mexican children. The court noted, however, that the increase in school enrollment was primarily attributable to the admission of children who were legal residents. *Id.*, at 575–576. It also found that while the "exclusion of all undocumented children from the public schools in Texas would eventually result in economies at some level," *id.*, at 576, funding from both the State and Federal Governments was based primarily on the number of children enrolled. In net effect then, barring undocumented children from the schools would save money, but it would "not necessarily" improve "the quality of education." *Id.*, at 577. The court further observed that the impact of § 21.031

was borne primarily by a very small subclass of illegal aliens, "entire families who have migrated illegally and—for all practical purposes—permanently to the United States." *Id.*, at 578. Finally, the court noted that under current laws and practices "the illegal alien of today may well be the legal alien of tomorrow," and that without an education, these undocumented children, "[already] disadvantaged as a result of poverty, lack of English-speaking ability, and undeniable racial prejudices . . . will become permanently locked into the lowest socio-economic class." *Id.*, at 577.

II

The Fourteenth Amendment provides that "[no] State shall . . . deprive any person of life, liberty, or property, without due process of law; nor deny to *any person within its jurisdiction* the equal protection of the laws." [Emphasis added.] Appellants argue at the outset that undocumented aliens, because of their immigration status, are not "persons within the jurisdiction" of the State of Texas, and that they therefore have no right to the equal protection of Texas law. We reject this argument. Whatever his status under the immigration laws, an alien is surely a "person" in any ordinary sense of that term. Aliens, even aliens whose presence in this country is unlawful, have long been recognized as "persons" guaranteed due process of law by the Fifth and Fourteenth Amendments. *Shaughnessy* v. *Mezei*, 345 U.S. 206, 212 (1953); *Wong Wing* v. *United States*, 163 U.S. 228, 238 (1896); *Yick Wo* v. *Hopkins*, 118 U.S. 356, 369 (1886). Indeed, we have clearly held that the Fifth Amendment protects aliens whose presence in this country is unlawful from invidious discrimination by the Federal Government. *Mathews* v. *Diaz*, 426 U.S. 67, 77 (1976).

. . .

Use of the phrase "within its jurisdiction" thus does not detract from, but rather confirms, the understanding that the protection of the Fourteenth Amendment extends to anyone, citizen or stranger, who *is* subject to the laws of a State, and reaches into every corner of a State's territory. That a person's initial entry into a State, or into the United States, was unlawful, and that he may for that reason be expelled, cannot negate the simple fact of his presence within the State's territorial perimeter. Given such presence, he is subject to the full range of obligations imposed by the State's civil and criminal laws. And until he leaves the jurisdiction—either voluntarily, or involuntarily in accordance with the Constitution and laws of the United States—he is entitled to the equal protection of the laws that a State may choose to establish.

Our conclusion that the illegal aliens who are plaintiffs in these cases may claim the benefit of the Fourteenth Amendment's guarantee of equal protection only begins the inquiry. The more difficult question is whether the Equal Protection Clause has been violated by the refusal of the State of Texas to reimburse local school boards for the education of children who cannot demonstrate that their presence within the United States is lawful, or by the imposition by those school boards of the burden of tuition on those children. It is to this question that we now turn.

III

The Equal Protection Clause directs that "all persons similarly circumstanced shall be treated alike." *F.S. Royster Guano Co.* v. *Virginia*, 253 U.S. 412, 415 (1920) . . .

We turn to a consideration of the standard appropriate for the evaluation of § 21.031.

A

Sheer incapability or lax enforcement of the laws barring entry into this country, coupled with the failure to establish an effective bar to the employment of undocumented aliens, has resulted in the creation of a substantial "shadow population" of illegal migrants—numbering in the millions—within our borders. This situation raises the specter of a permanent caste of undocumented resident aliens, encouraged by some to remain here as a source of cheap labor, but nevertheless denied the benefits that our society makes available to citizens and lawful residents. The existence of such an underclass presents most difficult problems for a Nation that prides itself on adherence to principles of equality under law.

The children who are plaintiffs in these cases are special members of this underclass. Persuasive arguments support the view that a State may withhold its beneficence from those whose very presence within the United States is the product of their own unlawful conduct. These arguments do not apply with the same force to classifications imposing disabilities on the minor *children* of such illegal entrants. At the least, those who elect to enter our territory by stealth and in violation of our law should be prepared to bear the consequences, including, but not limited to, deportation. But the children of those illegal entrants are not comparably situated. Their "parents have the ability to conform their conduct to societal norms," and presumably the ability to remove themselves from the State's jurisdiction; but the children who are plaintiffs in these cases "can affect neither their parents' conduct nor their own status." *Trimble* v. *Gordon*, 430 U.S. 762, 770 (1977). Even if the State found it expedient to control the conduct of adults by acting against their children, legislation directing the onus of a parent's misconduct against his children does not comport with fundamental conceptions of justice.

"[Visiting] . . . condemnation on the head of an infant is illogical and unjust. Moreover, imposing disabilities on the . . . child is contrary to the basic concept of our system that legal burdens should bear some relationship to individual responsibility or wrongdoing. Obviously, no child is responsible for his birth and penalizing the . . . child is an ineffectual—as well as unjust— way of deterring the parent." *Weber* v. *Aetna Casualty & Surety Co.*, 406 U.S. 164, 175 (1972) (footnote omitted).

Of course, undocumented status is not irrelevant to any proper legislative goal. Nor is undocumented status an absolutely immutable characteristic since it is the product of conscious, indeed unlawful, action. But §21.031 is directed against children, and imposes its discriminatory burden on the basis of a legal characteristic over which children can have little control. It is thus difficult to conceive of a rational justification for penalizing these children for their presence within the United States. Yet that appears to be precisely the effect of § 21.031.

Public education is not a "right" granted to individuals by the Constitution. *San Antonio Independent School Dist.* v. *Rodriguez*, 411 U.S. 1, 35 (1973). But neither is it merely some governmental "benefit" indistinguishable from other forms of social welfare legislation. Both the importance of education in maintaining our basic institutions, and the lasting impact of its deprivation on the life of the child, mark the distinction. The "American people have always regarded education and [the] acquisition of knowledge as matters of supreme importance." *Meyer* v. *Nebraska*, 262 U.S. 390, 400 (1923). We have recognized "the public schools as a most vital civic institution for the preservation of a democratic system of government," *Abington School District* v. *Schempp*, 374 U.S. 203, 230 (1963) (BRENNAN, J., concurring), and as the primary vehicle for transmitting "the values on which our society rests." *Ambach* v. *Norwick*, 441 U.S. 68, 76 (1979). "[As] . . . pointed out early in our history, . . . some degree of education is necessary to prepare citizens to participate effectively and intelligently in our open political system if we are to preserve freedom and independence." *Wisconsin* v. *Yoder*, 406 U.S. 205, 221 (1972). And these historic "perceptions of the public schools as inculcating fundamental values necessary to the maintenance of a democratic political system have been confirmed by the observations of social scientists." *Ambach* v. *Norwick, supra*, at 77. In addition, education provides the basic tools by which individuals might lead economically productive lives to the benefit of us all. In sum, education has a fundamental role in maintaining the fabric of our society. We cannot ignore the significant social costs borne by our Nation when select groups are denied the means to absorb the values and skills upon which our social order rests.

In addition to the pivotal role of education in sustaining our political and cultural heritage, denial of education to some isolated group of children poses an affront to one of the goals of the Equal Protection Clause: the abolition of governmental barriers presenting unreasonable obstacles to advancement on the basis of individual merit. Paradoxically, by depriving the children of any disfavored group of an education, we foreclose the means by which that group might raise the level of esteem in which it is held by the majority. But more directly, "education prepares individuals to be self-reliant and self-sufficient participants in society." *Wisconsin* v. *Yoder, supra,* at 221. Illiteracy is an enduring disability. The inability to read and write will handicap the individual deprived of a basic education each and every day of his life. The inestimable toll of that deprivation on the social, economic, intellectual, and psychological well-being of the individual, and the obstacle it poses to individual achievement, make it most difficult to reconcile the cost or the principle of a status-based denial of basic education with the framework of equality embodied in the Equal Protection Clause. What we said 28 years ago in *Brown* v. *Board of Education*, 347 U.S. 483 (1954), still holds true:

"Today, education is perhaps the most important function of state and local governments. Compulsory school attendance laws and the great expenditures for education both demonstrate our recognition of the importance of education to our democratic society. It is required in the performance of our most basic public responsibilities, even service in the armed forces. It is the very foundation of good citizenship. Today it is a principal instrument in awakening the child to cultural values, in preparing him for later professional training, and in helping him to adjust normally to his environment. In these days, it is doubtful that any child may reasonably be expected to succeed in life if he is denied the opportunity of an education. Such an opportunity, where the state has undertaken to provide it, is a right which must be made available to all on equal terms." *Id.,* at 493.

B

These well-settled principles allow us to determine the proper level of deference to be afforded § 21.031. Undocumented aliens cannot be treated as a suspect class because their presence in this country in violation of federal law is not a "constitutional irrelevancy." Nor is education a fundamental right; a State need not justify by compelling necessity every variation in the manner in which education is provided to its population. See *San Antonio Independent School Dist.* v. *Rodriguez, supra,* at 28–39. But more is involved in these cases than the abstract question whether § 21.031 discriminates against a suspect class, or whether education is a fundamental right. Section § 21.031 imposes a lifetime hardship on a discrete class of children not accountable for their disabling status. The stigma of illiteracy will mark them for the rest of their lives. By denying these children a basic education, we deny them the ability to live within the structure of our civic institutions, and foreclose any realistic possibility that they will contribute in even the smallest way to the progress of our Nation. In determining the rationality of § 21.031, we may appropriately take into account its costs to the Nation and to the innocent children who are its victims. In light of these countervailing costs, the discrimination contained in § 21.031 can hardly be considered rational unless it furthers some substantial goal of the State.

. . .

Finally, appellants suggest that undocumented children are appropriately singled out because their unlawful presence within the United States renders them less likely than other children to remain within the boundaries of the State, and to put their education to productive social or political use within the State. Even assuming that such an interest is legitimate, it is an interest that is most difficult to quantify. The State has no assurance that any child, citizen or not, will employ the education provided by the State within the confines of the State's borders. In any event, the record is clear that many of the undocumented children disabled by this classification will remain in this country indefinitely, and that some will become lawful residents or citizens of the United States. It is difficult to understand precisely what the State hopes to achieve by promoting the creation and perpetuation of a subclass of illiterates within our boundaries, surely adding to the problems and costs of unemployment, welfare, and crime. It is thus clear that whatever savings might be achieved by denying these children an education, they are wholly insubstantial in light of the costs involved to these children, the State, and the Nation.

VI

If the State is to deny a discrete group of innocent children the free public education that it offers to other children residing within its borders, that denial must be justified by a showing that it furthers some substantial state interest. No such showing was made here. Accordingly, the judgment of the Court of Appeals in each of these cases is

Affirmed.

DOCUMENT 8.2

Executive Memorandum, Oregon Board of Education, 2001

During the 1990s, Latino immigrants increasingly moved to the Southern states, the rural Midwest, and the Far West in search of employment. School districts in those areas, unaccustomed to large influxes of English language learners, faced numerous challenges in meeting federal, state, and district rules, including provisions for bilingual education and the legal status of children. In this document from 2001, the Oregon State Board of Education informs local districts of laws regarding undocumented students' access to public education.

From: Oregon Department of Education. Executive Memorandum 119–2000-01. February 15, 2001. www.ode.state.or.us/supportservices/memos/2000_01/119–01.htm. Accessed May 17, 2004.

Oregon Department of Education
Executive Memorandum 119–2000-01
February 15, 2001
TO: All Education Service District and School District Superintendents
RE: Undocumented Students' Rights To Attend Public Schools

Questions continue to arise from parents and school personnel as to educating undocumented students. Both Oregon law and the US Supreme Court affirm the right of undocumented youth to education in Oregon without regard to the citizenship of the child or parent.

Under ORS 339.115 the resident school district is obligated to provide a free education for all children between the ages of 5 and 19 residing in the district. Whether or not a student is a resident must be determined by ORS 339.133, which contains no provision for citizenship. Although districts may ask for the student's residence and age, care should be taken to not suggest that the student must have immigration papers in order to be enrolled in school. Also, all children between the ages of 7 and 18, regardless of their immigration status, are required to regularly attend school under Oregon's Compulsory Attendance Law. Under that law, all parents, regardless of their immigration status, have the legal duty, for any child who is under their control and who is between the age of 7 and 18, to send the child to school and maintain the child in regular attendance at the school.

In 1982 the US Supreme Court in *Plyer v. Doe* ruled that undocumented children have the same right to attend public primary and secondary schools as do U.S. citizens and permanent residents.

As a result of *Plyer v. Doe*, school districts must not:

- Deny admission to a student during initial enrollment or at any other time on the basis of undocumented status, (this could also be considered impermissible discrimination under ORS 659.150)
- Treat a student disparately to determine residency,
- Engage in any practices to deter or discourage the right of access of school,
- Require students or parents to disclose or document their immigration status,
- Make inquiries of students or parents that may expose their undocumented status,
- Require social security numbers from all students, as this may expose undocumented status (this is also impermissible under the Privacy Act of 1974).

Under the Family Education Rights and Privacy Act (FERPA), the Immigration and Naturalization Service, like any other outside agency, cannot obtain student record information from the public schools unless the school has written permission from the parent.

School personnel should be aware that they have no legal obligation or authority to enforce US immigration laws.

Simply stated, every child residing in an Oregon school district is entitled to a free public education without regard to immigration status of the child or the child's parents.

Please share this information with your building administrators and particularly those staff members that have initial contact with students . . .

Clark S. Brody
Deputy Superintendent of Public Instruction
Oregon Department of Education

Salem, Oregon

DOCUMENT 8.3

Proposition 227, the Unz Initiative of California, 1998

California is viewed as a bellwether state; other states look to the Golden State for trends in policies and politics. During the 1990s California voters passed Proposition 227 by a majority of 61 percent to 39 percent. The measure, often called the Unz Initiative because of its sponsorship by millionaire Ron Unz, effectively made it illegal to provide any public instruction for English language learners in any other language than English. Unz and his backers believed that the underachievement of English language learners indicated a failure of bilingual education to achieve its goals. For instance, Unz and associates contended that less than 5 percent of English language learners were reclassified to mainstream classrooms per year. Although opponents and academics questioned the data utilized to advocate for the measure, it passed by a two thirds vote and remains in place in 2004. Other states have followed suit, passing similar measures limiting bilingual education in Arizona and Massachusetts.

This initiative measure is submitted to the people in accordance with the provisions of Article II, Section 8 of the Constitution.

Proposed Law

SECTION 1. Chapter 3 (commencing with Section 300) is added to Part 1 of the Education Code, to read:

Chapter 3. English Language Education for Immigrant Children
Article 1. Findings and Declarations

300. The People of California find and declare as follows:

(a) Whereas, The English language is the national public language of the United States of America and of the State of California, is spoken by the vast majority of California residents, and is also the leading world language for science, technology, and international business, thereby being the language of economic opportunity; and

(b) Whereas, Immigrant parents are eager to have their children acquire a good knowledge of English, thereby allowing them to fully participate in the American Dream of economic and social advancement; and

(c) Whereas, The government and the public schools of California have a moral obligation and a constitutional duty to provide all of California's children, regardless of their ethnicity or national origins, with the skills necessary to become productive members of our society, and of these skills, literacy in the English language is among the most important; and

(d) Whereas, The public schools of California currently do a poor job of educating immigrant children, wasting financial resources on costly experimental language programs whose failure over the past two decades is demonstrated by the current high drop-out rates and low English literacy levels of many immigrant children; and

(e) Whereas, Young immigrant children can easily acquire full fluency in a new language, such as English, if they are heavily exposed to that language in the classroom at an early age.

(f) Therefore, It is resolved that: all children in California public schools shall be taught English as rapidly and effectively as possible.

Article 2. English Language Education

305. Subject to the exceptions provided in Article 3 (commencing with Section 310), all children in California public schools shall be taught English by being taught in English. In particular, this shall require that all children be placed in English language classrooms. Children who are English learners shall be educated through sheltered English immersion during a temporary transition period not normally intended to exceed one year. Local schools shall be permitted to place in the same classroom English learners of different ages but whose degree of English proficiency is similar. Local schools shall be encouraged to mix together in the same classroom English learners from different native-language groups but with the same degree of English fluency. Once English learners have acquired a good working knowledge of English, they shall be transferred to English language mainstream classrooms. As much as possible, current supplemental funding for English learners shall be maintained, subject to possible modification under Article 8 (commencing with Section 335) below.

306. The definitions of the terms used in this article and in Article 3 (commencing with Section 310) are as follows:

(a) "English learner" means a child who does not speak English or whose native language is not English and who is not currently able to perform ordinary classroom work in English, also known as a Limited English Proficiency or LEP child.

(b) "English language classroom" means a classroom in which the language of instruction used by the teaching personnel is overwhelmingly the English language, and in which such teaching personnel possess a good knowledge of the English language.

(c) "English language mainstream classroom" means a classroom in which the pupils either are native English language speakers or already have acquired reasonable fluency in English.

(d) "Sheltered English immersion" or "structured English immersion" means an English language acquisition process for young children in which nearly all classroom instruction is in English but with the curriculum and presentation designed for children who are learning the language.

(e) "Bilingual education/native language instruction" means a language acquisition process for pupils in which much or all instruction, textbooks, and teaching materials are in the child's native language.

Article 3. Parental Exceptions

310. The requirements of Section 305 may be waived with the prior written informed consent, to be provided annually, of the child's parents or legal guardian under the circumstances specified below and in Section 311. Such informed consent shall require that said parents or legal guardian personally visit the school to apply for the waiver and that they there be provided a full description of the educational materials to be used in the different educational program choices and all the educational opportunities available to the child. Under such parental waiver conditions, children may be transferred to classes where they are taught English and other subjects through bilingual education techniques or other generally recognized educational methodologies permitted by law. Individual schools in which 20 pupils or more of a given grade level receive a waiver shall be required to offer such a class; otherwise, they must allow the pupils to transfer to a public school in which such a class is offered.

311. The circumstances in which a parental exception waiver may be granted under Section 310 are as follows:

(a) Children who already know English: the child already possesses good English language skills, as measured by standardized tests of English vocabulary comprehension, reading, and writing, in which the child scores at or above the state average for his or her grade level or at or above the 5th grade average, whichever is lower; or

(b) Older children: the child is age 10 years or older, and it is the informed belief of the school principal and educational staff that an alternate course of educational study would be better suited to the child's rapid acquisition of basic English language skills; or

(c) Children with special needs: the child already has been placed for a period of not less than thirty days during that school year in an English language classroom and it is subsequently the informed belief of the school principal and educational staff that the child has such special physical, emotional, psychological, or educational needs that an alternate course of educational study would be better suited to the child's overall educational development. A written description of these special needs must be provided and any such decision is to be made subject to the examination and approval of the local school superintendent, under guidelines established by and subject to the review of the local Board of Education and ultimately the State Board of Education. The existence of such special needs shall not compel issuance of a waiver, and the parents shall be fully informed of their right to refuse to agree to a waiver.

Article 4. Community-Based English Tutoring

315. In furtherance of its constitutional and legal requirement to offer special language assistance to children coming from backgrounds of limited English proficiency, the state shall encourage family members and others to provide personal English language tutoring to such children, and support these efforts by raising the general level of English language knowledge in the community. Commencing with the fiscal year in which this initiative is enacted and for each of the nine fiscal years following thereafter, a sum of fifty million dollars ($50,000,000) per year is hereby appropriated from the General Fund for the purpose of providing additional funding for free or subsidized programs of adult English language instruction to parents or other members of the community who pledge to provide personal English language tutoring to California school children with limited English proficiency.

316. Programs funded pursuant to this section shall be provided through schools or community organizations. Funding for these programs shall be administered by the Office of the Superintendent of Public Instruction, and shall be disbursed at the discretion of the local school boards, under reasonable guidelines established by, and subject to the review of, the State Board of Education.

Article 5. Legal Standing and Parental Enforcement

320. As detailed in Article 2 (commencing with Section 305) and Article 3 (commencing with Section 310), all California school children have the right to be provided with an English language public education. If a California

school child has been denied the option of an English language instructional curriculum in public school, the child's parent or legal guardian shall have legal standing to sue for enforcement of the provisions of this statute, and if successful shall be awarded normal and customary attorney's fees and actual damages, but not punitive or consequential damages. Any school board member or other elected official or public school teacher or administrator who willfully and repeatedly refuses to implement the terms of this statute by providing such an English language educational option at an available public school to a California school child may be held personally liable for fees and actual damages by the child's parents or legal guardian.

Article 6. Severability

325. If any part or parts of this statute are found to be in conflict with federal law or the United States or the California State Constitution, the statute shall be implemented to the maximum extent that federal law, and the United States and the California State Constitution permit. Any provision held invalid shall be severed from the remaining portions of this statute.

Article 7. Operative Date

330. This initiative shall become operative for all school terms which begin more than sixty days following the date on which it becomes effective.

Article 8. Amendment

335. The provisions of this act may be amended by a statute that becomes effective upon approval by the electorate or by a statute to further the act's purpose passed by a two-thirds vote of each house of the Legislature and signed by the Governor.

Article 9. Interpretation

340. Under circumstances in which portions of this statute are subject to conflicting interpretations, Section 300 shall be assumed to contain the governing intent of the statute.

DOCUMENT 8.4a

Presidential Executive Order 12900

Although the 1980s had been called the "Decade of the Hispanics," Latino children and youth were still struggling on several indicators of academic achievement. Using his power as President, William Jefferson Clinton issued an executive order (a rule having the force of law) on February 22, 1994, to establish the White House Commission on Educational Excellence for Hispanic Americans. The commission's charge included finding ways to "eliminate educational inequities and disadvantages faced by Hispanic Americans," and "promote and publicize educational opportunities and programs of interest to Hispanic Americans." The presidential commission issued the comprehensive report, Our Nation on the Fault Line: Hispanic American Education, in 1996.

The White House
February 22, 1994

EDUCATIONAL EXCELLENCE FOR HISPANIC AMERICANS

By the authority vested in me as President by the Constitution and the laws of the United States of America, and in order to advance the development of human potential, to strengthen the Nation's capacity to provide high-quality education, and to increase opportunities for Hispanic Americans to participate in and benefit from Federal education programs, it is hereby ordered as follows:

Section 1. There shall be established in the Department of Education the President's Advisory Commission on Educational Excellence for Hispanic American (Commission). The Commission shall consist of not more than 25 members, who shall be appointed by the President and shall report to the Secretary of Education (Secretary). The Commission shall comprise representatives who: (a) have a history of involvement with the Hispanic community; (b) are from the education, civil rights, and business communities; or (c) are from civic associations representing the diversity within the Hispanic community. In addition, the President may appoint other representatives as he deems appropriate.

Sec. 2. The Commission shall provide advice to the President and the Secretary on: (a) the progress of Hispanic Americans toward achievement of the National Education Goals and other standards of educational accomplishment; (b) the development, monitoring, and coordination of Federal efforts to promote high-quality education for Hispanic Americans; (c) ways to increase State, private sector, and community involvement in improving education and (d) ways to expand and complement Federal education initiatives. The Commission shall provide advice to the President through the Secretary.

Sec. 3. There shall be established in the Department of Education the White House Initiative on Educational Excellence for Hispanic Americans (Initiative). The Initiative shall be an interagency working group coordinated by the Department of Education and shall be headed by a Director, who shall be a senior level Federal official. It shall provide the staff, resources, and assistance for the Commission and shall serve the Secretary in carrying out his or her responsibilities under this order. The Initiative is authorized to utilize the services, personnel, information, and facilities of other Federal, State, and local agencies with their consent, and with or without reimbursement, consistent with applicable law. To the extent permitted b [*sic*] law and regulations, each Federal agency shall cooperate in providing resources including personnel detailed to the Initiative, to meet the objectives of this order. The Initiative shall include both career civil service and appointed staff with expertise in the area of education, and shall provide advice to the Secretary on the implementation and coordination of education and related programs across Executive agencies.

Sec. 4. Each Executive department and each agency designated by the Secretary shall appoint a senior official, who is a full-time officer of the Federal Government and responsible for management or program administration, to report directly to the agency head on activity under this Executive order and to serve as liaison to the Commission and the Initiative. To the extend [*sic*] permitted by law and to the extent practicable, each Executive department and designated agency shall provide any appropriate information requested by the Commission or the staff of the Initiative, including data relating to the eligibility for and participation by Hispanic Americans in Federal education programs and the progress of Hispanic Americans in relation to the National Education Goals. Where adequate data is not available, the Commission shall suggest the means of collecting the data.

Sec. 5. The Secretary, in consultation with the Commission, shall: submit to the President an Annual Federal Plan to Promote Hispanic American Educational Excellence (Annual Federal Plan, or Plan). All actions described in the Plan shall [be] designed to help Hispanic Americans attain the educational improvement targets set forth in the National Education Goals and any standards established by the National Education Standards and Improvement Council. The Plan shall include data on eligibility for, and participation by, Hispanic Americans in Federal education programs, and such other aspects of the educational status of Hispanic Americans as the Secretary considers appropriate. This Plan also shall include, as an appendix, the text of the agency plans described in section 6 of this order. The Secretary, in consultation with the Commission and with the assistance of the Initiative staff, shall ensure that superintendents of Hispanic-serving school districts, presidents of Hispanic-serving institutions of higher education, directors of educational programs for Hispanic Americans, and other appropriate individuals are given the opportunity to comment on the proposed Annual Federal Plan. For purposes of this order, a "Hispanic-serving" school district or institution of higher education is any local education agency or institution of higher education, respectively, whose student population is more than 25 percent Hispanic.

Sec. 6. As part of the development of the Annual Federal Plan, each Executive department and each designated agency (hereinafter in this section referred to collectively as "agency") shall prepare a plan for, and shall document, both that agency's effort to increase Hispanic American participation in Federal education programs where Hispanic Americans currently are underserved, and that the agency's effort to improve educational outcomes for Hispanic Americans participating in Federal education programs. This plan shall address, among other relevant issues: (a) the elimination of unintended regulatory barriers to Hispanics American participation's in Federal programs; (b) the adequacy of announcements of program opportunities of interest to Hispanic-serving school districts, institutions of higher education, and agencies; and (c) ways of eliminating educational inequalities and disadvantages faced by Hispanic Americans. It also shall emphasize the facilitation of technical, planning, and development advice to Hispanic-serving school districts and institutions of higher education. Each agency's plan shall provide appropriate measurable objectives for proposed actions aimed at increasing Hispanic American participation in Federal education programs where Hispanic Americans currently are underserved. After the first years, each agency's plan also shall assess that agency's performance on the goals set in the previous year's annual plan. These plans shall be submitted by a date and time to be established by the Secretary.

Sec. 7. The Director of the Office of Personnel Management, in consultation with the Secretary and the Secretary of Labor, to the extent permitted by law, shall develop a program to promote recruitment of Hispanic students for part-time, summer, and permanent positions in the Federal Government.

Sec. 8. I have determined that the Commission shall be established in compliance with the Federal Advisory Committee Act, as amended (5, U.S.C. App. 2). Notwithstanding any other Executive order, the responsibilities of the President under the Federal Advisory Committee Act, as amended, shall be performed by the Secretary, in accordance with the guidelines and procedures established by the Administrator of General Services.

Sec. 9. Administration. **(a)** Members of the Commission shall serve without compensation, but shall be allowed travel expenses, including per diem in lieu of subsistence, as authorized by law for persons serving intermittently in the Government service (5 U.S.C. 5701–5707).

(b) The Commission and the Initiative shall obtain funding for their activities from the Department of Education.

(c) The Department of Education shall provide such administrative services for the Commission as may be required.

Sec. 10. Executive Order No. 12729 is revoked.

William J. Clinton

February 22, 1994

DOCUMENT 8.4b

"Our Nation on the Fault Line: Hispanic American Education," 1996

In this landmark report of 1996, the president's commission concluded that the U.S. educational system was failing its students of Hispanic background and called for renewed efforts to improve educational opportunity from the pre-K levels through the university. Interested readers may access a complete copy of the report through the U.S. Department of Education, http://www.ed.gov/pubs/FaultLine/index.html accessed May 10, 2004.

PRESIDENT'S ADVISORY COMMISSION ON EDUCATIONAL EXCELLENCE FOR HISPANIC AMERICANS

White House Initiative on Educational Excellence for Hispanic Americans

The Honorable William J. Clinton
The President of the United States
The Honorable Richard W. Riley
Secretary of Education

President's Advisory Commission on Educational Excellence for Hispanic Americans:

Honorable Ana "Cha" Margarita Guzmán, Chair; Linda G. Alvarado; Erlinda Paiz Archuleta; Cecilia Preciado Burciaga; George Castro; Darlene Chavira Chávez; David Cortiella; Miriam Cruz; Juliet Villareal Garcia; José Gonzalez; María Hernandez; Sonia Hernandez; Martin J. Koldyke; Guillermo Linares; Cipriano Muñoz; Eduardo Padrón; Janice Petrovich; Gloria Rodríguez; Waldemar Rojas; Isaura Santiago; John Phillip Santos; Samuel Vigil; Diana Wasserman; Rubén Zacarías.

Call to Action: An Executive Summary

The Commission calls upon the nation to improve education for Hispanic Americans. This Call to Action goes out to Hispanics and non-Hispanics alike—rich, middle-class, and poor—to work in partnership with the leadership and resources of government and the private sector.

The nature of the problem with the education of Hispanic Americans is rooted in a refusal to accept, to recognize, and to value the central role of Hispanics in the past, present, and future of this nation. The education of Hispanic Americans is characterized by a history of neglect, oppression, and periods of wanton denial of opportunity.

The successful resolution of what has become nothing less than a crisis is embedded in the collective and collaborative response of the nation; and it must be characterized by the affirmation of the value and dignity of Hispanic communities, families, and individuals.

DOCUMENT 8.5

The U.S. Census and Latinos, 2000

The increase in the Latino population between 1990 and 2000—58 percent—far exceeded the census bureau's predictions. In certain states such as Georgia and North Carolina, the population of Latinos tripled and quadrupled. The definition of Latino was further refined in the 2000 census questionnaire. For the first time in the history of the decennial census, Latinos could specify a racial category. The tables below capture pictorially key characteristics of this diverse group.

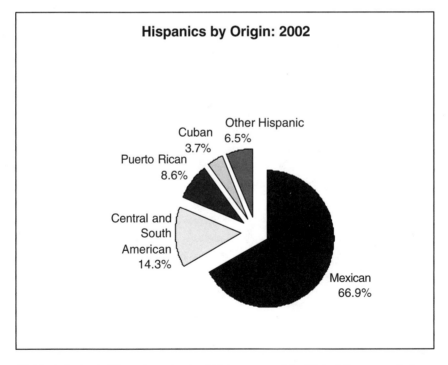

Table 1. In 2002, Hispanics comprised 13.3 percent of the United States population. Two-thirds of the 37.4 million are identified as Mexican. Source: U.S. Census Bureau, Annual Demographic Supplement to the March 2002 Current Population Survey.

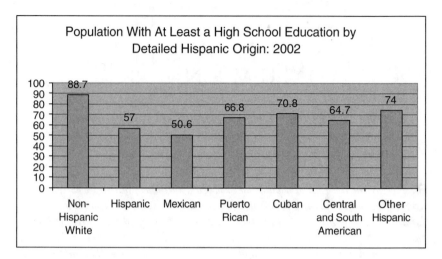

Table 2. No subgroup of Hispanics in 2002 graduated High School at the same rate as Non Hispanic Whites. Among the subgroups, Cubans have the highest graduation rate and Mexicans the lowest—the average among all Hispanics is more than 30 percent lower than among non-Hispanic Whites. Source: U.S. Census Bureau, Annual Demographic Supplement to the March 2002 Current Population Survey.

Reproduction of Questions on Race and Hispanic Origin From Census 2000

→ **NOTE: Please answer BOTH Questions 5 and 6.**

5. Is this person Spanish/Hispanic/Latino? Mark ☒ *the "No" box if not Spanish/Hispanic/Latino.*

☐ No, not Spanish/Hispanic/Latino   ☐ Yes, Puerto Rican
☐ Yes, Mexican, Mexican Am., Chicano   ☐ Yes, Cuban
☐ Yes, other Spanish/Hispanic/Latino — *Print group.* ↙

6. What is this person's race? *Mark* ☒ *one or more races to indicate what this person considers himself/herself to be.*

☐ White
☐ Black, African Am., or Negro
☐ American Indian or Alaska Native — *Print name of enrolled or principal tribe.* ↙

☐ Asian Indian   ☐ Japanese   ☐ Native Hawaiian
☐ Chinese   ☐ Korean   ☐ Guamanian or Chamorro
☐ Filipino   ☐ Vietnamese   ☐ Samoan
☐ Other Asian — *Print race.* ↙   ☐ Other Pacific Islander — *Print race.* ↙

☐ Some other race — *Print race.* ↙

Source: U.S. Census Bureau, Census 2000 questionnaire.

Table 3.

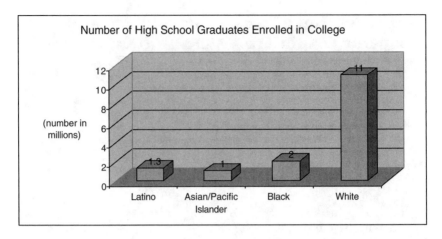

Table 4. In 2001, 1.3 million Latinos were enrolled in college. A higher proportion of Latino high school graduates attend college than any other racial/ethnic group, but their representation is disproportionately high in two-year colleges and disproportionately low in graduate and professional schools. Source: http://www.pewhispanic. org/site/docs/pdf/latinosinhighereducation-sept5–02.pdf (October Current Population Survey)

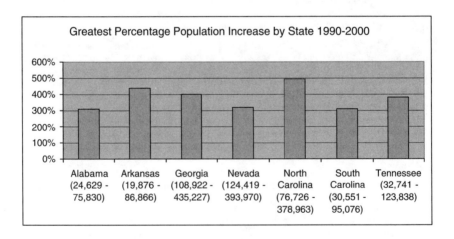

Table 5. Southern states experienced dramatic increases in the Latino population between 1990 and 2000. A region of the country more accustomed to black and white relations, the new Latino immigrants presented challenges to deeply rooted historical understandings of race and ethnicity.

DOCUMENT 8.6a

The Politics of Latino Identity

As this book has attempted to show, today's Latinos originate from a diverse group of sub-populations, with varying political views, historical and cultural experiences, and adaptations of the Spanish language. In the following documents the politics of modern day Latino identity are explored. To some, the word "Hispanic" is fraught with Iberian snobbism. Others believe the experience of Mexican Americans or Chicanos should not be subsumed under a pan-Latino umbrella, but be maintained as a separate name and history. Some citizens urge Latinos to assimilate quickly. In one response to these articles in The Washington Post, *an angry writer declared, "I have a solution for those Hispanic Latinos or Latino Hispanics who are divided over what to call themselves. If you were born in the United States or a naturalized citizen, call yourselves Americans, because that's really all that counts. The sooner we forget about where we all came from individually and think about where we are going collectively, the better off everyone in this country will be."[49] Struggles over identity are certain to continue as Latinos continue to increase in number and move to new areas of the country.*

LATINOS OR HISPANICS? A DEBATE ABOUT IDENTITY

*by **Darryl Fears**, Washington Post **staff writer***

From: *The Washington Post*, August 25, 2003, page A01. **Copyright 2003, The Washington Post, reprinted with permission.**

On a recent summer's day, Sandra Cisneros walked into Valenzuela's Latino Bookstore and thought she had discovered a treasure. It was one of the few independent book sellers in her home town of San Antonio, and on top of that, she said, its name appealed directly to her.

But within minutes, her mood changed. A clerk innocently used a word to describe a section of books that made Cisneros's skin crawl. "She used the word Hispanic," Cisneros said, her voice dripping with indignation. "I wanted to ask her, 'Why are you using that word?'"

"People who use that word don't know why they're using it," said Cisneros, a Mexican American poet and novelist. "To me, it's like a slave name. I'm a Latina."

That declaration—"I'm a Latina"—is resounding more and more through the vast and diverse Spanish-speaking population that dethroned African Americans as the nation's largest ethnic group a few months ago.

It is also deepening a somewhat hidden but contentious debate over how the group should identify itself—as Hispanics or Latinos. The debate is increasingly popping up wherever Spanish speakers gather.

It was raised last month at the National Council of La Raza's convention in Austin. The Internet is littered with articles and position papers on the issue. Civic organizations with Hispanic in their titles have withstood revolts by activist members seeking to replace it with the word Latino.

Cisneros refused to appear on the cover of *Hispanic* magazine earlier this year because of its name. She relented only after editors allowed her to wear a huge faux tattoo on her biceps that read "Pura Latina," or Pure Latina.

Another Mexican American writer, Luis J. Rodriguez, only reluctantly accepted an award from a Hispanic organization "because I'm not Hispanic," he said.

Some have called the argument an insignificant disagreement over words that is being blown out of proportion. But others believe such labels can change the course of a people, as advocates of "black power" showed when they cast aside the term Negro during their crusade for self-determination amid the 1960s civil rights movement.

"I think the debate reflects the flux this community is in right now," said Angelo Falcon, a senior policy executive for the Puerto Rican Legal and Education Fund. "It's almost like a story where you ask, 'Where might this community be going?'"

Although the terms "Latino" and "Hispanic" have been used interchangeably for decades, experts who have studied their meanings say the words trace the original bloodlines of Spanish speakers to different populations in opposite parts of the world.

Hispanics derive from the mostly white Iberian peninsula that includes Spain and Portugal, while Latinos are descended from the brown indigenous Indians of the Americas south of the United States and in the Caribbean, conquered by Spain centuries ago.

"Latino-Hispanic" is an ethnic category in which people can be of any race. They are white, like the Mexican American boxer Oscar de la Hoya, and black, like the Dominican baseball slugger Sammy Sosa.

They can also be Ameri-Indian and Asian. A great many are mixtures of several races. More than 90 percent of those who said they are of "some other race" on the 2000 Census identified themselves as Hispanic or Latino.

"As a poet, I'm especially sensitive to the power a word has," said Cisneros, who wrote the books "Caramelo" and "The House on Mango Street." "It's not a word. It's a way of looking at the world. It's a way of looking at meaning."

Duard Bradshaw has a different opinion. "I'll tell you why I like the word Hispanic," said the Panamanian president of the Hispanic National Bar Association. "If we use the word Latino, it excludes the Iberian peninsula and the Spaniards. The Iberian peninsula is where we came from. We all have that little thread that's from Spain."

A survey of the community conducted last year by the Pew Hispanic Center of Washington found that nearly all people from Spanish-speaking backgrounds identify themselves primarily by their place of national origin.

When asked to describe the wider community, more than half, 53 percent, said both Hispanic and Latino define them. A substantial but smaller group, 34 percent, favored the term Hispanic. The smallest group, 13 percent, said they preferred Latino. A survey by *Hispanic Trends* magazine produced a similar finding.

But advocates for the term Latino were unfazed.

"The very fact that it's called the Pew Hispanic Center tells you something," said Fernando Guerra, the Mexican American director of the Center for the Study of Los Angeles at Loyola Marymount University. "The fact that Hispanic is in the name of the organization . . . biased the question."

The term Hispanic was given prominence by the Nixon administration more than 30 years ago when it was added to the census questionnaire in 1970. Although that year's count of the large Mexican American, Puerto Rican and Cuban American populations was a disappointment, a seed had been planted.

By the 1980 Census, Hispanic had become fixed as the official government term. It appeared not only on census forms, but also on all other federal, state and municipal applications for employment, general assistance and school enrollment.

"It's a great gift that the government of the United States gave us," said Vincent Pinzon, the Colombian president and founder of the Americas Foundation. "If you want to acquire political muscle in this country, and you say you're just Argentinian or Colombian, then you have none."

But Mexican American activists in California and Puerto Rican activists in New York were not pleased. They favored a term that included the brown indigenous Indians who they believe are the source of their bloodline.

"Hispanic doesn't work for me because it's about people from Spain," said Rodriguez, author of the book "The Republic of East L.A." "I'm Mexican, and we were conquered by people from Spain, so it's kind of an insult."

Rodriguez's views are typical of Mexican Americans in Los Angeles, the epicenter of immigrants from that country, and the Chicano rights movement.

The term Chicano is thought to have originated as slang that described immigrants and refugees from the Mexican revolution. The term later evolved to define the uprising of Mexican American reformers and rights activists as well as farm laborers and other workers who lived in squalor while toiling for low pay.

As activists from other Latin countries joined the movement, Latino was adopted as an umbrella term for all groups.

"In L.A., if someone says he's Hispanic, and he's not from the East Coast, you begin to question this guy," said Guerra, the Loyola Marymount professor. "It means he didn't grow up in a Latino neighborhood."

In Washington, where the Pew Center is located, Salvadorans who dominate the area's large Central American population say "somos Latinos"—

we are Latinos—according to Jos [*sic*, José] Ramos, director of the United Salvadoran American Civic Committee.

"Hispanic is a category for the U.S. Census," he said. "It's a formality. For me, the correct term is Latino. It identifies people who speak the same language, people who share a vision of the historical meaning of our community. I am Salvadoran, and I am Latino."

But Cuban immigrants in Miami, conservative Mexican Americans in Texas and a group of Spanish descendants in New Mexico are among the groups that strongly identify themselves as Hispanic.

The word Latin dates to an 18th century spat between England and France, according to a historical resource guide written by journalist Frank del Olmo for the National Association of Hispanic Journalists.

Latin was used to distinguish Italy, France, Spain and their conquered territories in the Americas from the British empire and its colonies. Latino was popularized during the social movements of the 1960s, Guerra and other historians said.

The disagreement over the pair of ancient terms is an annoyance to some. When the subject came up at the National Council of La Raza's annual meeting, Lisa Navarette, the group's Cuban American spokeswoman, dismissed it. "We've got so many real important issues to work on, we can't be bothered with this nit-picking."

The community indeed faces daunting challenges: high unemployment, a skyrocketing high school dropout rate, widespread opposition to immigration reform and crowded communities.

But the issue isn't apt to disappear. A few years ago, Bradshaw's group, the Hispanic National Bar Association in Washington, had to fight off a resolution by a group of members to remove the word Hispanic from its name and replace it with Latino.

Last semester, students at Southern Methodist University in Dallas talked about changing the name Hispanic Student Services. And earlier this year, Cisneros, the author who abhors the word Hispanic, refused to accept an award from a Hispanic organization.

At the Latino bookstore Cisneros visited, owner Richard Martinez didn't know what to think. "I don't know which is correct," he said. "I'm a Mexican, a Latino, a Hispanic, whatever. Be who you are. Be proud, like everyone else."

DOCUMENT 8.6b

Readers of The Washington Post *responded to the previous article on Latino identity from various perspectives. In this article from the Ambassador of Spain, Ruperez points out that the United States is the third largest Spanish speaking country in the world and the Latino community can play a unique role in brokering relationships between Spain, Latin America, and the United States.*

"HISPANIC OR LATINO: ROOTS ARE THE SAME"

by Javier Ruperez, Ambassador of Spain to the United States

From: *The Washington Post*, September 6, 2003, page A17. **Copyright 2003, The Washington Post, reprinted with permission.**

Rejecting the term "Hispanic" and embracing the term "Latino" implies a huge contradiction that impairs reasoning ["Latinos or Hispanics? A Debate About Identity," front page, Aug. 25]. Is a Hispanic different from a Latino? Can we oppose the terms Latino and Hispanic in a legitimate fashion? Does that which is Latino identify only indigenous people and that which is Hispanic only a "conquistador"?

We Latinos or Hispanics, without distinction because such distinction is not possible, are a community with a history, language, cultural heritage and character with well-identified roots.

It is not only artificial to oppose the "Hispanic" to the "Latino" but dangerous as well. There are no confronting historical and cultural origins in our community. The MesoAmerican indigenous person is as Hispanic as the citizen from Buenos Aires or the AfroCaribbean person. Ethnic features are not the base for "Hispanic." Our nation is our language.

Hispanics are defined by their mixture of races—and that may be their greatest feature. The Hispanic universe is extremely diverse and includes races, religions and various cultural heritages. Cervantes, Velazquez and Picasso are as much a part of me as Botero, Neruda, Chichen Iza or Our Lady of Guadalupe.

A few weeks ago, the prime minister of the Spanish government, Jose Maria Aznar, pointed out at the annual meeting of the National Council of La Raza in Austin that the reason for seeking our cultural heritage is neither confrontation nor exclusion. Today the Hispanic community is composed of more than 400 million people. Those people use mostly one language, Spanish, which continues to grow and expand. The Spanish language belongs to all of us, not just one group.

In this changing world, U.S. Hispanics are going to be more relevant than ever. Spain looks at the advancement and growth of the Hispanic minority in the United States with a positive curiosity and somewhat pleasant surprise. The United States is the third largest Spanish-speaking nation in the world. And Spanish has become the second language in the country. The Hispanic community can be instrumental in the improvement of U.S. relations with Spain and Latin America.

When President Bush visited Spain in June 2001, he remarked that the American people are proud descendants of Hispanic heritage. And today we have added the vitality and dynamic forces of new generations of Hispanic Americans. There is no doubt that their strength and ability to transform will increase as their awareness as a community grows. Division among Hispanics will not help.

—Javier Ruperez, Washington

The writer is ambassador of Spain to the United States.

Notes

INTRODUCTION

1. Aurora Levins Morales, *The Historian as Curandera*, JSRI Working Paper no. 40, Julian Samora Research Institute, Michigan State University, East Lansing, Michigan, 1997, p. 1.
2. Suzanne Oboler, *Ethnic Labels, Latino Lives: Identity and the Politics of (Re) Presentation in the United States* (Minneapolis: University of Minnesota Press, 1995); and Clara E. Rodríguez, *Changing Race: Latinos, the Census, and the History of Ethnicity in the United States* (New York: New York University Press, 2000).
3. Victoria-María MacDonald, "Hispanic, Latino, Chicano, or 'Other'? Deconstructing the Relationship between Historians and Hispanic-American Educational History," *History of Education Quarterly* 41, no. 3 (fall 2001): 365–413.
4. The analysis in this section is drawn from MacDonald, "Hispanic, Latino, Chicano, or 'Other'?", 2001.
5. Peter Novick, *That Noble Dream: The "Objectivity Question" and the American Historical Profession* (Cambridge: Cambridge University Press, 1994).

CHAPTER 1

1. Frederick E. Eby, compiler, *Education in Texas: Source Materials*, University of Texas Bulletin no. 1824 (Austin: University of Texas, April 1918), pp. 5–6.
2. David J. Weber, *The Spanish Frontier in Northern New Spain* (New Haven and London: Yale University Press, 1992), p. 23.
3. Weber, *The Spanish Frontier*, p. 15.
4. Weber, *The Spanish Frontier*, pp. 125–128.
5. David Sweet, "The Ibero-American Frontier Mission in Native American History," in *The New Latin American Mission History*, ed. Erick Langer and Robert H. Jackson (Lincoln, NE and London: University of Nebraska Press, 1995), 1–48.
6. Martha Menchaca, *Recovering History, Constructing Race: The Indian, Black, and White Roots of Mexican Americans* (Austin: University of Texas Press, 2001).
7. Menchaca, *Recovering History, Constructing Race*, p. 81.
8. Lisbeth Haas, *Conquests and Historical Identities in California, 1769–1936* (Berkeley: University of California Press, 1995), p. 29.
9. Magnus Mörner, *Race Mixture In The History of Latin America* (Boston: Little, Brown and Company, 1967).
10. Rev. J. A. Burns, *The Catholic School System in the United States: Its Principles, Origin, and Establishment* (New York: Benziger Brothers, 1908), p. 507.
11. Martha Menchaca, "The Treaty of Guadalupe Hidalgo and the Racialization of the Mexican Population," in *The Elusive Quest for Equality: 150 Years of Chicano/*

Chicana Education, ed. José F. Moreno (Cambridge, MA: Harvard Educational Review, 1999), pp. 3–29.

12. Irving A. Leonard, *Books of the Brave: Being an Account of Books and of Men in the Spanish Conquest and Settlement of the Sixteenth-Century New World* (Cambridge, MA: Harvard University Press, 1949).

13. Lawrence Cremin, *American Education: The Colonial Experience, 1607–1783* (New York: Harper and Row, 1970), p. 181.

14. Cremin, *American Education: The Colonial Experience, 1607–1783*, pp. 235–248.

15. "Royal Orders Establishing Schools in Spanish America," trans. Carlos Castañeda and Mattie Austin Hatcher, in Frederick E. Eby, *Education in Texas: Source Materials* (Austin: University of Texas Bulletin, no. 1824, April 1918), 4–7; and Menchaca, "The Treaty of Guadalupe Hidalgo," pp. 12–15.

16. Burns, *The Catholic School System in the United States*, p. 49.

17. "Cedula of March 12, 1634," in Edgar Knight, ed. *Documentary History of Education in the South Before 1860*, vol. 1 (Chapel Hill: University of North Carolina Press, 1949), p. 666. Cedula of March 12, 1634 from *Statutes Relating to Florida in the Diocesan Synod Held by his Majesty's Command*, by the Right Rev. Dr. John Gareidae Palacios, Bishop of Cuba, June 1864. Translated by John Dawson Gilmary Shea.

18. Burns, *The Catholic School System in the United States*, p. 50; John Gilmary Shea, *History of the Catholic Church in the United States, Vol. 1, Colonial Days* (Akron, OH: D. H. McBride and Co.,1886), pp.469–470; "Rules of a School in St. Augustine, 1786," in Knight, *Documentary History of Education*, pp.728–732. Originals may be accessed in the East Florida Papers, 41 B4 (Library of Congress; Translated from the Spanish in Michael J. Curley, C. SS. R., *Church and State in the Spanish Floridas (1783–1822)* Studies in American History, the Catholic University of America. (Washington, D.C. Catholic University Press, 1940), pp.78–82; and Charles W. Arnade, "Raids, Sieges, and International Wars," in *The New History of Florida*, ed. Michael Gannon (Gainesville, FL: University Press of Florida, 1996), pp.100–116.

19. James McClellan, *An English School in Spanish St. Augustine, 1805*. Florida State University Studies No. 15. Published under the auspices of the Research Council of Florida State University. Tallahassee, FL 1954.

20. Michael Gannon, "First European Contacts," in *The New History of Florida*, ed. Michael Gannon (Gainesville, FL: University Press of Florida, 1996), pp.16–39; and Eugene Lyon, "Settlement and Survival," in *The New History of Florida*, ed. Michael Gannon (Gainesville, FL: University Press of Florida, 1996), pp.40–61; and Amy Turner Bushnell, "Republic of Spaniards, Republic of Indians," in *The New History of Florida*, ed. Michael Gannon (Gainesville, FL: University Press of Florida, 1996), pp.62–77.

21. David J. Weber, *What Caused the Pueblo Revolt of 1680?* (Boston: Bedford/St. Martin's Press, 1999), pp. 3–18.

22. Bernardo P. Gallegos, *Literacy, Education and Society in New Mexico, 1693–1821* (Albuquerque: University of New Mexico Press, 1992), p. 23.

23. Gallegos, *Literacy, Education and Society in New Mexico, 1693–1821*, pp. 22, 38–9.

24. Gallegos, *Literacy, Education and Society in New Mexico, 1693–1821*, pp. 40–41.

25. Gallegos, *Literacy, Education and Society in New Mexico, 1693–1821*, pp. 31–32.

26. Gallegos, *Literacy, Education and Society in New Mexico, 1693–1821*, p. 32.

27. Hubert Howe Bancroft, *The Works of Hubert Howe Bancroft*, vol. 18, History of California, vol. I, 1542–1800 (San Francisco: A.L. Bancroft & Co., Publishers, 1884), pp. 642–643.

28. Warren A. Beck, *New Mexico: A History of Four Centuries* (Norman: University of Oklahoma Press, 1962), p. 207 and Gallegos, *Literacy, Education and Society in New Mexico, 1693–1821*, pp. 41–42.

29. Perkin, "History of Universities," p. 20.

30. Max Berger, "Education in Texas during the Spanish and Mexican Periods," *Southwestern Historical Quarterly* 51 (July 1947): 41–53; and Eby, *Education in Texas*, pp. 8–10.

31. I. J. Cox, "Educational Efforts in San Fernando de Bexar," *Texas Historical Association Quarterly* 6 (July 1902): 28–29.

32. Cox, "Educational Efforts in San Fernando de Bexar," pp. 32–33.

33. Cox, "Educational Efforts in San Fernando de Bexar," p. 33.

34. Carl Kaestle, *Pillars of the Republic: Common Schools and American Society, 1780–1860* (New York: Hill and Wang, 1983), pp. 13–29.

35. Zephyrin Engelhardt, O.F.M., *The Missions and Missionaries of California*, vol. 1 (San Francisco: The James H. Barry Company, 1908), 88.

36. Manuel Patricio Servín, "California's Hispanic Heritage: A View into the Spanish Myth," in *New Spain's Far Northern Frontier: Essays on Spain in the American West, 1540–1821*, David J. Weber, ed. (Dallas, TX: Southern Methodist University Press, 1979), 120.

37. Engelhardt, *The Missions and Missionaries of California*, 163.

38. Weber, *The Spanish Frontier*, 241–2.

39. Hubert Howe Bancroft, *History of California*, vol. 1 *1542–1800*, vol. 18 of *The Works of Hubert Howe Bancroft* (Santa Barbara, CA: A. L. Bancroft and Co., Publishers, 1884), p. 642.

40. Bancroft, *History of California*, p. 643.

41. Servín, "California's Hispanic Heritage."

42. Bancroft, *History of California*, n. 42, p. 643.

43. Arizona, known in the colonial era as Pimería Alta, was too isolated to establish formal schools despite modest attempts. Menchaca, "The Treaty of Guadalupe Hidalgo," pp. 14–15.

44. Gallegos, *Literacy, Education, and Society in New Mexico, 1693–1821*, p. 95.

45. Document in Francis J. Weber, *Documents of California Catholic History (1784–1963)* (Los Angeles: Dawson's Book Shop, 1965), pp. 26–32.

46. Robert H. Jackson, "Introduction," in *The New Latin American Mission History*, ed. Erick Langer and Robert H. Jackson (Lincoln, NE and London: University of Nebraska Press, 1995), pp. vii–xviii.

47. David Sweet, "The Ibero-American Frontier Mission in Native American History," in *The New Latin American Mission History*, edited by Erick Langer and Robert H. Jackson (Lincoln, NE: University of Nebraska Press, 1995), pp. 1–48.

48. Weber, *The Spanish Frontier*, p. 108.

49. Ramón A. Gutiérrez, *When Jesus Came, the Corn Mothers Went Away: Marriage, Sexuality, and Power in New Mexico, 1500–1846* (Stanford, CA: Stanford University Press, 1991), pp. 127–130.

50. Carlos E. Castañeda, *The Mission Era: The Missions at Work, 1731–1761*, vol. 3, *Our Catholic Heritage in Texas, 1519–1936* (Austin, TX: Boeckmann-Jones, 1938), p. 27.

51. Castañeda, *The Mission Era*, p. 32.

52. Shea, *History of the Catholic Church*, vol. 1, p. 144.

53. Weber, *The Spanish Frontier*, n. 76, p. 401.

54. Weber, *The Spanish Frontier*, pp. 109–110.

55. Castañeda, *The Mission Era*, p. 33.

56. Gutiérrez, *When Jesus Came, the Corn Mothers Went Away*, p. 191.

57. Shea, *History of the Catholic Church in the United States*, vol. 4, p. 345.

58. Burns, *The Catholic School System in the United States*, p. 57. The diary is referred to as "Garce's Diary" and covers the year 1776 (n. 1, p. 56).

59. Gutiérrez, *When Jesus Came, the Corn Mothers Went Away*, p. 77.

60. Haas, *Conquests and Historical Identities in California, 1769–1936*, p. 27.

61. Richard C. Trexler, "From the Mouths of Babes: Christianization by Children in 16th Century New Spain," in *Religious Organization and Religious Experience*, ed. J. Davis (London: Academic Press, 1982), pp. 122–123.

62. Trexler, "From the Mouths of Babes," pp. 117–118.

63. Gallegos, *Literacy, Education, and Society in New Mexico*, pp. 65–66.

64. James F. Brooks, *Captives and Cousins: Slavery, Kinship, and Community in the Southwest Borderlands* (Chapel Hill: UNC Press, 2002).

65. Michael C. Meyer and William L. Sherman, *The Course of Mexican History*, 5th ed. (New York: Oxford University Press, 1995), esp. chapter 15, "Society and Stress in the Late Colonial Period."

66. Menchaca, *Recovering History, Constructing Race*, p. 159.

67. See example of the youth named Frasquillo who, after being trained to read and write Spanish and Latin, turned on his mentor in the 1680 revolt. "When the conspiracy was formed and the day for the massacre was fixed, this precocious boy entered ardently into it." Burns, *The Catholic School System in the United States*, pp. 209–210.

68. Bernard Bailyn, *Education in the Forming of American Society: Needs and Opportunities for Study* (New York: Vintage Books, 1960).

69. Harvey J. Graff, *The Labyrinths of Literacy: Reflections on Literacy Past and Present*, rev. ed. (Pittsburgh, PA: University of Pittsburgh Press, 1995).

70. Harold Perkin, "History of Universities," in *The History of Higher Education*, 2nd ed., ed. Lester F. Goodchild and Harold S. Wechsler (New York: Simon and Schuster, 1997), p. 9; and Doyce B. Nunis, Jr., *Books in Their Sea Chests: Reading along the Early California Coast* (California Library Association, 1964).

71. Perkin, "History of Universities," p. 9; and Nunis, *Books in Their Sea Chests*.

72. Leonard, *Books of the Brave*, p. 177.

73. David J. Weber, *The Mexican Frontier, 1821–1846: The American Southwest under Mexico* (Albuquerque: University of New Mexico Press, 1982); and Manuel G. Gonzales, *Mexicanos: A History of Mexicans in the United States* (Bloomington and Indianapolis: Indiana University Press, 1999).

74. Weber, *The Mexican Frontier*, pp. 16–17.

75. Weber, *The Mexican Frontier*, p. 17.

76. For a fuller account of his travels see Margaret Kress, "Diary of a Visit of Inspection of the Texas Missions Made by Fray Gasper José de Solís in the year 1767–8," *Southwestern Historical Quarterly* 35, no. 1 (1931–32): 28–76.

CHAPTER 2

1. Frederick E. Eby, compiler, *Education in Texas: Source Materials*, Austin: University of Texas Bulletin. No. 1824. April 1918, pp. 76–77.

2. Gordon Lee, ed., *Crusade against Ignorance: Thomas Jefferson on Education* (New York: Teachers College Press, 1961).

3. Report of the Mexican Secretary of State, November 1823, quoted in A Citizen of the United States (author), *A View of South America and Mexico, Comprising their history, the political condition, geography, agriculture, commerce, etc. of the Republics of Mexico, Guatemala, Colombia, Peru, the United Provinces of South America and Chili, with a complete history of the revolution, in each of these independent states*, 2 vols. in 1 (New York: Published for Subscribers, 1827), 125.

4. David N. Plank and Rick Ginsberg, eds., *Southern Cities, Southern Schools: Public Education in the Urban South* (New York, Greenwood Press, 1990); Victoria-María MacDonald, *The Persistence of Segregation: Race, Class, and Public Education in Columbus, Georgia, 1828–1998*, unpublished book manuscript, 2003; Edgar Knight, ed., *A Documentary History of Education in the South Before 1860*, vol. 2 (Chapel Hill: University of North Carolina Press, 1949–1953).

5. Manuel G. Gonzales, *Mexicanos: A History of Mexicans in the United States* (Bloomington: Indiana University Press, 1999).

6. Michael C. Meyer and William L. Sherman, *The Course of Mexican History* (New York: Oxford University Press, 1995).

7. Meyer and Sherman, *The Course of Mexican History*, p. 324.

8. David J. Weber, *The Mexican Frontier, 1821 to 1846: The American Southwest under Mexico* (Albuquerque: University of New Mexico Press, 1982).

9. Weber, *The Mexican Frontier*, p. 47.

10. Martha Menchaca, *Recovering History, Constructing Race: The Indian, Black, and White Roots of Mexican Americans* (Austin, TX: University of Texas Press, 2001), p. 167.

11. Menchaca, *Recovering History*, pp. 161–186.

12. Menchaca, *Recovering History*, p. 163.

13. Weber, *The Mexican Frontier*, p. 146.

14. Gonzales, *Mexicanos*, p. 65; Weber, *The Mexican Frontier*, pp. 208–210.

15. Raymond S. Brandes, trans. and annotator, "Times Gone By in Alta California: Recollections of Señora Dona Juana Machada Alipaz de Ridington (Wrightington)," *The Historical Society of Southern California Quarterly* 41 (September 1959): 195–240. Provides a first-person account of the constant fear (and reality) of Indians kidnapping women and assaulting Mexican settlers.

16. Fr. Zephyyrin Engelhardt, O. F. M., *The Missions and Missionaries of California*, vol. 20 (San Francisco: The James H. Barry Company, 1908), p. 549.

17. Meyer and Sherman, *The Course of Mexican History*, p. 184.

18. Weber, *The Mexican Frontier*, pp. 45–6.

19. Irving Berdine Richman, *California under Spain and Mexico, 1535–1847* (Boston: Houghton Miffling Co., 1911), p. 222.

20. Matt S. Meier and Feliciano Ribera, *Mexican Americans—American Mexicans: From Conquistadores to Chicanos* (New York: Hill and Wang, 1993), pp. 32–33; and Weber, *The Mexican Frontier*, p. 44.

21. Weber, *The Mexican Frontier*, p. 44.

22. George Harwood Phillips, "Indians and the Breakdown of the Spanish Mission System in California," in David J. Weber, ed., *New Spain's Far Northern Frontier: Essays on Spain in the American West, 1540–1821* (Dallas: Southern Methodist University Press, 1979), p. 264.

23. Menchaca, *Recovering History*, p. 171.

24. Gonzales, *Mexicanos*, p. 63.

25. Fr. Zephyrin Engelhardt, O.F.M., *The Missions and Missionaries of California*, vol. 1 (San Francisco: The James H. Barry Company, 1908), p. 88.

26. Engelhardt, *The Missions and Missionaries*, vol. 1, pp. 131–133.

27. Phillips, "Indians and the Breakdown of the Spanish Mission," pp. 267–268.
28. Phillips, "Indians and the Breakdown of the Spanish Mission," p. 265.
29. Paul Farnsworth and Robert H. Jackson, "Cultural, Economic, and Demographic Change in the Missions of Alta California: the Case of Nuestra Señora de la Soledad," in *The New Latin American Mission History*, ed. Erick Langer and Robert H. Jackson (Lincoln: University of Nebraska Press, 1995), pp. 124–5.
30. Francis J. Weber, *Documents of California Catholic History, 1784–1963* (Los Angeles: Dawson's Book Shop, 1965), pp. 33–36.
31. Wayne Moquin and Charles Van Doren, eds, *A Documentary History of the Mexican Americans* (New York: Praeger, 1971), pp. 130–136.
32. Albert Camarillo, *Chicanos in a Changing Society: From Mexican Pueblos to American Barrios in Santa Barbara and Southern California, 1848–1930* (Cambridge: Harvard University Press, 1979), pp. 9–10; and Menchaca, *Recovering History*, p. 181.
33. Weber, *The Mexican Frontier*, p. 211.
34. Farnsworth and Jackson, "Cultural, Economic and Demographic Change," p. 118.
35. Regulations reproduced in Daniel Tyler, "The Mexican Teacher," *Red River Valley Historical Review* 1 (1974): 214.
36. Eby, *Education in Texas*, p. 6.
37. Richard J. Altenbaugh, *The American People and Their Education: A Social History* (Upper Saddle River, NJ: Pearson Education, 2003), p. 45.
38. Carl F. Kaestle, *Pillars of the Republic: Common Schools and American Society, 1780–1860* (New York: Hill and Wang, 1983); Lawrence A. Cremin, *American Education: The National Experience, 1783–1876* (New York: Harper and Row, 1980); David B. Tyack, *Turning Points in American Educational History* (University Microfilms International, 1967); and Joel Perlmann and Robert A. Margo, *Women's Work?: American Schoolteachers, 1650–1920* (Chicago: University of Chicago Press, 2001).
39. Report of the Mexican secretary of state, November 1823, p. 125.
40. Eby, *Education in Texas*, p. 27.
41. Eby, *Education in Texas*, pp. 85–88.
42. Tyler, *The Mexican Teacher*, p. 209.
43. Eby, *Education in Texas*, p. 30.
44. Eby, *Education in Texas*, pp. 32–33.
45. Lawrence A. Cremin, *American Education: The National Experience, 1783–1876* (New York: Harper and Row, 1980), p. 171.
46. "Statement of the Ayuntamiento on the Purpose of Public Education, 1828," reproduced in I. J. Cox, "Educational Efforts in San Fernando," *Texas Historical Association Quarterly* 6 (July 1902): 62–63.
47. Tyler, "The Mexican Teacher," p. 219.
48. For more discussion of this educational plan see Rodney Hessinger, "Lancaster System," in Richard J. Altenbaugh, ed., *Historical Dictionary of American Education* (Westport, CT: Greenwood Press, 1999), pp. 208–9.
49. "Congress of the State of Coahuila and Texas—Decree No. 92," reproduced in Eby, *Education in Texas*, pp. 34–35.
50. Eby, *Education in Texas*, p. 36. The State Department papers of the Republic of Mexico most likely contain more documentation on Mexican-Era schools, research that urgently awaits scholarly attention.
51. Eby, *Education in Texas*, p. 37.
52. "Decree No. 240," reproduced in Eby, *Education in Texas*, p. 48–49.
53. Hubert Howe Bancroft, *The Works of Hubert Howe Bancroft, vol. 19, History of California. Vol.2. 1801–1824* (Santa Barbara, CA: Wallace Hebberd, 1966), p. 680.

54. Bancroft, pp. 603.
55. Tyler, "The Mexican Teacher," p. 220.
56. Bancroft, p. 613, 680.
57. Bancroft, p. 613.
58. Bancroft, p. 548.
59. Bancroft, n. 48, p. 680.
60. Ibid.
61. Bancroft, n. 48, p. 680.
62. Bancroft, B. n. 48, p. 680.
63. Hutchinson, cited in Menchaca, 1999, p. 18.
64. Tyler, "The Mexican Teacher," pp. 209–210; Lynn Marie Getz, *Schools of Their Own: The Education of Hispanos in New Mexico, 1850–1940* (Albuquerque: University of New Mexico Press, 1997), p. 5.
65. Rules reproduced in Tyler, "The Mexican Teacher," p. 215.
66. Kaestle, *Pillars of the Republic*.
67. Weber, *The Mexican Frontier*, 232.
68. Eby, *Education in Texas*, pp. 29–30.
69. Weber, *The Mexican Frontier*, pp. 232–3; and Doyce B. Nunis, Jr. *Books in Their Sea Chests: Reading along the Early California Coast* (California Library Association, 1964), pp. 6–7.
70. Quote from Nunis, *Books in Their Sea Chests*, p. 5.
71. Nunis, *Books in Their Sea Chests*, p. 4.
72. Lisbeth Haas, *Conquests and Historical Identities in California, 1769–1936* (Berkeley: University of California Press, 1995), p. 80.
73. Weber, *The Mexican Frontier*, p. 230.
74. Tyler, "The Mexican Teacher," p. 209.
75. For information on children sent abroad, see Tyler, "The Mexican Teacher," p. 211; and Richman, *California under Spain and Mexico*, pp. 345–346.
76. Tyler, "The Mexican Teacher," p. 209.

CHAPTER 3

1. David J. Weber, *The Mexican Frontier, 1821 to 1846: The American Southwest under Mexico* (Albuquerque: University of New Mexico Press, 1982), p. 213.
2. Manuel G. Gonzales, *Mexicanos: A History of Mexicans in the United States* (Bloomington and Indianapolis: Indiana University Press, 1999), p. 70.
3. Historians have offered multiple interpretations of the motives of the Mexican War. See for example, John D. Eisenhower, *So Far from God: The War with Mexico, 1846–1848* (New York: Random House, 1989).
4. Richard Griswold del Castillo, *The Treaty of Guadalupe Hidalgo: A Legacy of Conflict* (Norman and London: University of Oklahoma Press, 1990), p. 66.
5. Martha Menchaca, *Recovering History, Constructing Race: The Indian, Black, and White Roots of Mexican Americans* (Austin: University of Texas Press, 2001).
6. Charles Wollenberg, *All Deliberate Speed: Segregation and Exclusion in California Schools, 1855–1975* (Berkeley: University of California Press, 1978), p. 13.
7. Kenneth N. Owens, ed. *Riches For All: The California Gold Rush and the World* (Lincoln, NE: University of Nebraska Press, 2002). Leonard Pitt, *The Decline of the Californios: A Social History of the Spanish-Speaking Californios, 1846–1890* (Berkeley, Los Angeles, London: University of California Press, 1971).
8. Del Castillo, *Treaty of Guadalupe Hidalgo*, p. 190.

9. Del Castillo, *Treaty of Guadalupe Hidalgo*, p. 73.

10. Albert Camarillo, *Chicanos in a Changing Society: From Mexican Pueblos to American Barrios in Santa Barbara and Southern California, 1848–1930* (Cambridge, MA and London: Harvard University Press, 1979), p. 114.

11. Armando C. Alonzo, *Tejano Legacy: Rancheros and Settlers in South Texas, 1734–1900* (Albuquerque: University of New Mexico Press, 1998), p. 268.

12. Alonzo, *Tejano Legacy*, p. 9.

13. Pitt, *Decline of the Californios*; Del Castillo, *Treaty of Guadalupe*; Camarillo, *Chicanos*.

14. Arnoldo De León, *They Called Them Greasers: Anglo Attitudes toward Mexicans in Texas, 1821–1900* (Austin: University of Texas Press, 1983); Reginald Horsman, *Race and Manifest Destiny: The Origins of American Racial Anglo-Saxonism* (Cambridge, MA and London: Harvard University Press, 1981), pp. 208–248.

15. Horsman, *Race and Manifest Destiny*, p. 210.

16. De León, *They Called Them Greasers*, pp. 14–23.

17. Menchaca, *Recovering History*; Camarillo, *Chicanos*.

18. Jay P. Dolan, *The American Catholic Experience: A History from Colonial Times to the Present* (Garden City, NY: Image Books, 1987), p. 202 and 295.

19. See document 16, "Response to know-nothing attacks, San Antonio, 1855," in Timothy Matovina and Gerald E. Poyo, eds., *¡Presente!: U.S. Latino Catholics from Colonial Origins to the Present* (Maryknoll, NY: Orbis Books, 2000).

20. Robert J. Rosenbaum, *Mexicano Resistance in the Southwest: "The Sacred Right of Self-Preservation"* (Austin and London: University of Texas Press, 1981); Deena J. González, *Refusing the Favor: The Spanish-Mexican Women of Santa Fe, 1820–1880* (New York and Oxford: Oxford University Press, 1999).

21. *Tejano* was a term Hispanics in Texas utilized to identify themselves as early as 1833. See "TEJANO," the Handbook of Texas Online. Available at http://www.tsha.utexas.edu/handbook/online/articles/view/TT/pft7.html. (accessed August 15, 2003).

22. Frederick E. Eby, compiler. *Education in Texas: Source Materials*. Austin, TX: University of Texas Bulletin, no. 1824, April 1918, p. 130.

23. See Eby, *Education in Texas*, pp. 130–199 for information on Texas during the Republic

24. Arnoldo De León, *The Tejano Community, 1836–1900* (Albuquerque: University of New Mexico Press, 1982), pp. 1–22.

25. Michael R. Olneck and Marvin F. Lazerson, "The School Achievement of Immigrant Children, 1900–1930," *History of Education Quarterly* 14 (winter 1974): 453–482.

26. Eby, *Education in Texas*, p. 336.

27. H. P. N. Gammel, compiler and arranger, *The Laws of Texas, 1822–1897* (Austin: The Gammel book company, 1898), pp. 998–999.

28. De León, *The Tejano Community*, p. 197.

29. *First Annual Report of the Superintendent of Public Instruction of the State of Texas, 1871* (Austin: J.G. Tracy, State Printer, 1871), p. 10. In State Government Publications, States Other Than Wisconsin, Wisconsin Historical Society, Madison, Wisconsin.

30. Joel Perlmann, "Historical Legacies: 1840–1920," *The Annals of the American Academy of Political and Social Science*, vol. 508 (March 1990): 27–37.

31. De León, *The Tejano Community*, pp. 188–192.

32. *Second Annual Report of the Superintendent of Public Instruction of the State of Texas for the Year 1872* (Austin: James P. Newcomb and Co., 1873), p. 25. In State

Government Publications, States Other Than Wisconsin, Wisconsin Historical Society, Madison, Wisconsin.

33. De León, *The Tejano Community*, p. 191.
34. Carlos E. Castañeda, *The Church in Texas Since Independence, 1836–1950*, vol. 7 of *Our Catholic Heritage In Texas, 1519–1936* (Austin: Von Boeckmann-Jones Co., 1958), p. 307.
35. Guadalupe San Miguel, Jr., *"Let All of Them Take Heed": Mexican Americans and the Campaign for Education Equality in Texas, 1910–1981* (Austin: University of Texas Press, 1988), p. 11.
36. De León, *The Tejano Community*, p. 193.
37. Guadalupe San Miguel, Jr., and Richard R. Valencia, "From the Treaty of Guadalupe Hidalgo to Hopwood: The Educational Plight and Struggle of Mexican Americans in the Southwest." *Harvard Educational Review* 68, no. 3 (fall 1998): 358.
38. Guadalupe San Miguel, Jr., "The Schooling of Mexicanos in the Southwest, 1848–1891," in José F. Moreno, ed., *The Elusive Quest for Equality: 150 Years of Chicano/Chicana Education* (Cambridge: Harvard Educational Review, 1999): 31–51.
39. For a detailed history of these schools, see Carlos E. Castañeda, *The Church in Texas Since Independence*, pp. 285–358.
40. San Miguel, Jr. *"Let All of Them Take Heed,"* p. 9.
41. Castañeda, *The Church in Texas Since Independence*, p. 292.
42. Castañeda, *The Church in Texas Since Independence*, p. 294.
43. Castañeda, *The Church in Texas Since Independence*, pp. 299–300.
44. Castañeda, *The Church in Texas Since Independence*, pp. 329–331.
45. Biographical information on Melinda Rankin can be found in John C. Rayburn, "Introduction," *Texas in 1850* (Boston: Camrell and Moore, 1850), rpt. (Waco: Texian Press, 1966), iii–xxii.
46. Melinda Rankin, *Twenty Years among the Mexicans: A Narrative of Missionary Labor* (Cincinnati: Chase and Hall, Publishers, 1875), pp. 57–58.
47. Rankin, *Twenty Years among the Mexicans*, p. 65.
48. Rayburn, *Texas in 1850*, p. XI.
49. San Miguel, Jr., *"Let All of Them Take Heed,"* pp. 9–10.
50. Cinthia Salinas, "El Colegio Altamirano (1897–1958): The Educational Opportunities of Tejanos in Jim Hogg County," paper presented at the American Educational Research Association, Seattle, WA, April 2001; San Miguel, Jr., *"Let All of Them Take Heed,"* p. 10.
51. *Report of the Superintendent of Public Instruction of the State of California* (J. B. Devoe, State Printer, April 10, 1851), document in the Rare Book Collection, Library of Congress, Washington, D.C.
52. Pitt, *Decline of the Californios*, p. 7.
53. Menchaca, *Recovering History*, p. 265.
54. *First Annual Report of the Superintendent of Public Instruction, California, 1852* (Eugene Cassely, State Printer, 1872), p. 49. State Government Publications, States Other Than Wisconsin, Wisconsin Historical Society, Madison, Wisconsin.
55. Camarillo, *Chicanos in a Changing Society*, p. 17.
56. William Warren Ferrier, *Ninety Years of Education in California, 1846–1936* (Oakland: West Coast Printing, 1937), pp. 49–50.
57. Pitt, *Decline of the Californios*, p. 226.
58. Pitt, *Decline of the Californios*, p. 228.
59. "A Historical Sketch of the Public School System of California," *First Biennial Report of the Superintendent of Public Instruction of the State of California, for the*

School Years 1864 and 1865 (O. M. Clayes, State Printer), p. 244. Special Collections, Monroe C. Gutman Library, Harvard Graduate School of Education.

60. Appendix C, "Communication and Report of Bishop Alemany, concerning the Catholic Schools in California," in *Second Annual Report of the Superintendent of Public Instruction to the Legislature of the State of California* (George Kerr, State Printer, 1853), n.p.
61. "A Historical Sketch," p. 252.
62. "A Historical Sketch," p. 252.
63. See for example, Diane Ravitch, *The Great School Wars: A History of the New York City Public Schools* (Baltimore, MD: Johns Hopkins University Press, 2000).
64. Menchaca, *Recovering History*, pp. 256–267.
65. "A Historical Sketch," p. 258.
66. *Revised School Law, State of California*, approved March 24, 1866 (Sacramento: O.J. Clayes, State Printer, 1866), p. 18.
67. "John Swett on Schools in Southern California (1865)," in Sol Cohen, compiler, *Education in the United States: A Documentary History*, vol. 2. (New York: Random House, 1965), p. 1034.
68. "John Swett on Schools in Southern California," p.1038.
69. "John Swett on Schools in Southern California," p.1039.
70. Camarillo, *Chicanos in a Changing*; Del Castillo, *Treaty of Guadalupe Hidalgo*.
71. Frank C. Lockwood, *Pioneer Days in Arizona: From the Spanish Occupation to Statehood* (New York: Macmillan Company, 1932), p. 250.
72. Gonzales, *Mexicanos*, pp. 92–93.
73. Samuel Pressly McCrea, "Establishment of the Arizona School System," in *Biennial Report of the Superintendent of Public Instruction of the Territory of Arizona* (Phoenix: The H.H. McNeil Company, 1908), p. 82. State Government Publications, States Other Than Wisconsin, Wisconsin Historical Society, Madison.
74. McCrea, "Establishment of the Arizona," p. 86; Lockwood, *Pioneer Days*, p. 246.
75. Lockwood, *Pioneer Days*, p. 249.
76. Lockwood, *Pioneer Days*, p. 246.
77. *Report of the U.S. Commissioner of Education for the Year 1876* (Washington: GPO, 1878), p. 432. Government Documents Division, Florida State University.
78. McCrea, "Establishment of the Arizona," p. 99.
79. *Biennial Report of the Superintendent of Public Instruction of the Territory of Arizona, 1899–1900* (Phoenix: The H. H. M'Neil Co. Print, 1900), p. 12. Government Publications, States Other Than Wisconsin, Wisconsin Historical Society, Madison.
80. Thomas Sheridan, *Arizona: A History* (Tucson: University of Arizona Press, 1995), p. 123
81. Gonzales, *Mexicanos*, p. 97.
82. McCrea, "Establishment of the Arizona."
83. *Annual Report of the U.S. Commissioner of Education, 1870* (Washington, GPO, 1870), pp. 326–327. Government Documents Division, Florida State University.
84. Lynn Marie Getz, *Schools of Their Own: the Education of Hispanos in New Mexico, 1850–1940* (Albuquerque: University of New Mexico Press, 1997), pp. 13–28.
85. Getz, *Schools of Their Own*, p. 14.
86. "List of Superintendents of Public Instruction," *Twenty-Ninth and Thirtieth Annual Reports of the State Superintendent of Public Instruction, New Mexico, 1919–1920* (Albuquerque: Central Ptg. Co., 1918), p. 7. Government Publications, States Other Than Wisconsin, Wisconsin Historical Society, Madison.
87. Gonzales, *Mexicanos*, p. 104.

88. *Compilation of the School Laws of New Mexico/Leyes de Escuelas de Nuevo Mexico* (East Las Vegas: J.A. Carruth, printer, 1889), p. 14. Rare Book Room, Library of Congress, Washington, D.C. (LB2529 .N6 1889).

89. *Report of the Superintendent of Public Instruction of the Territory of New Mexico, 1891* (Santa Fe, NM: New Mexican Printing Co., 1892), p. 6. Government Publications, States Other Than Wisconsin, Wisconsin Historical Society, Madison.

90. Getz, *Schools of Their Own*, p. 21.

91. *Education in New Mexico. Third Annual Report of Hon. W.G. Ritch, Secretary of the Territory, to the National Bureau of Education, 1875* (Northwestern Steam Printing House, 1876), p. 3. Government Publications, States other than Wisconsin, Wisconsin Historical Society, Madison.

92. *Education in New Mexico, 1875*, p.3.

93. Getz, *Schools of Their Own*, p. 15.

94. *Education in New Mexico, 1875*, p. 12.

95. *Education in New Mexico. Report of Hon. W. G. Ritch to the Commissioner of Education for the Year 1874.* (Santa Fe, NM: Manderfield and Tucker, printers, 1875), p. 12. Government Publications, States Other Than Wisconsin, Wisconsin Historical Society, Madison.

96. *Report of the Superintendent of Public Instruction of the Territory of New Mexico for the Year Ending 1892* (Santa Fe, NM: New Mexican Printing Co., 1892), p. 16. Government Publications, States Other Than Wisconsin, Wisconsin Historical Society, Madison.

97. *Report of the Superintendent of Public Instruction of the Territory of New Mexico for the Year Ending, 1891* (Santa Fe, NM: New Mexican Printing Co., 1892), p. 16. Government Publications, States other than Wisconsin, Wisconsin Historical Society, Madison.

98. Data for this section from *Report of the Superintendent of Public Instruction of the Territory of New Mexico for the Year Ending, 1891*, p. 22.

99. Victoria-María MacDonald, "Hispanic, Latino, Chicano, or 'Other'?: Deconstructing the Relationship between Historians and Hispanic-American Educational History," *History of Education Quarterly* 41, no.3 (fall 2001): 395.

100. Examples of works on this topic include Susan M. Yohn, *A Contest of Faiths: Missionary Women and Pluralism in the American Southwest* (Ithaca: Cornell University Press, 1995); Mark Banker, *Presbyterian Missions and Cultural Interaction in the Far Southwest, 1850–1950* (Urbana: University of Illinois Press, 1993); Jerry A. David, "Matilda Allison on the Anglo-Hispanic Frontier: Prebyterian Schooling in New Mexico, 1880–1910," *American Presbyterians* 74 (fall 1996): 171–182; Norman J. Bender, *Winning the West for Christ: Sheldon Jackson and Presbyterianism on the Rocky Mountain Frontier, 1869–1880* (Albuquerque: University of New Mexico Press, 1996); and Sara Deutsch, *No Separate Refuge: Culture, Class, and Gender on the Anglo-Hispanic Frontier in the American Southwest, 1880–1940* (New York: Oxford Press, 1987).

101. *Report of the Superintendent of Public Instruction of the Territory of New Mexico for the Year Ending, 1891*, p. 18.

102. Ibid., p. 19–20.

103. MacDonald, "Hispanic, Latino, Chicano, or 'Other,'" p. 396.

104. Yohn, *A Contest of Faiths*, pp.171–172.

105. This section has been included with permission from Victoria-María MacDonald and Teresa García, "Historical Perspectives on Latino Access to Higher Education, 1848–1990," in Jeanett Castellanos and Lee Jones, eds., *The Majority in the*

Minority: *Expanding the Representation of Latina/o Faculty, Administrators and Students in Higher Education* (Sterling, VA: Stylus Press, 2003), pp. 19–22.

106. Roger Geiger, "New themes in the history of nineteenth-century colleges," in R. Geiger, ed., *The American College in the Nineteenth Century* (Nashville: Vanderbilt University Press, 2000), pp. 1–36.
107. Gerald McKevitt, "Hispanic Californians and Catholic higher education: the diary of Jesús María Estudillo, 1857–1864," *California History* 69 (winter 1990–1991): 320–331.
108. D. J. León and D. McNeil, "A precursor to affirmative action: Californios and Mexicans in the University of California, 1870–72," *Perspectives in Mexican American Studies* 3 (1992): 179–206.
109. León and McNeil, "A precursor to affirmative action," p. 194.
110. Nicolas Kanellos, *Hispanic Firsts: 500 Years of Extraordinary Achievement* (Detroit: Visible Ink Press, 1997).
111. Thomas P. Carter, *Mexican Americans in School: A History of Educational Neglect* (General Education Board, 1970), p. 30.
112. Kanellos, *Hispanic Firsts*, p. 45; Gonzales, *Mexicanos*, p. 95.
113. Thomas E. Sheridan, *Arizona, A History* (Tucson: University of Arizona Press, 1995); Gonzales, *Mexicanos*, p. 94.
114. Ferrier, *Ninety Years of Education in California*, p. 202.
115. McKevitt, "Hispanic Californians and Catholic higher education," p. 322.
116. Ferrier, *Ninety Years of Education in California*, p. 208.
117. Kanellos, *Hispanic Firsts*, p. 44; De León, *The Tejano Community, 1836–1900* (Albuquerque: University of New Mexico Press, 1982), p. 190.
118. Pitt, *Decline of the Californios*; Camarillo, *Chicanos in a Changing Society*, p.69.
119. Biographical information from John C. Rayburn, "Introduction, " in Melinda Rankin, *Texas in 1850* (rpt. Waco, TX: Texian Press, 1966), iii–xxxiii. For profiles of other women teachers of this era see Polly S. Kaufman, *Women Teachers on the Frontier* (New Haven: Yale University Press, 1980).

CHAPTER 4

1. Quoted in Louis A. Pérez, *The War of 1898: The U.S. and Cuba in History and Historiography* (Chapel Hill: University of North Carolina Press, 1998), p. 112.
2. Frederick Merk, *Manifest Destiny and Mission in American History: A Reinterpretation* (New York: Vintage Books, 1966).
3. Thomas G. Paterson and Stephen G. Rabe, eds., *Imperial Surge: The United States Abroad, The 1890s-Early 1900s* (Lexington, MA: D.C. Heath, 1992), p. 156.
4. Merk, *Manifest Destiny*, p. 254.
5. Juan Gonzalez, *Harvest of Empire: A History of Latinos in America* (New York: Viking Press, 2000), p. 63.
6. José Trías Monge, *Puerto Rico: The Trials of the Oldest Colony in the World* (New Haven: Yale University Press, 1997), p. 43.
7. Monge, *Puerto Rico*.
8. Quoted in Louis A. Pérez, Jr., "The Imperial Design: Politics and Pedagogy in Occupied Cuba, 1899–1902," *Cuban Studies/Estudios Cubanos* 12 (summer 1982): 8.
9. Louis A. Pérez, *The War of 1898*.
10. Pérez, Jr., "The Imperial Design," p. 12.

11. For further discussion of the Harvard program see Edwards D. Fitchen, "The United States Military Government, Alexis E. Frye and Cuban Education, 1898–1902," *Revista/Review Interamericana* 2 (summer 1972): 123–159 and Fitchen, "The Cuban Teachers and Harvard, 1900: An Early Experiment in Inter-American Cultural Exchange," *Horizontes: Revista de la Universidad Católica de Puerto Rico* 26 (1973).

12. Pérez, Jr., "The Imperial Design," p. 12.

13. Pérez, Jr., "The Imperial Design," p. 13.

14. Quoted in Pérez, Jr., "The Imperial Design," pp. 13–14.

15. Quoted in Pérez, Jr., "The Imperial Design," p. 18, n. 53.

16. "Education in Porto Rico," in *Report of the Commissioner of Education*, vol. 2., in *Annual Report of the Department of the Interior* (Washington: Government Printing Office, 1903), p. 1204.

17. Aida Negron de Montilla, *Americanization in Puerto Rico and the Public School System, 1900–1930* (Rio de Piedras, P.R.: Universidad de Puerto Rico, 1975), pp. 9–10.

18. "Education in Porto Rico," in *Report of the Commissioner of Education for the Year 1899–1900*, vol. 1 (Washington: Government Printing Office, 1901), p. 254.

19. De Montilla, *Americanization*, p. 10.

20. President McKinley's special commissioner Henry K. Carroll is identified as one such advisor. See Arturo Morales Carrión, *Puerto Rico: A Political and Cultural History* (New York: W.W. Norton and Co., 1983), pp. 147–157.

21. Letter to English supervisors from John Eaton, May 13, 1899, in "Education in Porto Rico," 1:236–8.

22. "Education in Porto Rico," 1:236–8.

23. *Report of the Commissioner of Education*, vol. 1 in *Annual Reports of the Department of the Interior for the fiscal year ended June 30, 1901* (Washington: Government Printing Office, 1902), p. 62.

24. *Report of the Commissioner of Education for the year ended June 30, 1907*, vol. 1 (Washington: Government Printing Office, 1908), 331.

25. Donal F. Lindsey, *Indians at Hampton Institute, 1877–1923* (Urbana: University of Illinois Press, 1995).

26. Samuel McCune Lindsay, "Inauguration of the American School System in Porto Rico," chapter 15 of *Report of the Commissioner of Education for the year ending June 30, 1905*. Vol. 1 (Washington: Government Printing Office, 1907), p. 332.

27. Carlos Fraticelli-Rodríguez, *Education and Imperialism: The Puerto Rican Experience in Higher Education, 1898–1986*. Centro de Estudios Puertorriqueños Working Paper Series. Higher Education Task Force (New York: Hunter College, 1986).

28. De Montilla, *Americanization*, p. 27.

29. The appointees of the president during these years were the following: Mr. Martin G. Brumbaugh, August 8, 1900 to November 15, 1902. Mr. Samuel M. Lindsay, February 12, 1902 to October 1, 1904. Mr. Roland P. Falkner, October 2, 1904 to August 8, 1907. Mr. Edwin G. Dexter, August 9, 1907 to June 30, 1912. Edward M. Bainter, July 1, 1912 to May 15, 1915. Mr. Paul G. Miller, August 19, 1915 to September 29, 1921. From De Montilla, *Americanization*, p. 35.

30. Juan José Osuna, *A History of Education in Puerto Rico* (Rio Piedras, P.R.: Editorial de la Universidad de Puerto Rico, 1949), p. 282.

31. Osuna, *A History of Education in Puerto Rico*, pp. 347–350.

32. Quoted in Osuna, *A History of Education in Puerto Rico*, pp. 359–360.

33. Ellen Condliffe Lagemann, *An Elusive Science: The Troubling History of Educational Research*, and Osuna, *A History of Education in Puerto Rico*, p. 360.
34. Both reports, *A Survey of the Public Educational System of Puerto Rico*, Teachers College, Columbia University (1926) and Victor S. Clark, *Porto Rico and Its Problems* (1928), Brookings Institute, Washington, D.C. are quoted from extensively in Osuna, *A History of Education in Puerto Rico*, pp. 341–364.

CHAPTER 5

1. Manuel Gamio, *The Mexican Immigrant, His Life-Story: Autobiographic Documents Collected by Manuel Gamio* (Chicago: University of Chicago Press, 1931), pp. 222–223.
2. Juan Gonzalez, *Harvest of Empire: A History of Latinos in America* (New York: Viking Press, 2000); and Virginia Sánchez- Korrol, From *Colonia to Community: The History of Puerto Ricans in New York City* (Berkeley: University of California Press, 1994).
3. Manuel G. Gonzales, *Mexicanos: A History of Mexicans in the United States* (Bloomington, IN: Indiana University Press, 1999), pp. 139–160.
4. Carey McWilliams, *North from Mexico: The Spanish-Speaking People of the United States* (rpt., New York: Greenwood Press, 1968).
5. For a comprehensive listing see Sylvia Alicia Gonzales, *Hispanic American Voluntary Organizations* (Westport, CT: Greenwood Press, 1985).
6. George J. Sánchez, *Becoming Mexican American: Ethnicity, Culture and Identity in Chicano Los Angeles, 1900–1945* (New York: Oxford Press, 1993).
7. Gilbert G. Gonzales, *Chicano Education in the Era of Segregation* (Philadelphia: Balch Institute Press, 1990).
8. Gonzales, *Chicano Education;* Guadalupe San Miguel and Richard R. Valencia, "From the Treaty of Guadalupe Hidalgo to Hopwood: The Educational Plight and Struggle of Mexican Americans in the Southwest," *Harvard Educational Review* 68 (fall 1998): 353–412.
9. Ruben Donato, *The Other Struggle for Equal Schools: Mexican Americans during the Civil Rights Era* (Albany: State University Press of New York, 1997), p. 8.
10. Donato, *The Other Struggle;* Gonzales, *Chicano Education.*
11. Guadalupe San Miguel, Jr., *"Let All of Them Take Heed": Mexican Americans and the Campaign for Educational Equality in Texas, 1910–1981* (Austin: University of Texas Press, 1987).
12. Delgado V. Bastrop Independent School District, the Handbook of Texas Online, available at www.tsha.utexas.edu/handbook/online/articles/view/DD/jrd1.html.
13. Donato, *The Other Struggle.*
14. Victoria-María MacDonald and Scott Beck, "Educational History in Black and Brown: Paths of Divergence and Convergence, 1900–1990," paper presented at the History of Education Society Annual Meeting, San Antonio, Texas, October 2000.
15. Delgado V. Bastrop Independent School District.
16. San Miguel, Jr., *"Let All of Them Take Heed."*
17. San Miguel, Jr., *"Brown, Not White": School Integration and the Chicano Movement in Houston* (College Station: Texas A and M University Press, 2001).

18. Rubén Donato, "Hispano Education and the Implications of Autonomy: Four School Systems in Southern Colorado, 1920–1963," *Harvard Educational Review* 69 (summer 1999): 117–149; and Lynne Marie Getz, *Schools of Their Own: The Education of Hispanos in New Mexico, 1850–1940* (Albuquerque: University of New Mexico Press, 1997).

19. *Nineteenth and Twentieth Annual Reports of the Territorial Superintendent of Public Instruction to the Governor of New Mexico for the Years 1909–1910* (Santa Fe: The New Mexican Printing Company, 1911), p. 19 and p. 144; *Twenty-seventh and Twenty-Eighth Annual Reports of the State Superintendent of Public Instruction to the Governor of New Mexico for the Years 1917–1918* (Albuquerque, NM: Central Printing Co., 1918), quote on p. 30.

20. Getz, *Schools of Their Own.*

21. Information on higher education in this section is from Victoria-María MacDonald and Teresa García, "Historical Perspectives on Latino Access to Higher Education, 1848–1990," in Jeannette Castellanos and Lee Jones, *The Majority in the Minority: Expanding the Representation of Latina/a Faculty, Administrators and Students in Higher Education* (Sterling, VA: Stylus Press, 2003), pp. 24–27.

22. Dennis Valdés, *Barrios Norteños: St. Paul and Midwestern Mexican Communities in the Twentieth Century* (Austin: University of Texas Press, 2000).

23. Carlos Muñoz, *Youth, Identity, Power: The Chicano Movement* (London: Verso Books, 1989).

24. Armando Navarro, *Mexican American Youth Organization: Avant-garde of the Chicano Movement in Texas* (Austin: University of Texas Press, 1995), pp. 49–50.

25. Muñoz, *Youth, Identity, Power*, pp. 24–25.

26. Esquivel Twyoniak and Mario García, *Migrant Daughter: Coming of Age as a Mexican American Woman* (Berkeley: University of California Press, 2000).

27. Muñoz, *Youth, Identity, Power.*

28. Muñoz, *Youth, Identity, Power*, pp. 48–49.

29. Carl V. Allsup, *The American G.I. Forum: Origins and Evolution* (Austin: University of Texas, Center for Mexican American Studies, no. 6, 1982).

30. Donato, "Hispano Education and the Implications of Autonomy," pp. 138–139.

31. Twyoniak and García, *Migrant Daughter.*

32. Mario T. García, *Mexican Americans: Leadership, Ideology and Identity, 1930–1960* (New Haven: Yale University Press, 1989).

33. Sánchez-Korrol, *From Colonia to Community*, p. 21.

34. Ibid.

35. Sánchez-Korroll, *From Colonia to Community*; and Meyer Weinberg, *A Chance to Learn.*

36. Meyer Weinberg, *A Chance to Learn: The History of Race and Education in the United States.* (Cambridge: Cambridge University Press, 1977).

37. Virginia Sánchez-Korroll, "Toward Bilingual Education: Puerto Rican Women Teachers in New York City Schools, 1947–1967," in Altagarcia Ortiz, ed., *Puerto Rican Women and Work: Bridges in Transnational Labor* (Philadelphia: Temple University Press, 1996), pp. 82–104.

38. Twyoniak and García, *Migrant Daughter*; Dennis Nodín Valdés, *Al Norte: Agricultural Workers in the Great Lakes Region, 1917–1970* (Austin: University of Texas Press, 1991); Elva Treviño Hart, *Barefoot Heart: Stories of a Migrant Child* (Tempe: Bilingual Press, 1999).

39. *Report of the Public Schools of El Paso, Texas, 1903–1904*, p. 13. *Report of the Public Schools of El Paso, Texas, 1904–05*, Library of Congress, Washington, D.C.

CHAPTER 6

1. Mrs. Sylvia Carothers, director, Florida Children's Commission, *Report to the Governor of Florida on the Cuban Refugee Problem in Miami*, February 21, 1961, Florida State Archives, R.A. Gray Building, Department of State, Tallahassee, FL.
2. Juan Gonzalez, *Harvest of Empire: A History of Latinos in America* (New York: Viking Press, 2000), p. 110.
3. James S. Olson and Judith E. Olson, *Cuban Americans: From Trauma to Triumph* (New York: Twayne Publishers, 1995), p. 24.
4. Alex Stepick and Carol Dutton Stepick, "Power and Identity: Miami Cubans," in Marcelo M. Suárez-Orozco and Mariela M. Páez, *Latinos Remaking America* (Berkeley: University of California Press, 2002), pp. 75–92.
5. Victoria-María MacDonald, "Hispanic, Latino, Chicano, or 'Other'?: Deconstructing the Relationship between Historians and Hispanic-American Educational History," *History of Education Quarterly* 41 (fall 2001): 410–411.
6. María Cristina García, *Havana USA: Cuba Exiles and Cuban Americans in South Florida, 1959–1994* (Berkeley: University of California Press, 1996), p. 43.
7. García, *Havana USA.*
8. Letter from Joe Hall to Tracy Voorhees, November 23, 1960, Rockefeller Foundation Archives, Cuban Refugees, Box 67, Folder 565, Rockefeller Archives Center (RAC), Sleepy Hollow, New York.
9. Carothers, *Report to the Governor of Florida*, 1961, p.5.
10. Ibid., p. 5.
11. Thomas D. Bailey, superintendent, Florida State Department of Education, *Report on Cuban Non-Immigrant Children Enrolled in Schools of Dade County, Florida*, State Department of Education, Tallahassee, Florida. January 27, 1961, p. 1. Florida State Archives, Tallahassee, FL.
12. Tracy S. Voorhees, *Report to the President of the United States on the Cuban Refugee Problem*, January 18, 1961 (Washington: United States Government Printing Office, 1961), Rockefeller Foundation Archives, Cuban Refugees, Box 67, Folder 565, Rockefeller Archives Center (RAC), Sleepy Hollow, New York.
13. Stepick and Stepick, "Power and Identity," p. 76.
14. Bailey, *Report on Cuban*, p. 3.
15. Bailey, *Report on Cuban*, p. 4.
16. Bailey, *Report on Cuban*, p. 4.
17. Bailey, *Report on Cuban*, p. 4.
18. Bailey, *Report on Cuban*, p. 6.
19. Bailey, *Report on Cuban*, p. 7.
20. Bailey, *Report on Cuban*, p. 7.
21. Olson and Olson, *Cuban Americans*, p. 99.
22. Ibid.
23. Virginia Sánchez-Karroll, *From Colonia to Community: The History of Puerto Ricans in New York City* (Berkeley: University of California Press, 1994).
24. Letter from Thomas D. Bailey to James L. Burnsted, February 2, 1961, Cubans, Series 1126, Carton 5, Florida State Archives, Department of State, Tallahassee, FL.
25. Letter from James L. Burnsted to Thomas D. Bailey, January 26, 1961, Cubans, Series 1126, Carton 5, Florida State Archives, Department of State, Tallahassee, FL.
26. Bailey to Burnsted, February 2, 1961.
27. García, *Havana USA*, pp. 26–29.

28. Quoted in García, *Havana USA*, pp. 26–27.

29. Ibid.

30. Ibid., p. 28.

31. Olson and Olson, *Cuban Americans*, p. 64.

32. Alejandro Portes and Alex Stepick, *City on the Edge: The Transformation of Miami* (Berkeley: University of California Press, 1993).

33. Marjorie L. Fillyaw, "A New Kind of Refugee," *Ave Maria National Catholic Weekly* n.d., circa 1960, Cuban Refugees, Series 276, Carton 2, Florida State Archives, Department of State, Tallahassee, FL.

34. Yvonne M. Conde, *Operation Pedro Pan: The Untold Exodus of 14,048 Cuban Children* (New York: Routledge Press, 1999); Victor Andres Triay, *Fleeing Castro: Operation Pedro Pan and the Cuban Children's Program* (Gainesville: University Press of Florida, 1998); and Marilyn Gilroy, "Operation Pedro Pan: Memories and Accolades 40 Years Later," *The Hispanic Outlook in Higher Education* (July 15, 2002): 7–10.

35. Gilroy, "Operation Pedro Pan," p. 8.

36. Conde, *Operation Pedro Pan*.

37. Stepick and Stepick, "Power and Identity," p. 81. The author was also a resident of Miami during this turbulent era and a witness to the social and political conditions.

38. García, *Havana USA*, p. 65.

39. Silva, *Children of Mariel*, p. 28.

40. Helga Silva, *The Children of Mariel from Shock to Integration: Cuban Refugee Children in South Florida Schools* (Miami: Cuban American National Foundation, Inc. 1985), p. 14.

41. Silva, *Children of Mariel*, quotation from page 22.

42. Silva, *Children of Mariel*, pp. 36–42.

43. "Milwaukee Public Schools Recent Immigrant and Refugee Assessment and Bilingual Academic Support Program, 1981," Hispanics in Wisconsin, M89–116, folder, "Next edition—Hispanic Bibliography," Wisconsin Historical Society, Madison, Wisconsin.

44. Cuban Refugees at Camp McCoy photographs, ca. 1980–ca. 1981. Photographs by Dave Archibald, M94–198, Wisconsin State Historical Society, Madison, Wisconsin.

45. For a comparison of educational levels between Cubans, Mexicans, Puerto Ricans, and other Hispanics see table 1.5, "Selected Characteristics of Hispanics by Type of Origin," in Marcelo M. Suárez-Orozco and Mariela M. Páez, "Introduction: The Research Agenda," in Suárez-Orozco and Páez, *Latinos Remaking America*, p. 27.

CHAPTER 7

1. Letter from Students of Claremont Men's College to Mr. Edwin H. Corbin. March 3, 1969, Rockefeller Foundation Archives, Record Group 1.2, series 200, box 19, folder 160, Claremont Colleges, Mexican-American Students, Rockefeller Archives Center (RAC), Sleepy Hollow, NY.

2. Ysidro Ramon Macias, "The Chicano Movement," in *A Documentary History of the Mexican Americans*, edited by Wayne Moquin and Charles Van Doren (New York: Praeger Publishers, 1971), p. 391. Originally appeared in *Wilson Library Bulletin*, March 1970.

3. Manuel G. Gonzales, *Mexicanos: A History of Mexicans in the United States* (Bloomington: Indiana University Press, 1999), pp. 191–222; Juan Gonzalez, *Harvest of Empire: A History of Latinos in America* (New York: Viking Press, 2000), pp. 167–182; Andrés Torres and José E. Velázquez, eds., *The Puerto Rican Movement: Voices from the Diaspora* (Philadelphia: Temple University Press, 1998); and Miguel Melendez, *We Took the Streets: Fighting for Latino Rights with the Young Lords* (New York: St. Martin's Press, 2003).

4. Gonzalez, *Harvest of Empire*, p. 174.

5. Gonzales, *Mexicanos*, p. 194.

6. Ian F. Haney López, *Racism on Trial: The Chicano Fight for Justice* (Cambridge: Belknap Press of Harvard University Press, 2003), pp. 16–17.

7. Testimony of Patricia Delgado, *Hearings before the United States Commission on Civil Rights*. San Francisco, CA, May 1–3, 1967 and Oakland, CA, May 4–6, 1967 (Washington: U.S. Government Printing Office, 1967), p. 408.

8. López, *Racism on Trial*, p. 17.

9. López, *Racism on Trial*, p. 20.

10. F. Arturo Rosales, *Chicano!: The History of the Mexican American Civil Rights Movement* (Houston, TX: Arte Público Press, 1996), pp. 184–151 and Gonzales, *Mexicanos*, pp. 209–210.

11. Rosales, *Chicano!*, p. 192.

12. López, *Racism on Trial*, pp. 20–23.

13. Guadalupe San Miguel, Jr., *"Let All of Them Take Heed": Mexican Americans and the Campaign for Educational Equality in Texas, 1910–1981* (Austin: University of Texas Press, 1987), p. 171.

14. San Miguel, Jr., *"Let All of Them Take Heed,"* p. 178.

15. Guadalupe San Miguel, Jr., *Brown, Not White: School Integration and the Chicano Movement in Houston* (College Station: Texas A and M Press, 2001).

16. San Miguel, Jr., *"Let All of Them Take Heed,"* p. 192.

17. Laws of 90th Cong.—1st session, P.L. 90–247, January 2, 1968, p. 919.

18. Sandra Del Valle, "Bilingual Education for Puerto Ricans in New York City: From Hope to Compromise," *Harvard Educational Review* 68 (summer 1998): 193–217.

19. Herbert J. Teitelbaum and Richard J. Hiller, "Bilingual Education: The Legal Mandate," *Harvard Educational Review* 47 (May 1977): 139.

20. Iris C. Rotberg, "Some Legal and Research Considerations in Establishing Federal Policy in Bilingual Education," *Harvard Educational Review* 52 (May 1982): 151.

21. Alberto M. Ochoa, "Language Policy and Social Implications for Addressing the Bicultural Immigrant Experience in the United States," in *Culture and Difference: Critical Perspectives on the Bicultural Experience in the United States*, edited by Antonia Darder (Westport, CT: Bergin and Garvey, 1995), pp. 227–253; Carlos J. Ovando, "Politics and Pedagogy: The Case of Bilingual Education," *Harvard Educational Review* 60, no. 3 (1990): pp. 341–356; James Crawford, *Bilingual Education: History, Politics, Theory and Practice* (Trenton, NJ: Crane, 1989); Guadalupe Valdés, "Dual-Language Immersion Programs: A Cautionary Note Concerning the Education of Language-Minority Students," *Harvard Educational Review* 67 (fall 1997): 391–429; and Rosalie Pedalino Porter, *Forked Tongue: The Politics of Bilingual Education* (New York: Basic Books, 1990).

22. Richard J. Margolis, *The Losers: A Report on Puerto Ricans and the Public Schools*, p. 1. Commissioned by Aspira, Inc. May, 1968. Vertical File—Education—Elementary School. Center for Puerto Rican Studies, Hunter College, New York City.

23. Francesco Cordasco and Eugene Bucchioni, eds., *Puerto Rican Children in Mainland Schools* (Metuchen, NJ: The Scarecrow Press, Inc., 1968), pp. 353–4.
24. Gonzalez, *Harvest of Empire*, p. 93; and Sonia Nieto, "Puerto Rican Students in the U.S. Schools: A Brief History," Sonia Nieto, ed., in *Puerto Rican Students in U.S. Schools* (Mahwah, NJ: Lawrence Erlbaum Associates, 2000), p. 16.
25. Madeleine E. López, "New York, Puerto Ricans, and the Dilemmas of Integration," unpublished paper, Lehman College, 2003; and Nieto, "A Brief History," p. 16.
26. Jorge Duany, *The Puerto Rican Nation on the Move: Identities on the Island and in the United States* (Chapel Hill: University of North Carolina Press, 2002), p. 202.
27. Duany, *Puerto Rican Nation*, pp. 202–206; and Gonzalez, *Harvest of Empire*, p. 94.
28. Virginia E. Sanchéz-Korrol, *From Colonia to Community: The History of Puerto Ricans in New York City* (Berkeley: University of California Press, 1983, rpt. 1994).
29. Duany, *Puerto Rican Nation*, p. 202.
30. López, "New York, Puerto Ricans, and the Dilemmas of Integration."
31. Sylvia Alicia Gonzales, *Hispanic American Voluntary Organizations* (Westport, CT: Greenwood Press, 1985).
32. Duany, *Puerto Rican Nation*, p. 203; Miguel "Mickey" Melendez, *We Took the Streets: Fighting for Latino Rights with the Young Lords* (New York: St. Martin's Press, 2003); Felix M. Padilla, *Latino Ethnic Consciousness: The Case of Mexican Americans and Puerto Ricans in Chicago* (Notre Dame Press: University of Notre Dame Press, 1985), p. 53; and Gonzalez, *Harvest of Empire*, p. 93.
33. For a description of ASPIRA's founder, Antonia Pantoja, see "Memorias de una Vida de Obra (Memories of a Life of Work): An Interview with Antonia Pantoja," *Harvard Educational Review* 68 (summer 1998): 244–258.
34. Nieto, "A Brief History," p. 20.
35. Nieto, "A Brief History," p. 16; and Richard Margolis, *The Losers: A Report on Puerto Ricans and the Public Schools*. Commissioned by ASPIRA Inc. May, 1968 (Newark, NJ: ASPIRA Inc.) Vertical file, Center for Puerto Rican Studies, Hunter College, New York, NY.
36. Nieto, "A Brief History," p. 21.
37. Muñoz, p. 200.
38. From Rockefeller Foundation Archives officer interview. "Interviews: JEB" with Edgar C. Reckard Jr., chaplain, Claremont Colleges. Dated June 23, 1967, Rockefeller Foundation Archives, Record Group. 1.2, series 200, box 19, folder 158, Rockefeller Archives Center, Sleepy Hollow, NY
39. Navarro, 1995.
40. Valdez, 2000, p. 159.
41. De León, 2001.
42. Carter, 1970, p. 31.
43. Muñoz, 1989, p. 142.
44. Muñoz, 1989, p. 58.
45. Navarro, 1995, p. 86.
46. Olivas; Muñoz, 1989, p. 85.
47. Dolores Delgado Bernal 1998.
48. Navarro, 1995; Munoz, 1989; Valdez, 2000.
49. "Resolution to Grant Funds to Claremont College," Rockefeller Foundation 67071, June 30, 1968, Rockefeller Foundation Archives, Record Group. 1.2, series 200, box 19, folder 158, Rockefeller Archives Center, Sleepy Hollow, NY, From Rockefeller Foundation Archives officer interview. "Interviews: JEB" with Edgar C. Reckard Jr., chaplain, Claremont Colleges.

50. Cordasco and Bucchioni, *Puerto Rican Children in Mainland Schools*, pp. 353–354.
51. Traub, 1994, p. 44.
52. Sánchez-Korrol, 1994.
53. Santiago, 1995, p. xviii.
54. Serrano, 1998.
55. Serrano, 1998.
56. "Despierta Boricua," and "An Open Letter to President Brewster," vertical file—higher education—Yale College, Center for Puerto Rican Studies, Hunter College, New York, NY.
57. "Catholic High Schools in New York City," vertical file—higher education, Center for Puerto Rican Studies, Hunter College, New York, NY.
58. Rodriguez-Fraticelli, 1986.
59. Serrano, 1998.
60. Valdez, 2000, p. 100.
61. Traub, 1994.
62. Keller, Deneed, and Magallán, 1991.
63. Richard Rodriguez, *Brown, The Last Discovery of America* (New York: Penguin, 2002), p. 95.

CHAPTER 8

1. Victoria-María MacDonald, *Returning to María: A Journey through Cultures and Self*, unpublished manuscript, 1994.
2. Racial and Ethnic Tensions in American Communities: Poverty, Inequality, and Discrimination.—A National Perspective. Executive Summary and Transcript of Hearing Held in Washington, D.C. May 21–22, 1992 (Washington: U.S. Commission on Civil Rights, 1992), p. 8.
3. Alejandro Portes and Rubén G. Rumbaut, *Immigrant America: A Portrait*, 2nd ed. (Berkeley: University of California Press, 1996).
4. Juan Gonzalez, *Harvest of Empire: A History of Latinos in America* (New York: Viking Press, 2000), p. 129; Alejandro Portes and Alex Stepick, *City on the Edge: The Transformation of Miami* (Berkeley: University of California Press, 1993), pp. 206–207.
5. Stanton Wortham, Enrique G. Murillo, Jr., and Edmund T. Hamann, *Education in the New Latino Diaspora: Policy and the Politics of Identity* (Westport, CT: Ablex, 2002).
6. Leonard Dinnerstein, Roger L. Nichols, and David M. Reimers, *Natives and Strangers: A Multicultural History of Americans* (New York: Oxford University Press, 1996).
7. Data from the 2000 Census U.S. Bureau of the Census. Available online at www.census.gov
8. For example, scholars have increasingly turned to examining the urban concentration of blacks and Latinos. See for example, Roger Waldinger, *Still the Promised City?: African-Americans and New Immigrants in Postindustrial New York* (Cambridge: Harvard University Press, 1996) and James Jennings, ed., *Blacks, Latinos and Asians in Urban America: Status and Prospects for Politics and Activism* (Westport, CT: Praeger Publishing, 1994).
9. Clara E. Rodríguez, *Changing Race: Latinos, the Census, and the History of Ethnicity in the United States*. (New York: New York University Press, 2000).

10. Gary Orfield, "Commentary," in *Latinos Remaking America*, edited by Marcelo M. Suárez-Orozco and Mariela M. Páez (Berkeley: University of California Press and David Rockefeller Center for Latin American Studies, Harvard University, 2002), p. 389.

11. Rubén G. Rumbaut and Alejandro Portes, *Ethnicities: Children of Immigrants in America* (Berkeley: University of California Press and New York: Russell Sage Foundation, 2001).

12. Silvia Torres-Saillant and Ramona Hernández, *The Dominican Americans* (Westport, CT: Greenwood Press, 1998), p. 38; and Alejandro Portes and Alex Stepick, *City on the Edge: The Transformation of Miami* (Berkeley: University of California Press, 1993), pp. 206–207.

13. Gonzalez, *Harvest of Empire*, p. 118.

14. Julia Alvarez, *How the García Girls Lost Their Accents* (New York: Plume Books, 1992).

15. Torres-Saillant and Hernández, *The Dominican Americans*, p. 34.

16. Torres-Saillant and Hernández, *The Dominican Americans*, p. 157.

17. Gonzalez, *Harvest of Empire*, p. 117.

18. Marcelo M. Suárez-Orozco, *Central American Refugees and U.S. High Schools: A Psychosocial Study of Motivation and Achievement* (Stanford, CA: Stanford University Press, 1989), p. 59.

19. Gonzalez, *Harvest of Empire*, p. 129.

20. Suárez-Orozco, *Central American Refugees*, p. 71.

21. *Racial and Ethnic Tensions in American Communities: Poverty, Inequality, and Discrimination*, vol. 1: *The Mount Pleasant Report* (Washington, D.C.: Report of the United States Commission on Civil Rights, January 1993), p. 124.

22. Carola Suárez-Orozco and Marcelo M. Suárez-Orozco, *Children of Immigrants* (Cambridge: Harvard University Press, 2001).

23. Portes and Stepick, *City on the Edge*, p. 151.

24. *Racial and Ethnic Tensions in American Communities: Poverty, Inequality, and Discrimination*, vol. 1: *The Mount Pleasant Report*, p. 124.

25. Ibid., p. 134.

26. Ibid.

27. Suárez-Orozco and Suárez-Orozco, *Children of Immigrants*.

28. Eugene E. García, *Hispanic Education in the United States: Raíces y Alas* (Lanham, MD: Rowman and Littlefield, 2001), pp. 92–93.

29. García, *Hispanic Education*, p. 93.

30. Portes and Stepick, *City on the Edge*, p. 161.

31. *Racial and Ethnic Tensions in American Communities: Poverty, Inequality, and Discrimination*, vol. IV: *The Miami Report* (Washington, D.C.: Report of the United States Commission on Civil Rights, October 1997), p. 31.

32. Ibid.

33. Angela Valenzuela, *Subtractive Schooling: U.S.-Mexican Youth and the Politics of Caring* (Albany: State University of New York Press, 1999), p. 19.

34. National Center for Education Statistics [NCES]. *The Condition of Education for Hispanic Americans* (Washington, D.C.: Government Printing Office, 1980).

35. Gary Orfield, Susan E. Eaton, and The Harvard Project on School Desegregation, *Dismantling Desegregation: The Quiet Reversal of Brown v. Board of Education* (New Press: New York, 1996).

36. Marcelo M. Suárez-Orozco and Mariela M. Páez, eds., *Latinos Remaking America* (Berkeley: University of California Press and David Rockefeller Center for Latin

American Studies, Harvard University, 2002), is an excellent example of cutting-edge research on this topic.

37. Scholars, politicians, and policymakers disagree on whether a correct enumeration has been conducted, particularly regarding undocumented Latino immigrants. In addition, the 2000 Census allowed individuals for the first time to check boxes for more than one racial category, thus complicating agreement on the accuracy of the data. Based on data from the United States Census Bureau Table Population and Housing Table 1 (PHC-T-1). See www.census.gov/population/www/cen2000/phc-t1.html

38. Scott A. L. Beck and Martha Allexsaht-Snider, "Recent Language Minority Education Policy in Georgia: Appropriation, Assimilation, and Americanization," in *Education, Policy, and the Politics of Identity in the New Latino Diaspora*, edited by Stanton Wortham, Ted Hamann, and Enrique Murillo, Jr. (Westport, CT: Ablex Press, 2001); Ted Hamann, *The Georgia Project: A Binational Attempt to Reinvent a School District in Response to Latino Newcomers*, Ph.D. diss., University of Pennsylvania, Ann Arbor, University Microfilms, 1999; Lynn Schnaiberg, "Immigration's Final Frontier," *Education Week*, 23 February 1994; Kathleen Kennedy Manzo, "Rural N.C. to Get Aid for LEP Student Influx," *Education Week*, 27 January 1999; Doug Cumming, "State School Board Wants Classes for Immigrants to Expand," *The Atlanta Journal-Constitution*, March, 12, 1999, C10.

39. U.S. Congress, Committee on Education and Labor. Staff Report on Hispanics' Access to Higher Education. 99th Congress, 1st Session, Serial No. 99-K, (Washington, D.C.: Government Printing Office, 1985) p. 3.

40. U.S. Congress, 1985, p. 4.

41. Michael A. Olivas, "Research on Latino College Students: A Theoretical Framework and Inquiry," in Michael A. Olivas, ed. *Latino College Students* (New York: Teachers College Press, 1986), p. 2.

42. U.S. Congress, 1985, p. 1.

43. The name "TRIO" refers to a set of federal outreach programs funded under Title IV of the Higher Education Act of 1965. The programs expanded beyond the original three—Upward Bound, Talent Search, and Student Support Services—but the name TRIO persists.

44. Berta Vigil Laden, "Hispanic-Serving Institutions: Myths and Realities," *Peabody Journal of Education* 76, no. 1 (2001): 73–92.

45. Laden, "Hispanic-Serving," 2001.

46. Jorge Chapa, "Affirmative Action, X Percent Plans, and Latino Access to Higher Education in the Twenty-first Century," in Suárez-Orozco and Páez, eds., *Latinos Remaking America*.

47. Victor Torrejón, "Residency urged for some kids of entrants," *Arizona Daily Star*. July 18, 2002, p.B8.

48. Richard Fry, "Latinos in Higher Education: Many Enroll, Too Few Graduate." Pew Hispanic Center Research Report, September 5, 2002. Accessed from http://www.pewhispanic.org/site/docs/pdf/latinosinhighereducation-sept5–02.pdf. May 19, 2004.

49. Patrick Frazier, "What's in a Nombre?" Editorial, pg. A28. *The Washington Post*, August 30, 2003, Saturday, Final Edition. Accessed via washingtonpost.com, September 18, 2003.

Resources

BOOKS AND ARTICLES

Acuña, Rodolfo. *Occupied America: A History of Chicanos*, 3d. ed. New York: Harper Collins, 1988.

———. *Sometimes There Is No Other Side: Chicanos and the Myth of Equality*. Notre Dame: University of Notre Dame Press, 1998.

Allsup, Carl V. *The American G.I. Forum: Origins and Evolution*. Austin, TX: Center for Mexican American Studies, the University of Texas at Austin. Distributed by the University of Texas Press, Monograph/Center for Mexican American Studies, the University of Texas at Austin, no. 6, 1982.

Alonzo, Armando C. *Tejano Legacy: Rancheros and Settlers in South Texas, 1734–1900*. Albuquerque: University of New Mexico Press, 1998.

Altenbaugh, Richard J. *The American People and Their Education: A Social History*. Upper Saddle River, NJ: Pearson, 2003.

Bancroft, Hubert Howe. *History of California*, vol. 1, *1542–1800*, vol. 18 of *The Works of Hubert Howe Bancroft*. San Francisco: A.L. Bancroft & Co., Publishers, 1884.

Bancroft, Hubert Howe. *History of California*, vol. 2, *1801–1824*, vol. 19 of *The Works of Hubert Howe Bancroft*. Santa Barbara, CA: Wallace Hebberd, 1966.

Banker, Mark. *Presbyterian Missions and Cultural Interaction in the Far Southwest, 1850–1950*. Urbana: University of Illinois Press, 1993.

Bannon, John Francis. *Herbert Eugene Bolton: The Historian and the Man, 1870–1953*. Tucson: University of Arizona Press, 1978.

Beck, Warren A. *New Mexico: A History of Four Centuries*. Norman, OK: University of Oklahoma Press, 1962.

Bender, Norman J. *Winning the West for Christ: Sheldon Jackson and Presbyterianism on the Rocky Mountain Frontier, 1869–1880*. Albuquerque: University of New Mexico Press, 1996.

Berger, Max. "Education in Texas during the Spanish and Mexican Periods." *Southwestern Historical Quarterly* 51 (July 1947): 41–53.

Bogardus, Emory S. *The Mexican Immigrant: An Annotated Bibliography*. Los Angeles: The Council on International Relations, 1929.

———. *The Mexican in the United States*. Los Angeles: University of Southern California Press, 1934.

Bolton, Herbert E. *Wider Horizons of American History*. New York: D. Appleton-Century Co., 1939.

Boswell, Thomas D., and James R. Curtis. *The Cuban-American Experience: Culture, Images, and Perspectives*. Totowa, NJ: Rowman and Allaheld, 1983.

Bradford, James C., ed. *Crucible of Empire: The Spanish-American War and Its Aftermath*. Annapolis, MD: Naval Institute Press, 1993.

Brandes, Raymond S. "Times Gone by in Alta California: Recollections of Señora Dona Juana Machada Alipaz de Ridington (Wrightington)." *The Historical Society of Southern California Quarterly* 41 (September 1959).

Brooks, James F. *Captives and Cousins: Slavery, Kinship, and Community in the Southwest Borderlands.* Chapel Hill, NC: University of North Carolina Press, 2002.

Burns, Rev. J. A. *The Catholic School System in the United States: Its Principles, Origin, and Establishment.* New York: Benziger Brothers, 1908.

Bushnell, Amy Turner. "Republic of Spaniards, Republic of Indians." In *The New History of Florida*, edited by Michael Gannon. Gainesville, FL: University Press of Florida, 1996.

Camarillo, Albert. *Chicanos in a Changing Society: From Mexican Pueblos to American Barrios in Santa Barbara and Southern California, 1848–1930.* Cambridge, MA: Harvard University Press, 1979.

Carrión, Arturo Morales. *Puerto Rico: A Political and Cultural History.* New York: W.W. Norton and Co., 1983.

Carter, Thomas P. *Mexican Americans in School: A History of Educational Neglect.* New York: College Entrance Examination Board, 1970.

Carter, Thomas P., and Roberto Segura. *Mexican Americans in the Public Schools: A Decade of Change.* New York: College Entrance Examination Board, 1979.

Castañeda, Carlos E. *The Mission Era: The Missions at Work, 1731–1761.* Vol. 3 of *Our Catholic Heritage in Texas, 1519–1936.* Austin, TX: Boeckmann-Jones, 1938.

———. *The Church in Texas Since Independence, 1836–1950.* Vol. 7 of *Our Catholic Heritage In Texas, 1519–1936.* Austin: Von Boeckmann-Jones Co., 1958.

Cecelski, David. *Along Freedom Road: Hyde County, North Carolina and the Fate of Black Schools in the South.* Chapel Hill: University of North Carolina Press, 1994.

Clark, Victor S. *Porto Rico and Its Problems.* Washington, D.C.: Brookings Institute, 1928.

Cohen, Sol. *Education in the United States: A Documentary History.* Vol. 2. New York: Random House, 1974.

Conde, Yvonne M. *Operation Pedro Pan: The Untold Exodus of 14,048 Cuban Children.* New York: Routledge Press, 1999.

Contreras, A. Reynaldo, and Leonard A. Valverde. "The Impact of *Brown* on the Education of Latinos." In *Brown v. Board of Education at Forty: A Commemorative Issue Dedicated to the Late Thurgood Marshall, Journal of Negro Education* 63, no. 3 (summer 1994): 470–481.

Cook, Katherine, and Florence E. Reynolds. *The Education of Native and Minority Groups, A Bibliography, 1923–32.* Washington, D.C.: U.S. Dept. of Education, Bulletin No. 12, 1933.

Cordasco, Francesco, and Eugene Bucchioni, eds. *Puerto Rican Children in Mainland Schools.* Metuchen, NJ: The Scarecrow Press, 1968.

———. *The Puerto Rican Community and Its Children on the Mainland: a Sourcebook for Teachers, Social Workers, and Other Professionals.* 3d. rev. ed. Metuchen, NJ: The Scarecrow Press, 1982.

Cortés, Carlos E., ed. *The Cuban Experience in the United States.* New York: Arno Press, 1980.

———. *Cuban Refugee Programs.* New York: Arno Press, 1980.

Cox, I. J. "Educational Efforts in San Fernando de Bexar," *Texas Historical Association Quarterly* 6 (July 1902): 28–29.

Crahan, Margaret E. "Religious Penetration and Nationalism in Cuba: U.S. Methodist Activities, 1898–1958." *Revista/Review Interamericana* 8 (summer 1978): 204–224.

Crawford, James. *Bilingual Education: History, Politics, Theory and Practice.* Trenton, NJ: Crane, 1989.

Cremin, Lawrence A. *The Wonderful World of Ellwood Patterson Cubberley: An Essay on the Historiography of American Education.* New York: Bureau of Publications, Columbia University Teachers College, 1965.

———. *American Education: The Colonial Experience, 1607–1783.* New York: Harper and Row, 1970.

———. *American Education: The National Experience, 1783–1876.* New York: Harper and Row, 1980.

Curley, Michael J. *Church and state in the Spanish Floridas (1783–1822).* Washington, D.C., The Catholic University of America Press, 1940.

Darder, Antonia, et al., eds. *Latinos and Education: A Critical Reader.* New York: Routledge Press, 1997.

David, Jerry A. "Matilda Allison on the Anglo-Hispanic Frontier: Presbyterian Schooling in New Mexico, 1880–1910." *American Presbyterians* 74 (fall 1996): 171–182.

Davis, Matthew. "Herschel T. Manuel and Latino Educational Policy, 1925–1975." Ph.D. diss., University of Texas, 2000.

De La Torre, Adela, and B. M. Pesquera. *Building with Our Hands: New Directions in Chicana Studies.* Berkeley: University of California Press, 1993.

De León, Arnoldo. *The Tejano Community, 1836–1900.* Albuquerque: University of New Mexico Press, 1982.

———. *They Called Them Greasers: Anglo Attitudes toward Mexicans in Texas, 1821–1900.* Austin: University of Texas Press, 1983.

———. *Ethnicity in the Sunbelt: Mexican Americans in Houston.* College Station, TX: Texas A and M University Press, 2001.

de Montilla, Aida Negron. *Americanization in Puerto Rico and the Public School System, 1900–1930.* Rio de Piedras, P.R.: Universidad de Puerto Rico, 1975.

del Castillo, Richard Griswold. *La Familia: Chicano Families in the Urban Southwest, 1848 to the Present.* Notre Dame, IN: University of Notre Dame Press, 1984.

———. *The Treaty of Guadalupe Hidalgo: A Legacy of Conflict.* Norman, OK: University of Oklahoma Press, 1990.

———. "Chicano Historical Discourse: An Overview and Evaluation of the 1980s." *Perspectives in Mexican American Studies* 4 (1993): 1–25.

del Castillo, Richard, and Arnoldo De León. *North to Aztlán: A History of Mexican Americans in the United States.* New York: Twayne Publishers, 1996.

Del Valle, Sandra. "Bilingual Education for Puerto Ricans in New York City: From Hope to Compromise." *Harvard Educational Review* 68 (summer 1998): 193–217.

Delgado Bernal, D. "Grassroots Leadership Reconceptualized: Chicana Oral Histories and the 1968 East Los Angeles School Blowouts." *Frontiers: A Journal of Women Studies* 19, no. 2 (1998): 113–142.

DeSantis, Hugh. "The Imperialistic Impulse and American Innocence, 1865–1900." In *American Foreign Relations: A Historiographical Review*, edited by Gerald K. Haines and J. Samuel Walker, 65–90. Westport, CT: Greenwood Press, 1981.

Deutsch, Sara. *No Separate Refuge: Culture, Class, and Gender on the Anglo-Hispanic Frontier in the American Soutwest, 1880–1940.* New York: Oxford University Press, 1987.

Dinnerstein, Leonard, Nichols, Roger L., and David M. Reimers, eds. *Natives and Strangers: A Multicultural History of Americans.* 4th. ed. New York: Oxford University Press, 2003.

Dolan, Jay P. *The American Catholic Experience: A History From Colonial Times to the Present.* Garden City, NY: Image Books, 1987.

Donato, Rubén. *The Other Struggle for Equal Schools: Mexican Americans during the Civil Rights Era.* Albany: State University Press of New York, 1997.

———. "Hispano Education and the Implications of Autonomy: Four School Systems in Southern Colorado, 1920–1963." *Harvard Educational Review* 69 (summer 1999): 117–149.

Donato, R. and Marvin Lazerson. "New Directions in American Educational History: Problems and Prospects." *Educational Researcher* 29, no. 8 (2000): 4–15.

Donnelly, Thomas C. "Educational Progress in New Mexico and Some Present Problems." *New Mexico Quarterly Review* (autumn 1946): 305–17.

Duany, Jorge. *The Puerto Rican Nation on the Move: Identities on the Island and in the United States.* Chapel Hill: University of North Carolina Press, 2002.

Eby, Frederick E., comp. *Education in Texas: Source Materials.* Austin, TX: University of Texas Bulletin, no. 1824, April, 1918.

———. *The Development of Education in Texas.* New York: Macmillan Co., 1925.

Eisenhower, John D. *So Far from God: The War with Mexico, 1846–1848.* New York: Random House, 1989.

Elsasser, Ana, et al. *Las Mujeres: Conversations from a Hispanic Community.* Old Westbury, NY: Feminist Press, 1980.

Engelhardt, Fr. Zephyyrin, O. F. M. *The Missions and Missionaries of California.* Vol. 1. San Francisco: The James H. Barry Company, 1908.

Engelhardt, Fr. Zephyyrin, O. F. M. *The Missions and Missionaries of California.* Vol. 2. San Francisco: The James H. Barry Company, 1908.

Estrada, Jorge A. *The Depiction of Hispanic People in the California State–Adopted Sixth Grade Readings Texts in 1982.* Coral Gables, FL: La Torre de Papel Inc., 1994.

Fagen, Richard F., Richard A. Brody, and Thomas J. O'Leary. *Cubans in Exile: Disaffection and the Revolution.* Stanford: Stanford University Press, 1968.

Falcón, Luis Nieves, et al. *Stereotypes, Distortions and Omissions in U.S. History Textbooks.* New York: Council on Interracial Books for Children, 1977.

Farnsworth, Paul and Robert H. Jackson. "Cultural, Economic, and Demographic Change in the Missions of Alta California: The Case of Nuestra Señora de la Soledad." In *The New Latin American Mission History*, edited by Erick Langer and Robert H. Jackson, 109–129. Lincoln: University of Nebraska Press, 1995.

Ferrier, William Warren. *Ninety Years of Education in California, 1846–1936.* Oakland, CA: West Coast Printing, 1937.

Fitchen, Edwards D. "The United States Military Government, Alexis E. Frye and Cuban Education, 1898–1902." *Revista/Review Interamericana* 2 (summer 1972).

———. "The Cuban Teachers and Harvard, 1900: An Early Experiment in Inter-American Cultural Exchange." *Horizontes: Revista de la Universidad Católica de Puerto Rico* 26 (1973).

Fitzpatrick, J. P. *Puerto Rican Americans: The Meaning of Migration to the Mainland.* 2d. ed. Englewood Cliff, NJ: Prentice-Hall, 1987.

Foner, Philip S. *The Spanish-Cuban-American War and the Birth of American Imperialism 1895–1902.* Vol. 2. New York: Monthly Review Press, 1972.

Foner, Philip S., and Richard C. Winchester. *The Anti-Imperialist Reader: A Documentary History of Anti-Imperialism in the United States.* London: Holmes and Meier Publishers, Inc., 1984.

"Forum: 'Why the West Is Lost': Comments and Response." *William and Mary Quarterly* 51, 3d. ser. (October 1994): 717–754.

Fraticelli-Rodríguez, Carlos. *Education and Imperialism: The Puerto Rican Experience in Higher Education, 1898–1986.* Centro de Estudios Puertorriqueños Working Paper Series. Higher Education Task Force. Hunter College, New York, 1986.

Frazier, Patrick. "What's in a Nombre?" Editorial. Pg.A28. *The Washington Post.* August 30, 2003, Saturday, Final Edition. Accessed via washingtonpost.com September 18, 2003.

Fry, Richard. "Latinos in Higher Education: Many Enroll, Too Few Graduate." Pew Hispanic Center Research Report, 2002. Accessed from www.pewhispanic.org/ May 14, 2004.

Gaines, John Strother. "The Treatment of Mexican-American History in Contemporary American High School Textbooks." Ed.D. Diss., University of Southern California, 1971.

Gallegos, Bernardo P. *Literacy, Education and Society in New Mexico, 1693–1821.* Albuquerque: University of New Mexico Press, 1992.

Gammel, H. P. N., comp. and arr. *The Laws of Texas, 1822–1897.* 10 vols. Austin, TX: Gammel Book Company, 1898.

Gannon, Michael. "First European Contacts." In *The New History of Florida,* edited by Michael Gannon. Gainesville, FL: University Press of Florida, 1996.

García, Alma M. *Chicana Feminist Thought: The Basic Historical Writings.* New York: Routledge, 1997.

García, Eugene E. *Hispanic Education in the United States: Raíces y Alas.* Lanham, MD: Rowman and Littlefield Publishers, Inc., 2001.

García, María Cristina. *Havana USA: Cuba Exiles and Cuban Americans in South Florida, 1959–1994.* Berkeley: University of California Press, 1996.

García, Mario T. *Mexican Americans: Leadership, Ideology and Identity, 1930–1960.* New Haven: Yale University Press, 1989.

———. "The Hispanic in American History: Myth and Reality — A Review of L. H. Gann and Peter J. Duigan, *The Hispanic in the United States: A History.*" *Latino Studies Journal* (January 1991): 75–85.

Geiger, Roger. "New Themes in the History of Nineteenth-Century Colleges." In *The American College in the Nineteenth Century,* edited by Roger Geiger, 1–36. Nashville: Vanderbilt University Press, 2000.

Getz, Lynne Marie. "Extending the Helping Hand to Hispanics: The Role of the General Education Board in New Mexico in the 1930s." *Teachers College Record* 93 (spring 1992): 500–515.

———. *Schools of Their Own: The Education of Hispanos in New Mexico, 1850–1940.* Albuquerque: University of New Mexico Press, 1997.

Gibson, Charles. *The Black Legend: Anti-Spanish Attitudes in the Old World and the New.* New York: Alfred A. Knopf, 1971.

Gilroy, Marilyn. "Operation Pedro Pan: Memories and Accolades 40 Years Later." *The Hispanic Outlook in Higher Education* (July 15, 2002): 7–10.

Glasser, Ruth. *My Music is My Flag: Puerto Rican Musicians and Their New York Communities, 1917–1940.* Berkeley and Los Angeles: University of California Press, 1994.

Glazer, Nathan, and Daniel Patrick Moynihan. *Beyond the Melting Pot: The Negroes, Puerto Ricans, Jews, Italians, and Irish of New York City.* Cambridge, MA: MIT Press, 1963.

González, Gilbert G. *Chicano Education in the Era of Segregation.* Philadelphia: Balch Institute Press, 1990.

González, Gilbert G., and Raúl Fernández. "Chicano History: Transcending Cultural Models." *Pacific Historical Review* 63 (November 1994): 469–497.

Gonzales, Manuel G. *Mexicanos: A History of Mexicans in the United States.* Bloomington: Indiana University Press, 1999.

Gonzales, Sylvia Alicia. *Hispanic American Voluntary Organizations.* Westport, CT: Greenwood Press, 1985.

González, Deena J. *Refusing the Favor: The Spanish-Mexican Women of Santa Fe, 1820–1880.* New York: Oxford University Press, 1999.

Gonzalez, Juan. *Harvest of Empire: A History of Latinos in America.* New York: Viking Press, 2000.

Gordon, Linda. *The Great Arizona Orphan Abduction.* Cambridge: Harvard University Press, 1999.

Grose, Howard B. *Advance in the Antilles: the new era in Cuba and Porto Rico.* New York: Eaton and Mains, 1910.

Gutiérrez, Ramón A. *When Jesus Came, the Corn Mothers Went Away: Marriage, Sexuality, and Power in New Mexico, 1500–1846.* Stanford, CA: Stanford University Press, 1991.

———. "Community, Patriarchy and Individualism: The Politics of Chicano History and the Dream of Equality." *American Quarterly* 45 (March 1993): 44–72.

Haas, Lisbeth. *Conquests and Historical Identities in California, 1769–1936.* Berkeley, CA: University of California Press, 1995.

Handlin, Ocar. *The Newcomers: Negroes and Puerto Ricans in a Changing Metropolis.* Cambridge: Harvard University Press, 1959.

Haslip-Viera, Gabriel, and Sherrie Baver, eds. *Latinos in New York: A Community in Transition.* Notre Dame: University of Notre Dame Press, 1994.

Healy, David. *Drive to Hegemony: the United States in the Caribbean, 1898–1917.* Madison: University of Wisconsin Press, 1988.

Hoffman, Abraham. "The Writing of Chicano Urban History: From Bare Beginnings to Significant Studies." *Journal of Urban History* 12 (February 1986): 199–205.

Hoffmann, Nancy. *Woman's "True" Profession: Voices from the History of Teaching.* Old Westbury, NY: Feminist Press, 1982.

Hofstadter, Richard. *Social Darwinism in American Thought.* Boston: Beacon Press, 1955.

Horsman, Reginald. *Race and Manifest Destiny: The Origins of American Racial Anglo-Saxonism.* Cambridge, MA: Harvard University Press, 1981.

Hudson, Charles. "Research on the Eastern Spanish Borderlands." In *The Hispanic Experience in North America: Sources for Study in the United States,* edited by Lawrence A. Clayton, 81–95. Columbus: Ohio State University Press, 1992.

Hunt, Michael. *Ideology and U.S. Foreign Policy.* New Haven: Yale University Press, 1987.

Hurtado, Albert L. *Intimate Frontiers: Sex, Gender, and Culture in Old California.* Albuquerque: University of New Mexico Press, 1999.

Ibarra, Robert A. *Beyond Affirmative Action: Reframing the Context of Higher Education.* Madison: University of Wisconsin Press, 2001.

Jacobs, Wilbur R. *On Turner's Trail: 100 Years of Writing Western History.* Lawrence: University Press of Kansas, 1994.

Jameson, Elizabeth, and Susan Armitage. *Writing the Range: Race, Class, and Culture in the Women's West.* Norman: University of Oklahoma Press, 1997.

Jones, Sylvester. "Religious Conditions in Cuba." *The Missionary Review of the World* 3 (March 1907): 182–188.

Jordan, Winthrop D., Miriam Greenblatt, and John S. Bowes. *The Americans: A History.* Evanston, IL: McDouglas, Little and Co., 1994.

Jumonville, Neil. "Liberals and the Historical Past, 1948–1997." In *Henry Steele Commager: Midcentury Liberalism and the History of the Present,* 230–259. Chapel Hill, NC: University of North Carolina Press, 1999.

Kaestle, Carl. *Pillars of the Republic.* New York: Hill and Wang, 1983.

Kagan, Richard L. "Prescott's Paradigm: American Historical Scholarship and the Decline of Spain" *American Historical Review* 101, no. 2 (April 1996): 423–446.

Kanellos, Nicolas. *Hispanic Firsts: 500 Years of Extraordinary Achievement.* Detroit: Visible Ink Press, 1997.

Kanellos, Nicolas, and Claudio Esteva-Fabregat, eds. *The Handbook of Hispanic Cultures in the United States.* Houston: Arte Público Press, 1994.

Keller, Gary D., et al. *Assessment and Access: Hispanics in Higher Education.* Albany: State University of New York Press, 1991.

Knight, Edgar, ed. *Documentary History of Education in the South Before 1860.* Vol. 1. Chapel Hill: University of North Carolina Press, 1949.

Kress, Margaret. "Diary of a Visit of Inspection of the Texas Missions Made by Fray Gasper José de Solís in the Year 1767–8." *Southwestern Historical Quarterly* 35, no. 1 (1931–32): pp. 28–76.

Laden, Berta Vigil. "Hispanic-Serving Institutions: Myths and Realities." *Peabody Journal of Education* 76, no. 1 (2001): 73–92.

Langer, Erick, & Robert H. Jackson, eds. *The New Latin American Mission History.* Lincoln, NE: University of Nebraska Press, 1995.

Latino Education: Status and Prospects: State of Hispanic America 1998. Washington, D.C.: National Council of La Raza, 1998.

Lazerson, Marvin. "Revisionism and American Educational History." *Harvard Educational Review* 43 (spring 1973): 269–283.

———. "Understanding American Catholic Education History." *History of Education Quarterly* 17 (fall 1977): 297–317.

León, David J. "Manuel M. Corella: The Broken Trajectory of the First Latino Student and Teacher at the University of California, 1869–1874." *Aztlán* 26 (spring 2001): 171–79.

León, David J. and D. McNeil. "A Precursor to Affirmative Action: Californios and Mexicans in the University of California, 1870–72." *Perspectives in Mexican American Studies* 3 (1992): 179–206.

Leonard, Irving A. *Books of the Brave: Being an Account of Books and of Men in the Spanish Conquest and Settlement of the Sixteenth-Century New World.* Cambridge, MA: Harvard University Press, 1949.

Limerick, Patricia Nelson. *The Legacy of Conquest: The Unbroken Past of the American West.* New York: W.W. Norton and Co., 1987.

Lindsay, B., and M. J. Justiz, eds. *Quest for Equity in Higher Education: Toward New Paradigms in an Evolving Affirmative Action Era.* Albany, NY: State University Press, 2001.

Lindsay, Samuel McCune. "Inauguration of the American School System in Porto Rico," chapter 15 of *Report of the Commissioner of Education for the year ending June 30, 1905.* Vol. 1. Washington: Government Printing Office, 1907.

Lindsey, Donal F. *Indians at Hampton Institute, 1877–1923.* Urbana: University of Illinois Press, 1995.

Lockwood, Frank C. *Pioneer Days in Arizona: From the Spanish Occupation to Statehood.* New York: Macmillan Company, 1932.

López, Ian F. Haney. *Racism on Trial: The Chicano Fight for Justice.* Cambridge: Belknap Press of Harvard University Press, 2003.

López-Williams, Madeleine. "Speaking American: Puerto Ricans, Language, and the Formation of a New York City Community (1938–1968)." Ph.D. diss., Princeton University, 2003.

Mabry, Donald J. "Twentieth-Century Mexican Education: A Review." *History of Education Quarterly* (spring/summer 1985).

MacDonald, Victoria-María. "Hispanic American Women." In *Historical Dictionary of Women's Education in the United States*, edited by Linda Eisenmann, 201–203. Westport, CT: Greenwood Press, 1998.

MacDonald, Victoria-María. "Hispanic, Latino, Chicano, or 'Other'?: Deconstructing the Relationship between Historians and Hispanic-American Educational History." *History of Education Quarterly* 41, no. 3 (fall 2001): 365–413.

MacDonald, Victoria-María and Scott Beck. "Educational History in Black and Brown: Paths of Divergence and Convergence, 1900–1990." Paper presented at the History of Education Society Annual Meeting, San Antonio, Texas, October 2000.

MacDonald, Victoria-María, and Michael Craig. "Schooling the Other South: Cherokee Indian Education in 1820s and 1830s Georgia." Paper presented at the American Educational Research Association Annual Meeting, Chicago, Illinois, 1997.

MacDonald, Victoria-María, and Teresa García. "Historical Perspectives on Latino Access to Higher Education, 1848–1990." In *The Majority in the Minority: Expanding the Representation of Latina/o Faculty, Administrators and Students in Higher Education*, edited by Jeanett Castellanos and Lee Jones, 13–43. Sterling, VA: Stylus Press, 2003.

MacDonald, Victoria-María. *The Persistence of Segregation: Race, Class, and Public Education in Columbus, Georgia, 1828–1998*. Unpublished manuscript. In authors' possession, Florida State University, Tallahassee, FL. 2004.

Macias, Ysidro Ramon. "The Chicano Movement." In *A Documentary History of the Mexican Americans*, edited by Wayne Moquin and Charles Van Doren. New York: Praeger Publishers, 1971.

Mackey, William F., and Von Nieda Beebe. *Bilingual Schools for a Bicultural Community: Miami's Adaptation to the Cuban Refugees*. Rowley, MA: Newbury House, 1977.

Magnaghi, Russell M. *Herbert E. Bolton and the Historiography of the Americas*. Westport, CT: Greenwood Press, 1998.

Maltby, William S. *The Black Legend in England: The Development of anti-Spanish Sentiment, 1558–1660*. Durham, NC: Duke University Press, 1971.

Manuel, Hershel T. "The Educational Problem Presented by the Spanish-Speaking Child of the Southwest." *School and Society* 40 (1934): 692–695.

Manuel, Hershel T. *Spanish-Speaking Children of the Southwest: Their Education and the Public Welfare*. Austin: University of Texas Press, 1965.

Martínez, Elizabeth. "Brown David v. White Goliath." *Z Magazine* (January 1996): 57–62.

Martínez, Oscar J., ed. *U.S.-Mexico Borderlands: Historical and Contemporary Perspectives*. Wilmington, DE: Scholarly Resources, 1996.

Martínez y Alíre, Jerome J., J. C. L. "The Influence of the Roman Catholic Church in New Mexico under Mexican Administration, 1821–1848." In *Seeds of Struggle/Harvest of Faith: The Papers of the Archdiocese of Santa Fe Catholic Cuarto Centennial Conference: The History of the Catholic Church in New Mexico*, edited by Thomas J. Steele, Paul Rhetts, and Barbe Awalt, 329–344. Albuquerque, NM: LPD Press, 1998.

Masud-Piloto, Felix. *With Open Arms: The Political Dynamics of the Migration from Revolutionary Cuba to the United States*. New York: Rowman and Littlefield, 1988.

Matovina, Timothy, and Gerald E. Poyo, eds. *¡Presente!: U.S. Latino Catholics from Colonial Origins to the Present*. Maryknoll, NY: Orbis Books, 2000.

Matute-Bianchi, M. E. "Ethnic Identities and Patterns of School Success and Failure among Mexican-Descent and Japanese American Students in a California

High School: An Ethnographic Analysis." *American Journal of Education* 95 (1986): 233–255.

McClellan, James E. "English School in Spanish St. Augustine, 1805." In *Education in Florida Past and Present*. Florida State University Studies, No. 15. Tallahassee, FL: Florida State University, 1954.

McKevitt, Gerald. "Hispanic Californians and Catholic higher education: The Diary of Jesús María Estudillo, 1857–1864." *California History* 69 (winter 1990–1991): 320–331, 401–403.

McWilliams, Carey. *North from Mexico: The Spanish-Speaking People of the United States*. Reprint. New York: Greenwood Press, 1968.

Melendez, Miguel "Mickey." *We Took the Streets: Fighting for Latino Rights with the Young Lords*. New York: St. Martin's Press, 2003.

Menchaca, Martha. *The Mexican Outsiders: A Community History of Marginalization and Discrimination in California*. Austin: University of Texas Press, 1995.

———. "The Treaty of Guadalupe Hidalgo and the Racialization of the Mexican Population." In *The Elusive Quest for Equality: 150 Years of Chicano/Chicana Education*, edited by José F. Moreno, 3–30. Cambridge, MA: Harvard Educational Review, 1999.

———. *Recovering History, Constructing Race: The Indian, Black, and White Roots of Mexican Americans*. Austin: University of Texas Press, 2001.

Meier, Matt S. and Feliciano Ribera. *Mexican Americans — American Mexicans: From Conquistadores to Chicanos*. New York: Hill and Wang, 1993.

Merk, Frederick. *Manifest Destiny and Mission in American History: A Reinterpretation*. New York: Knopf, 1963.

Meyer, Michael C. and William L. Sherman. *The Course of Mexican History*. New York: Oxford University Press, 1995.

Mirandé, Alfredo, and Evangelina Enríquez. *La Chicana: The Mexican-American Woman*. Chicago: University of Chicago Press, 1979.

Monge, José Trías. *Puerto Rico: The Trials of the Oldest Colony in the World*. New Haven: Yale University Press, 1997.

Moquin, Wayne and Charles Van Doren, eds. *A Documentary History of the Mexican Americans*. New York: Praeger, 1971.

Mora, Magdalena, and Adelaida R. del Castillo, eds. *Mexican Women in the United States: Struggles Past and Present*. Los Angeles: UCLA, Chicano Studies Research Center Publications, 1979.

Moreno, José F., ed. *The Elusive Quest for Equality: 150 Years of Chicano/Chicana Education*. Cambridge: Harvard Educational Review, 1999.

Mormino, Gary R., and George E. Pozzetta. *The Immigrant World of Ybor City: Italians and Their Latin Neighbors in Tampa, 1885–1985*. Urbana: University of Illinois Press, 1990.

Mörner, Magnus. *Race Mixture in the History of Latin America*. Boston: Little, Brown and Company, 1967.

Muñoz, Carlos. *Youth, Identity, Power: The Chicano Movement*. London: Verso Books, 1989.

Nash, Gerald D. *Creating the West: Historical Interpretations 1890–1990*. Albuquerque: University of New Mexico Press, 1991.

Navarrette, R. *A Darker Shade of Crimson: Odyssey of a Harvard Chicano*. New York: Bantam Books, 1993.

Navarro, Armando. *Mexican American Youth Organization: Avant-garde of the Chicano Movement in Texas*. Austin: University of Texas Press, 1995.

Navarro, Jose-Manuel. *Creating Tropical Yankees: The "Spiritual Conquest" of Puerto Rico, 1898–1908.* Ph.D. diss., University of Chicago, 1995.

Neblett, Sterling A. *Methodism's First Fifty Years in Cuba.* Wilmore, KY: Asbury Press, 1976.

Nieto, Sonia. "A History of the Education of Puerto Rican Students in U.S. Schools: 'Losers,' 'Outsiders,' or 'Leaders'?" In *Handbook of Research on Multicultural Education,* edited by James A. Banks and Cherry Banks, 388–411. New York: Macmillan Press, 1995.

———. "Fact and Fiction: Stories of Puerto Rican students in U.S. Schools." *Harvard Educational Review* 68 (summer 1998): 133–163.

Nieto, Sonia, ed. *Puerto Rican Students in U.S. Schools.* Mahwah, NJ: Lawrence Erlbaum Associates, 2000.

Novick, Peter. *That Noble Dream: The "Objectivity Question" and the American Historical Profession.* Cambridge: Cambridge University Press, 1988.

Nunis, Doyce B., Jr. *Books in Their Sea Chests: Reading along the Early California Coast.* California Library Association, 1964.

Oboler, Suzanne. *Ethnic Labels, Latino Lives: Identity and the Politics of (Re)Presentation in the United States.* Minneapolis: University of Minnesota Press, 1995.

Ochoa, Alberto M. "Language Policy and Social Implications for Addressing the Bicultural Immigrant Experience in the United States." In *Culture and Difference: Critical Perspectives on the Bicultural Experience in the United States,* edited by Antonia Darder, 227–253. Westport, CT: Bergin and Garvey, 1995.

Ogbu, John. *Minority Education and Caste: The American in Cross-Cultural Perspective.* New York: Academic Press, 1978.

———. "Racial Stratification and Education: The Case of Stockton, California." *IRCD Bulletin* 12 (Summer 1977):1–27.

———. "Variability in Minority School Performance: A Problem in Search of an Explanation." *Anthropology and Education Quarterly* 18 (1987): 312–334.

———. "Class Stratification, Racial Stratification, and Schooling." In *Class, Race, and Gender in American Education,* edited by Lois Weis, 163–182. Albany: State University of New York Press, 1988.

———. "Understanding Cultural Diversity and Learning." In *Handbook of Research on Multicultural Education,* edited by James and Cherry McGee Banks, 582–593. New York: MacMillan Publishing, 1995.

Ogbu, John, and M. A. Gibson, eds. *Minority Status and Schooling: A Comparative Study of Immigrants and Involuntary Minorities.* New York: Garland Press, 1991.

Olivas, Michael A. "Research on Latino College Students: A Theoretical Framework and Inquiry." In *Latino College Students,* edited by Michael A. Olivas, 1–25. New York: Teachers College Press, 1986.

Olivas, Michael. "Indian, Chicano and Puerto Rican Colleges: Status and Issues." In *ASHE Reader in the History of Higher Education,* edited by L. F. Goodchild and H. Wechsler. 2nd ed. Needham Heights, MA: Ginn, 1997.

Olneck, Michael R., and Marvin F. Lazerson. "The School Achievement of Immigrant Children, 1900–1930." *History of Education Quarterly* 14 (winter 1974): 453–482.

Olson, James S., and Judith E. Olson. *Cuban Americans: From Trauma to Triumph.* New York: Twayne Publishers, 1995.

Orfield, Gary. "Commentary." In *Latinos Remaking America,* edited by Marcelo M. Suárez-Orozco and Mariela M. Páez. Berkeley: University of California Press and David Rockefeller Center for Latin American Studies, Harvard University, 2002.

Ortiz, Altagracia. "Introduction." In *Puerto Rican Women and Work: Bridges in Transnational Labor*, edited by Altagracia Ortiz. Philadelphia: Temple University Press, 1996.

Osuna, Juan José. *A History of Education in Puerto Rico*. Rio Piedras, P.R.: Editorial de la Universidad de Puerto Rico, 1949.

Ovando, Carlos J. "Politics and Pedagogy: The Case of Bilingual Education." *Harvard Educational Review* 60, no. 3 (1990): 341–356.

Owens, Kenneth N., ed. *Riches For All: The California Gold Rush and the World*. Lincoln, NE: University of Nebraska Press, 2002.

Padilla, Felix M. *Latino Ethnic Consciousness: The Case of Mexican Americans and Puerto Ricans in Chicago*. Notre Dame, IN: University of Notre Dame Press, 1985.

———. *Puerto Rican Chicago*. Notre Dame, IN: University of Notre Dame Press, 1987.

Pantoja, Antonia. "Memorias de una Vida de Obra (Memories of a Life of Work): An Interview with Antonia Pantoja." *Harvard Educational Review* 68, no. 2 (1998): 244–258.

Paterson, Thomas G. "United States Intervention in Cuba, 1898: Interpretations of the Spanish-American-Cuban-Filipino War." *The History Teacher* 29 (May 1996): 341–61.

Paterson, Thomas G., and Stephen G. Rabe, eds. *Imperial Surge: The United States Abroad, The 1890s-Early 1900s*. Lexington, MA: D.C. Heath, 1992.

Perea, Juan F. "The Black/White Binary Paradigm of Race: the 'Normal Science' of American Racial Thought." *California Law Review* 85 (October 1997): 1213–1258.

Pérez, Emma. *The Decolonial Imaginary: Writing Chicanas Into History*. Bloomington: Indiana University Press, 1999.

Pérez, Jr., Louis A. "The Imperial Design: Politics and Pedagogy in Occupied Cuba, 1899–1902." *Cuban Studies/Estudios Cubanos* 12 (summer 1982): 1–19.

———. "Cuba-U.S. Relations: A Survey of Twentieth Century Historiography." *Inter-American Review of Bibliography* 39 (1989): 311–328.

———. *Essays on Cuban History: Historiography and Research*. Gainesville, FL: University of Florida Press, 1995.

———. *The War of 1898: The United States and Cuba in History and Historiography*. Chapel Hill: University of North Carolina Press, 1998.

Perlmann, Joel. *Ethnic Differences: Schooling and Social Structure among the Irish, Italians, Jews and Blacks in an American City, 1880–1935*. Cambridge: Cambridge University Press, 1988.

———. "Historical Legacies: 1840–1920." *The Annals of the American Academy of Political and Social Science* 508 (March 1990): 27–37.

Perlmann, Joel, and Margo, Robert A. *Women's Work?: American Schoolteachers, 1650-1920*. Chicago: University of Chicago Press, 2001.

Perlmann, Joel, and Roger Waldinger. "Second Generation Decline?: Children of Immigrants, Past and Present — A Reconsideration." *International Migration Review* 3 (1997): 893–922.

Phillips, George Harwood. "Indians and the Breakdown of the Spanish Mission System in California." In *New Spain's Far Northern Frontier: Essays on Spain in the American West, 1540–1821*, edited by David J. Weber. Dallas: Southern Methodist University Press, 1979.

Pitt, Leonard. *The Decline of the Californios: A Social History of the Spanish-Speaking Californians, 1846–1890*. Berkeley and Los Angeles: University of California Press, 1966.

Plank, David N. and Rick Ginsberd, eds. *Southern Cities, Southern Schools: Public Education in the Urban South*. Westport, CT: Greenwood Press, 1990.

Porter, Rosalie Pedalino. *Forked Tongue: The Politics of Bilingual Education.* New York: Basic Books, 1990.

Portes, Alejandro. "Children of Immigrants: Segmented Assimilation and Its Determinants." In *The Economic Sociology of Immigration: Essays on Networks, Ethnicity, and Entrepreneurship,* edited by Alejandro Portes, 248–279. New York: Russell Sage Foundation, 1995.

———. *The New Second Generation.* New York: Russell Sage Foundation, 1996.

Portes, Alejandro, and Robert L. Bach. *Latin Journey: Cuban and Mexican Immigrants in the United States.* Berkeley: University of California Press, 1985.

Portes, Alejandro, and Rubén G. Rumbaut. *Immigrant America: A Portrait.* 2nd ed. Berkeley: University of California Press, 1996.

Portes, Alejandro, and Richard Schauffler. "Language and the Second Generation: Bilingualism Yesterday and Today." *International Migration Review* 28 (winter 1994): 640–661.

Portes, Alejandro, and Alex Stepick. *City on the Edge: The Transformation of Miami.* Berkeley: University of California Press, 1993.

Powell, Philip Wayne. *Tree of Hate: Propaganda and Prejudices Affecting United States Relations with the Hispanic World.* New York: Basic Books, 1971.

Poyo, Gerald E., and Gilberto M. Hinojosa. "Spanish Texas and Borderlands Historiography in Transition: Implications for United States History." *Journal of American History* 75 (September 1988): 393–416.

Prida, Dolores, and Dolores Ribner. "100 Books about Puerto Ricans: A Study in Racism, Sexism and Colonialism." *Interracial Digest* 1 (1972): 38–41.

Priestley, Herbert Ingram. *The Coming of the White Man, 1492–1848.* New York: Macmillan Co., 1929.

Raftery, Judith R. "Missing the Mark: Intelligence Testing in Los Angeles Public Schools, 1922–1932." *History of Education Quarterly* (spring 1988).

———. *Land of Fair Promise: Politics and Reform in Los Angeles Schools, 1885–1941.* Palo Alto, CA: Stanford University Press, 1992.

Ravitch, Diane. *The Great School Wars: A History of the New York City Public Schools.* Baltimore, MD: Johns Hopkins University Press, 2000.

Reynolds, Annie. *The Education of Spanish-Speaking Children in Five Southwestern States.* U.S. Department of the Interior, Office of Education, Bulletin 1933, no. 11. Washington, D.C.: Government Printing Office, 1933.

Richman, Irving Berdine. *California under Spain and Mexico, 1535–1847.* Boston: Houghton Mifflin, 1911.

Rochín, Refugio I., and Dennis N. Valdés, eds. *Voices of a New Chicana/o History.* East Lansing: Michigan State University Press, 2000.

Rodríguez, Clara E. *Changing Race: Latinos, the Census, and the History of Ethnicity in the United States.* New York: New York University Press, 2001.

Rodríguez, Félix V. Matos. "New Currents in Puerto Rican History: Legacy, Continuity, and Challenges of the 'Nueva Historia'." *Latin American Research Review* 32 (1997): 193–208.

Rodriguez, Richard. *Brown: The Last Discovery of America.* New York: Viking Press, 2002.

Rodriguez, Roberto. "Black/Latino Relations: An Unnecessary Conflict." *Black Issues in Higher Education* 11 (October 6, 1994): 40–2.

Rodríguez-Fraticelli, Carlos. "Education, Politics and Imperialism: The Reorganization of the Cuban Public Elementary School System during the First American Occupation, 1899–1902." Ph.D. diss., University of California, San Diego, 1984.

———. *Education and Imperialism: The Puerto Rican Experience in Higher Education, 1898–1986.* Centro de Estudios Puertorriqueños Working Paper Series. Higher Education Task Force. Hunter College, New York, 1986.

Rosales, F. Arturo. *Chicano!: The History of the Mexican American Civil Rights Movement.* Houston, Texas: University of Houston, Arte Público Press, 1996.

Rosenbaum, Robert J. *Mexicano Resistance in the Southwest: "The Sacred Right of Self-Preservation."* Austin: University of Texas Press, 1981.

Rotberg, Iris C. "Some Legal and Research Considerations in Establishing Federal Policy in Bilingual Education." *Harvard Educational Review* 52 (May 1982): 151.

Ruiz, Vicki, L. *Cannery Women, Cannery Lives: Mexican Women, Unionization, and the California Food Processing Industry, 1930–1950.* Albuquerque, NM: University of New Mexico Press, 1987.

———. "Star Struck: Acculturation, Adolescence, and the Mexican American Woman, 1920–1950." In *Between Two Worlds: Mexican Immigrants in the United States*, edited by David G. Gutiérrez, 125–47. Wilmington, DE: Scholarly Resources, 1996.

———. *From Out of the Shadows: Mexican Women in Twentieth-Century America.* New York: Oxford University Press, 1998.

Ruiz, Vicki L., and Ellen Carol DuBois. *Unequal Sisters: A Multicultural Reader in U.S. Women's History*, 3d. ed. New York: Routledge Press, 2000.

Rumbaut, Rubén G., and Alejandro Portes. *Ethnicities: Children of Immigrants in America.* Berkeley: University of California Press and New York: Russell Sage Foundation, 2001.

Rury, John L. "Who Became Teachers?: The Social Characteristics of Teachers in American History." In *American Teachers: Histories of a Profession at Work*, edited by Donald Warren, 9–48. New York: Macmillan, 1989.

San Miguel, Guadalupe, Jr. "The Struggle against Separate and Unequal Schools: Middle Class Mexican Americans and the Desegregation Campaign in Texas, 1929–1957." *History of Education Quarterly* (fall 1983).

———. "Status of the Historiography of Chicano Education: A Preliminary Analysis." *History of Education Quarterly* 26 (winter 1986): 523–536.

———. *"Let All of Them Take Heed": Mexican Americans and the Campaign for Education Equality in Texas, 1910–1981.* Austin, TX: University of Texas Press, 1988.

———. "The Schooling of Mexicanos in the Southwest, 1848–1891." In *The Elusive Quest for Equality: 150 Years of Chicano/Chicana Education*, edited by José F. Moreno, 31–51. Cambridge: Harvard Educational Review, 1999.

———. *"Brown, Not White": School Integration and the Chicano Movement in Houston.* College Station, TX: Texas A and M University Press, 2001.

San Miguel, Guadalupe, and Richard R. Valencia. "From the Treaty of Guadalupe Hidalgo to Hopwood: The Educational Plight and Struggle of Mexican Americans in the Southwest." *Harvard Educational Review* 68 (fall 1998): 353–412.

Sánchez, George I. *Forgotten People: A Study of New Mexicans.* Albuquerque: University of New Mexico Press, 1940.

Sánchez, George J. *Becoming Mexican American: Ethnicity, Culture and Identity in Chicano Los Angeles, 1900–1945.* New York: Oxford Press, 1993.

Sánchez-Korrol, Virginia. *From Colonia to Community: The History of Puerto Ricans in New York City.* Berkeley: University of California Press, 1994.

Sánchez-Korrol, Virginia. "Toward Bilingual Education: Puerto Rican Women Teachers in New York City Schools, 1947–1967." In *Puerto Rican Women and Work: Bridges in Transnational Labor*, edited by Altagarcía Ortiz, 82–104. Philadelphia: Temple University Press, 1996.

Sanders, James W. *The Education of an Urban Minority: Catholics in Chicago, 1833- 1965.* New York: Oxford University Press, 1977.

Santiago, Roberto, ed. *Boricuas: Influential Puerto Rican Writings — An Anthology.* New York: Ballantine Books, 1995.

Santiago-Valles, Kelvin A. *"Subject People" and Colonial Discourses: Economic Transformation and Social Disorder in Puerto Rico, 1898–1947.* Albany: SUNY Press, 1994.

———. "'Higher Womanhood' Among the 'Lower Races': Julia McNair Henry in Puerto Rico and the 'Burdens' of 1898." *Radical History Review* 73 (1999): 47–73.

Saragoza, Alex M. "Recent Chicano Historiography: An Interpretive Essay." *Aztlan* 19 (1990): 1–77.

Schlossman, Steven. "Self-Evident Remedy? George I. Sánchez, Segregation, and Enduring Dilemmas in Bilingual Education." *Teachers College Record* 84 (summer 1983): 871–907.

Sears, Jesse B., and Adin D. Henderson. *Cubberley of Stanford and His Contribution to American Education.* Stanford: Stanford University Press, 1957.

Seller, Maxine Schwartz. "The Education of the Immigrant Woman, 1900–1935." *Journal of Urban History* 4 (1978): 307–330.

———. "Immigrants in the Schools — Again: Historical and Contemporary Perspectives on the Education of Post-1965 Immigrants in the United States." *Educational Foundations* (spring 1989): 53–75.

———, ed. *Women Educators in the United States, 1820–1993: A Bio-bibliographic Sourcebook.* Westport, CT: Greenwood Press, 1994.

Serrano, Basilio. "¡Rifle, Cañón, y Escopeta! A Chronicle of the Puerto Rican Student Union." In *The Puerto Rican Movement: Voices from the Diaspora*, edited by A. Torres and J. Velásquez, 124–143. Philadelphia: Temple University Press, 1998.

Servín, Manuel Patricio. "California's Hispanic Heritage: A View into the Spanish Myth." In *New Spain's Far Northern Frontier: Essays on Spain in the American West, 1540–1821*, edited by David J. Weber. Dallas, TX: Southern Methodist University Press, 1979.

Sexton, Patricia Cayo. *Spanish Harlem, Anatomy of Poverty.* New York: Harper and Row, 1965.

Shea, John Gilmary. *History of the Catholic Church in the United States: Colonial Days, 1521–1763 With Portraits, Views, Maps and Fac-similes.* Vol.1. New York: D.H. McBride and Co., 1886.

———. *History of the Catholic Church in the United States, 1844–1866, With Portraits, Views, Maps and Fac-similes.* Vol. 4. New York: D.H. McBride and Co., 1886.

Sheridan, Thomas E. *Arizona: A History.* Tucson, AZ: University of Arizona Press, 1995.

Silva, Helga. *The Children of Mariel from Shock to Integration: Cuban Refugee Children in South Florida Schools.* Miami: Cuban American National Foundation, Inc., 1985.

Sleeter, Christine A., and Carl A. Grant. "Race, Class, Gender, and Disability in Current Textbooks." In *The Politics of the Textbook*, edited by Michael W. Apple and Linda K. Christian-Smith, 78–110. New York: Routledge Press, 1991.

Stefancic, Jean. "Latino and Latina Critical Theory: An Annotated Bibliography." *California Law Review* 85 (October 1997): 1509–1584.

Stepick, Alex, and Carol Dutton Stepick. "Power and Identity: Miami Cubans." In *Latinos Remaking America*, edited by Marcelo M. Suárez-Orozco and Mariela M. Páez, 75–92. Berkeley, CA: University of California Press, 2002.

Suárez-Orozco, Carola, & Marcelo M. Suárez-Orozco. *Transformations: Immigration, Family Life and Achievement Motivation among Latino Adolescents.* Stanford: Stanford University Press, 1995.

————. *Children of Immigrants*. Cambridge, MA: Harvard University Press, 2001.

Suárez-Orozco, Marcelo M. "Immigrant Adaptation to Schooling: A Hispanic Case." In *Minority Status and Schooling: A Comparative Study of Immigrant and Involuntary Minorities*, edited by M. A. Gibson and John Ogbu, 37–61. New York: Garland Press, 1991.

Suárez-Orozco, Marcelo M., and Mariela M. Páez, eds. *Latinos Remaking America*. Cambridge, MA: David Rockefeller Center for Latino American Studies, Harvard University and Berkeley: University of California Press, 2002.

Suárez-Orozco, Marcelo M., and Carola Suárez-Orozco. *Central American Refugees and U.S. High Schools: A Psychosocial Study of Motivation and Achievement*. Stanford: Stanford University Press, 1989.

————. *Transformations: Immigration, Family Life, and Achievement Motivation among Latino Adolescents*. Stanford: Stanford University Press, 1995.

Sweet, David. "The Ibero-American Frontier Mission in Native American History." In *The New Latin American Mission History*, edited by Erick Langer and Robert H. Jackson, 1–48. Lincoln, NE: University of Nebraska Press, 1995.

Teitelbaum, Herbert J., and Richard J. Hiller. "Bilingual Education: The Legal Mandate." *Harvard Educational Review* 47 (May 1977): 139.

Tentler, Leslie Woodcock. "On the Margins: The State of American Catholic History." *American Quarterly* 45 (March 1993): 104–127.

Thelen, David. "The Nation and Beyond: Transnational Perspectives on United States History." *Journal of American History* 86 (December 1999): 965–975.

Torre, Carlos Antonio, Hugo Rodríguez Vecchini, and William Burgos, eds. *The Commuter Nation: Perspectives on Puerto Rican Migration*. Ríos Pedras, P.R.: Universidad de Puerto Rico, 1994.

Torrejón, Victor. "Residency urged for some kids of entrants," *Arizona Daily Star*. July 18, 2002, p.B8.

Torres, Andrés, and José E Velázquez, eds. *The Puerto Rican Movement: Voices from the Diaspora*. Philadelphia: Temple University Press, 1998.

Torres-Saillant, Silvia, and Ramona Hernández. *The Dominican Americans*. Westport, CT: Greenwood Press, 1998.

Traub, J. *City on a Hill: Testing the American Dream at City College*. New York: Addison-Wesley Publishing, 1994.

Trexler, Richard C. "From the Mouths of Babes: Christianization by Children in 16th Century New Spain." In *Religious Organization and Religious Experience*, edited by J. Davis, 122–3. London: Academic Press, 1982.

Triay, Victor Andres. *Fleeing Castro: Operation Pedro Pan and the Cuban Children's Program*. Gainesville: University Press of Florida, 1998.

Twyoniak, F. Esquivel and Mario T. García. *Migrant Daughter: Coming of Age as a Mexican American Woman*. Berkeley: University of California Press, 2000.

Tyack, David B., comp. *Turning Points in American Educational History*. Waltham, MA: Blaisdell Pub. Co., 1967.

Tyler, Daniel. "The Mexican Teacher." *Red River Valley Historical Review* 1 (1974).

Valdés, Dennis. *Barrios Norteños: St. Paul and Midwestern Mexican Communities in the Twentieth Century*. Austin: University of Texas Press, 2000.

Valdés, Guadalupe. "Dual-Language Immersion Programs: A Cautionary Note Concerning the Education of Language-Minority Students." *Harvard Educational Review* 67 (fall 1997): 391–429.

Valdez, Francisco. "Under Construction: LatCrit Consciousness, Community, and Theory." *California Law Review and La Raza Law Journal* 85 (October 1997): 1087–1142.

Walker, Vanessa Siddle. *Their Highest Potential: An African-American School Community in the Segregated South.* Chapel Hill: University of North Carolina Press, 1996.

Walsh, Bryan O. "Cuban Refugee Children." *Journal of Inter-American Studies and World Affairs* 13 (July-October 1971): 378–414.

Weber, David J. " 'Scarce More Than Apes': Historical Roots of Anglo American Stereotypes of Mexicans in the Border Region." In *New Spain's Far Northern Frontier: Essays on Spain in the American West, 1540–1821*, edited by David J. Weber, 295–307. Albuquerque: University of New Mexico Press, 1979.

———. *The Mexican Frontier, 1821 to 1846: The American Southwest under Mexico.* Albuquerque: University of New Mexico Press, 1982.

———. *The Spanish Frontier in Northern New Spain.* New Haven: Yale University Press, 1992.

———. "The Spanish Legacy in North America and the Historical Imagination." *Western Historical Quarterly* 23 (February 1992): 4–24.

———. "John Francis Bannon and the Historiography of the Spanish Borderlands: Retrospect and Prospect." In *Establishing Exceptionalism: Historiography and the Colonial Americas*, edited by Amy Turner Bushnell, 297–330. Vol. 5 of *An Expanding World: The European Impact on World History.* Brookfield, VT: Variorum, 1995.

———. *What Caused the Pueblo Revolt of 1680?* Boston: Bedford/St. Martin's Press, 1999.

Weber, Francis J. *Documents of California Catholic History, 1784–1963.* Los Angeles: Dawson's Book Shop, 1965.

Weinberg, Meyer. *A Chance to Learn: The History of Race and Education in the United States.* Cambridge: Cambridge University Press, 1977.

Welsh, Michael. "A Prophet without Honor: George I. Sánchez and Bilingualism in New Mexico." *New Mexico Historical Review* 69 (January 1994):19–34.

Wollenberg, Charles. *All Deliberate Speed: Segregation and Exclusion in California Schools, 1855–1975.* Berkeley, CA: University of California Press, 1976.

Wortham, Stanton, Enrique G. Murillo, Jr., & Edmund T. Hamann. *Education in the New Latino Diaspora: Policy and the Politics of Identity.* Westport, CT: Ablex, 2002.

Yohn, Susan M. "An Education in the Validity of Pluralism: The Meeting between Presbyterian Mission Teachers and Hispanic Catholics in New Mexico, 1870–1912." *History of Education Quarterly* (fall 1991).

———. *A Contest of Faiths: Missionary Women and Pluralism in the American Southwest.* Ithaca, NY: Cornell University Press, 1995.

GOVERNMENT AND ARCHIVAL SOURCES

Annual Report of the U.S. Commissioner of Education, 1870. Washington: Government Printing Office, 1870.

Annual Report of the Commissioner of Education for the Fiscal Year Ended June 30, 1901. Vol. 1. Washington, D.C.: Government Printing Office, 1902.

Annual Report of the U.S. Commissioner of Education, 1903. Washington: Government Printing Office, 1903.

Annual Report of the U.S. Commissioner of Education, 1905. Washington: Government Printing Office, 1905.

Bailey, Thomas D. Report on Cuban Non-Immigrant Children Enrolled in Schools of Dade County, Florida. January 27, 1961. Florida State Archives and Library, Tallahassee, Florida.

Biennial Report of the Superintendent of Public Instruction of the Territory of Arizona, 1899–1900. Phoenix: The H.H. McNeil Co. Printer, 1900. State Government Publications, States Other Than Wisconsin, Wisconsin Historical Society, Madison, Wisconsin.

Carothers, Sylvia. Report to the Governor of Florida on the Cuban Refugee Problem in Miami. February 21, 1961. Florida State Archives and Library, Tallahassee, Florida.

A Citizen of the United States, A View of South America and Mexico, Comprising their history, the political condition, geography, agriculture, commerce, etc. of the Republics of Mexico, Guatemala, Colombia, Peru, the United Provinces of South America and Chili, with a complete history of the revolution, in each of these independent states. New York: Published for Subscribers, 1827.

Clark, Victor S. "Report to General John Eaton, Director of Public Instruction." San Juan, Puerto Rico, March 14, 1899. In Report of the Commissioner of Education for the Year 1899–1900. Vol. 1. Washington, D.C.: Government Printing Office, 1901.

"Communication and Report of Bishop Alemany, Concerning the Catholic Schools in California." In Second Annual Report of the Superintendent of Public Instruction to the Legislature of the State of California, n. p. George Kerr, State Printer, 1853. Rare Book Room, Library of Congress, Washington, D.C.

Compilation of the School Laws of New Mexico, Leyes de Escuelas de Nuevo Mexico. East Las Vegas, NM: J.A. Carruth, Printer, Binder and Blank Book Manufacturer, 1889. Rare Book Room, Library of Congress, Washington, D.C.

Delgado v. Bastrop Independent School District, 1948 (Texas). [Not recorded] No. 388 Civil. United States District Court for the Western District of Texas, Austin Division. National Archives and Records Administration, Southwest Region, Forth Worth, Texas.

Education in New Mexico. Report of Hon. W.G. Ritch to the commissioner of education for the year 1874. Santa Fe, NM: Manderfield and Tucker, Printers, 1875. State Government Publications, States Other Than Wisconsin, Wisconsin Historical Society, Madison, Wisconsin.

Education in New Mexico. Third Annual Report of Hon. W.G. Ritch, Secretary of the Territory, to the National Bureau of Education, 1875 (Northwestern Steam Printing House, 1876). Government Publications, States other than Wisconsin, Wisconsin Historical Society, Madison, Wisconsin.

"Education in Porto Rico." In Report of the Commissioner of Education for the Year 1899–1900. Vol. 1. Washington, D.C.: Government Printing Office, 1901, p. 254.

"Education in Porto Rico." In Report of the Commissioner of Education. Vol. 2. In Annual Report of the Department of the Interior. Washington, D.C.: Government Printing Office, 1903, p. 1204.

"Expedition of Cuban Teachers to Cambridge, Mass." In Report of the Commissioner of Education for the Year 1899–1900. Vol. 2. Washington, Government Printing Office, 1901, p. 1377–1385.

First Annual Report of the Superintendent of Public Instruction to the Legislature of the State of, California, 1852. Eugene Cassely, State Printer, 1872. State Government Publications, States Other Than Wisconsin, Wisconsin Historical Society, Madison, Wisconsin.

First Annual Report of the Superintendent of Public Instruction of the State of Texas, 1871. Austin: J.G. Tracy, State Printer, 1871. State Government Publications, States Other Than Wisconsin, Wisconsin Historical Society, Madison, Wisconsin.

Florida's Division of Child Welfare. "Cuban Refugee Assistance Program Administrative Files, 1962–1972." Florida State Archives and Library, Tallahassee, Florida. Florida's Division of Child Welfare. "Cuban Refugee Children Program Correspondence, 1961–1968." Florida State Archives and Library, Tallahassee, Florida.

Fourth Biennial Report of the Superintendent of Public Instruction of the State of California. Sacramento, CA: T.A. Springer, State Printer, 1871. Rare Book Room, Library of Congress, Washington D.C.

Gammel, H.P.N., compiled and arranger. The Laws of Texas, 1822–1897. 10 Vols. Austin: The Gammel Book Company, 1898.

"Historical Sketch of the Public School System in California." In First Biennial Report of the Superintendent of Public Instruction of the State of California for the School Years 1864 and 1865. Special Collections, Monroe C. Gutman Library, Harvard Graduate School of Education.

Independent School District v. Salvatierra, 33 S. W. 2d 790 (Tex. Civ. App. San Antonio, 1930), cert. Denied, 284 U.S. 580 (1931).

Laws of 90th Cong. — 1st session. P.L. 90–247. January 2, 1968, p. 919.

Lindsay, Samuel McCune. "Inauguration of the American School System in Porto Rico," chapter 15 of Report of the Commissioner of Education for the year ending June 30, 1905. Vol. 1. Washington: Government Printing Office, 1907.

"List of Superintendents of Public Instruction," Twenty-Ninth and Thirtieth Annual Reports of the State Superintendent of Public Instruction, New Mexico, 1919–1920, p. 7. State Government Publications, States Other Than Wisconsin, Wisconsin Historical Society, Madison, Wisconsin.

McCrea, Samuel Pressly. "Establishment of the Arizona School System." Biennial Report of the Superintendent of Public Instruction of the Territory of Arizona, 1908. Phoenix: the H.H. McNeil Company, 1908. State Government Publications, States Other Than Wisconsin, Wisconsin Historical Society, Madison.

Méndez et al. v. Westminister School District, 64, F. Supp. 544 (S. D. Cal 1946), 161 F.2D 744.

National Center for Education Statistics. The Condition of Education for Hispanic Americans. Washington, D.C. Government Printing Office, 1985.

Nineteenth and Twentieth Annual Reports of the Territorial Superintendent of Public Instruction to the Governor of New Mexico for the Years 1909–1910. Santa Fe: The New Mexican Printing Company, 1911. Rare Book Room, Library of Congress, Washington, D.C

President's Advisory Commission on Educational Excellence for Hispanic Americans. Our Nation on the Fault Line: Hispanic American Education. Government Printing Office: Washington, D.C., 1996.

Rankin, Melinda. Twenty Years among the Mexicans: A Narrative of Missionary Labor. Cincinnati: Chase and Hall, Publishers, 1875.

"Regulations for the Public Schools of the Island of Cuba, June 30, 1900." In Report of the Commissioner of Education for the Year 1899–1900. Vol. 2. Washington: Government Printing Office, 1901, 1643–1647.

Report of M.C. Monroe, Superintendent of Merced County Schools. In Fourth Biennial Report of the Superintendent of Public Instruction of the State of California. Sacramento, CA: T.A. Spring, State Printer, 1871, 148. Rare Book Room, Library of Congress, Washington D.C.

Report of the U.S. Commissioner of Education for the Year 1876. Washington: Government Printing Office, 1878.

Reports of the Commissioner of Education for the Year 1874. Washington: Government Printing Office, 1875.

Revised School Law, State of California. Approved March 24, 1866. Sacramento: O. J. Clayes, State Printer, 1866. Rare Book Room, Library of Congress, Washington, D.C

Second Annual Report of the Superintendent of Public Instruction of the State of Texas for the Year 1872. Austin: James P. Newcomb and Co., 1873. State Government Publications, States Other Than Wisconsin, Wisconsin Historical Society, Madison.

"Statement of the Condition and Progress of Education in the Territory of New Mexico." In Reports of the Commissioner of Education for the Year 1874. Washington, D.C.: Government Printing Office, 1875.

A Survey of the Public Educational System of Puerto Rico. New York: Teachers College, Columbia University, 1926.

Territory of New Mexico. Report of the Superintendent of Public Instruction of the Territory of New Mexico for the Year Ending, 1891. Santa Fe, NM: New Mexican Printing Co., 1892. Rare Book Room, Library of Congress, Washington D.C.

Testimony of Patricia Delgado, San Francisco, Hearings Before the United States Commission on Civil Rights. San Francisco, CA, May 1–3, 1967 and Oakland, CA, May 4–6, 1967. Washington D.C.: U.S. Government Printing Office, 1967.

Twenty-seventh and Twenty-Eighth Annual Reports of the State Superintendent of Public Instruction to the Governor of New Mexico for the Years 1917–1918. Albuquerque, NM: Central Printing Co., 1918. State Government Publications, States Other Than Wisconsin, Wisconsin Historical Society, Madison.

U.S. Commissioner of Education. Annual Reports of the Secretary of Education, Department of Porto Rico. Washington, D.C.: Government Printing Office, 1901–1910.

U.S. Congress. Committee on Education and Labor. Staff Report on Hispanics' Access to Higher Education. 99th Congress, 1st Session, Serial No. 99-K, Washington, D.C., 1985.

Voorhees, Tracy S. Report to the President of the United States on the Cuban Refugee Problem. January 18, 1961. Washington: United States Government Printing Office, 1961. Rockefeller Foundation Archives, Box 67, Folder 565, Rockefeller Archive Center, Sleepy Hollow, New York.

Whigham, E. L. The Cuban refugee in the public schools of Dade County, Florida. Miami, FL: Dept. of Administrative Research, 1970. Florida State Archives and Library, Tallahassee, Florida.

ARCHIVES

Benson Latin American Collection. University of Texas, Austin. Austin, Texas.

Bexar Archives, East Texas Research Center, Ralph W. Steen Library, Stephen F. Austin State University, Texas.

Center for Puerto Rican Studies/El Centro de Estudios Puertorriqueños, Hunter College, New York, NY.

Florida State Archives and Library, Tallahassee, FL.

Library of Congress, Washington D.C.

National Archives and Records Administration, Southwest Region, Forth Worth, Texas.

Rockefeller Archives Center (RAC), Sleepy Hollow, NY.
Special Collections, Monroe C. Gutman Library, Harvard Graduate School of
      Education, Cambridge, MA.
Texas A &M Archives, Cushing Memorial Library, Texas A&M University, College
      Station, Texas.
Wisconsin Historical Society, Madison, WI.

Index